# Emerging Topics in Metal Complexes: Pharmacological Activity

# Emerging Topics in Metal Complexes: Pharmacological Activity

Guest Editors

**Agnieszka Ścibior**
**Juan Llopis**
**Manuel Aureliano**

MDPI

Basel • Beijing • Wuhan • Barcelona • Belgrade • Novi Sad • Cluj • Manchester

*Guest Editors*
Agnieszka Ścibior
Department of Biomedicine and Environmental Research
The John Paul II Catholic University of Lublin
Lublin
Poland

Juan Llopis
Department of Physiology
Biomedical Research Center
University of Granada
Granada
Spain

Manuel Aureliano
Faculdade de Ciências e Tecnologia (FCT)
Universidade do Algarve
Faro
Portugal

*Editorial Office*
MDPI AG
Grosspeteranlage 5
4052 Basel, Switzerland

This is a reprint of the Special Issue, published open access by the journal *International Journal of Molecular Sciences* (ISSN 1422-0067), freely accessible at: https://www.mdpi.com/journal/ijms/special_issues/G300O371XX.

For citation purposes, cite each article independently as indicated on the article page online and as indicated below:

Lastname, A.A.; Lastname, B.B. Article Title. *Journal Name* **Year**, *Volume Number*, Page Range.

**ISBN 978-3-7258-4340-4 (Hbk)**
**ISBN 978-3-7258-4339-8 (PDF)**
**https://doi.org/10.3390/books978-3-7258-4339-8**

# Contents

# About the Editors

**Agnieszka Ścibior**

Agnieszka Ścibior, PhD, DSc, has been an Associate Professor since 2019. From 2014 to 2022, she was the Head of the Laboratory of Oxidative Stress in the Centre for Interdisciplinary Research at the John Paul II Catholic University of Lublin (Poland). Now, she works in the Department of Biomedicine and Environmental Research at the John Paul II Catholic University of Lublin (Poland). Her main fields of interest include the toxicity of selected elements, metal interactions, and mechanisms of metal toxicity (in vitro/in vivo experimental model). She is a member of the Polish Laboratory Animal Science Association, the Polish Society of Toxicology, and the Polish Magnesiological Society. She has been conferred awards of the Rector of the John Paul II Catholic University of Lublin for original and creative scientific achievements. She has served as Guest Editor in such journals as *IJMS*, *Biology*, and *Toxics*.

**Juan Llopis**

Juan Llopis is an Emeritus Professor, University of Granada, Spain. He is also a Research Director at University of Granada, Department of Physiology, Institute of Nutrition and Food Technology "José Mataix Verdú", Biomedical Research Centre. He is a Full Professor of "Fisiología Digestiva Y Nutrición" and has served in various editorial roles, such as Associate Editor and/or Guest Editor in several journals such as *IJMS*, *Biology*, *Toxics*, and *Pharmaceutics*.

**Manuel Aureliano**

Manuel Aureliano is a Full Professor of Biochemistry (Aggregation in Inorganic Biochemistry) with the Faculty of Sciences and Technology of the University of Algarve, Faro, Portugal, teaching biochemistry, inorganic biochemistry, muscle contraction, and advanced chemistry and biochemistry. He completed his undergraduate, master, and PhD studies at the University of Coimbra, Coimbra, Portugal. He was Director of the Biochemistry degree at the University of Algarve for about two decades, from 1998 to 2013 and from 2021 to 2025. At the Algarve Centre for Marine Sciences (CCMAR-Algarve), he investigates the applications of decavanadate and others polyoxometalates in the environment and biomedicine. He was selected as an "Outstanding Reviewer" (top 10 reviewers) for the journal *Metallomics* during 3 consecutive years (2017, 2018, and 2019). To date, he has participated in the revision of more than 900 papers from about 140 journals. Moreover, he has served in various editorial roles, including Editor, Associate Editor, and/or Guest Editor, with many journals. In 2021, 2022, 2023, and 2024, he was included in the "Worlds Top 2% Scientists list" (impact-career and year). Recently (December 2024), he was awarded the 3rd edition of the "UAlg Researcher Award" attributed by the University of Algarve.

# Preface

Due to their pharmacological potential, metal complexes and compounds have been arousing the interest of many research centers worldwide. They are used not only in the treatment of some modern-age diseases but also in clinical diagnostics. Currently, metal complexes and compounds (such as polyoxometalates) are prominent players in medicine and medicinal chemistry.

The results of certain studies have demonstrated that gold (Au), silver (Ag), platinum (Pt), copper (Cu), zinc (Zn), and ruthenium (Ru) coordination compounds exhibit anti-tumor, anti-microbial, and immunosuppressive activity. Some of them, i.e., Cu and Zn complexes, have also been studied in the context of neurodegenerative illnesses such as Alzheimer's disease, which is a constantly growing social concern. Numerous studies have also revealed the potential therapeutic benefits of vanadium (V) complexes and compounds, which are at the forefront among the different metals examined for their potential therapeutic use. Its multidirectional biological properties (i.e., antiviral, antibacterial, antiproliferative, anti-inflammatory, antihypertensive, anti-allergic, anti-oxidative, anticancer, antidiabetic, anti-ulcer, and anti-obesity activity), as well as its neuroprotective, nephroprotective, and cardioprotective effects, provide a basis for studies of the unique class of pharmacological agents containing this metal.

Despite the progress in the knowledge of the pharmacological activity of certain metals and their complexes, further studies are still needed, as the mechanisms of action of the potential drugs containing metal ions are not sufficiently recognized yet. Moreover, the side effects of the use of metal coordination compounds in the treatment of some diseases (mainly cancer) and resistance to treatment are a serious problem in modern medicine. Therefore, there is a constant search for new therapeutics based on metal complexes that would ensure better treatment results and contribute to a significant slowdown in the progression of the disease. The therapeutic targets, bioavailability, and pharmacokinetics of metal-containing compounds have to be clarified in detail with reference to metal-based drugs, which may guarantee potential clinical success.

We invited authors to submit original research papers or review articles on the pharmacological activity of metal coordination compounds with a focus on strategies improving their pharmacological features and on links between metal-containing complexes and potential therapies of human pathologies with mechanistic analysis.

We hope that the present reprint will be used to preserve the memory of future generations of students, researchers, and professors. We also hope that we can all find this reprint valuable to our own research work. We would like to thank the Editorial team and the reviewers for their priceless support and help throughout this process.

**Agnieszka Ścibior, Juan Llopis, and Manuel Aureliano**
*Guest Editors*

International Journal of
*Molecular Sciences*

MDPI

*Editorial*

# Emerging Topics in Metal Complexes: Pharmacological Activity

Agnieszka Ścibior [1,*], Manuel Aureliano [2,3,*] and Juan Llopis [4]

1 Laboratory of Oxidative Stress, Department of Biomedicine and Environmental Research, The John Paul II Catholic University of Lublin, 20-708 Lublin, Poland
2 Faculdade de Ciências e Tecnologia (FCT), Campus de Gambelas, Universidade do Algarve, 8005-139 Faro, Portugal
3 Centro de Ciências do Mar (CCMar), Campus de Gambelas, 8005-139 Faro, Portugal
4 Institute of Nutrition and Food Technology "José Mataix Verdú", Department of Physiology, Biomedical Research Center, University of Granada, Avda del Conocimiento s/n., 18100 Armilla, Granada, Spain; jllopis@ugr.es
* Correspondence: agnieszka.scibior@kul.pl (A.Ś.); maalves@ualg.pt (M.A.)

## 1. Introduction and Scope

This Special Issue (SI), "Emerging Topics in Metal Complexes: Pharmacological Activity", includes reports updating our knowledge on metals with multidirectional biological properties and metal-containing compounds/complexes for their potential therapeutic applications, with a focus on strategies improving their pharmacological features. The current SI also refers to the biofortification process, which, in the near future, may become a promising source of dietary enrichment with elements important for human health. The scientific articles making up the current SI, i.e., one review and eight original papers (nine in total), garnered a total of six citations and over 8600 visualizations (April 2024). A graphical summary of the content included in this Special Issue is presented in Figure 1, showing a central image with seven hexagons reflecting the seven papers on antitumor, antimicrobial, antioxidant, and CT-imaging properties, flanked by one hexagon each side for the biofortification of wheat sprouts with iron (Fe), zinc (Zn), magnesium (Mg), and chromium (Cr), which can effectively stimulate plant growth (right) and by the influence of vanadium (V) in human neurodegenerative disorders such as Alzheimer's disease (AD), Parkinson's disease (PD), and amyotrophic lateral sclerosis (ALS) (left).

Data collected in this SI clearly indicate that further studies are needed to better recognize the effects of the interaction between vanadium, of which its pharmacological activity has been repeatedly described [1,2], and other metals, especially those with antioxidant properties [3], as it may be helpful in optimizing the therapeutic potential of vanadium and contribute to the development of effective therapeutic approaches that use this metal in the future. The data also point to the need to better understand the relationships between the structure and properties of metal compounds/complexes, as it may help to explain the modes of their action, thereby making them therapeutically effective. Further studies should also be carried out to define the biological effects of metals/metal-containing species with a special focus on the relationship between the dose and response with the estimation of safe concentrations of the tested metals/metal-containing species at which the potential therapeutic effects will be observed. It is also important to recognize the pharmacokinetic and pharmacodynamic properties of metals and metal-containing compounds/complexes and to clarify the mechanisms of their action in detail. This is crucial in view of the development of metallodrugs in general.

Moreover, there is great interest in repurposing medicines, allowing 'older' medicines with marketing approval, such as metformin, to be made available to patients in a timely manner as 'new' treatments [4]. Recently, it was suggested that decavanadate and metformin-decavanadate decrease the proliferation of melanoma cells after treatment with these compounds [5]. It was also suggested that certain cell signaling pathways important in the development of cancer are altered by treatment with decavanadate, and how it alters the cell cycle of melanoma cells. Taken together, several studies demonstrate that these applications of polyoxometalates (POMs) in the treatment of cancer [6,7], and specifically vanadium and polyoxovanadates (POVs) in melanoma [5,8], are possible, suggesting a therapeutic opportunity, and strengthening the potential use of these potential metallopharmaceuticals in the near future as anticancer agents, among others biomedical applications such as the understanding of human neurodegenerative disorders [7–14], [contribution 1].

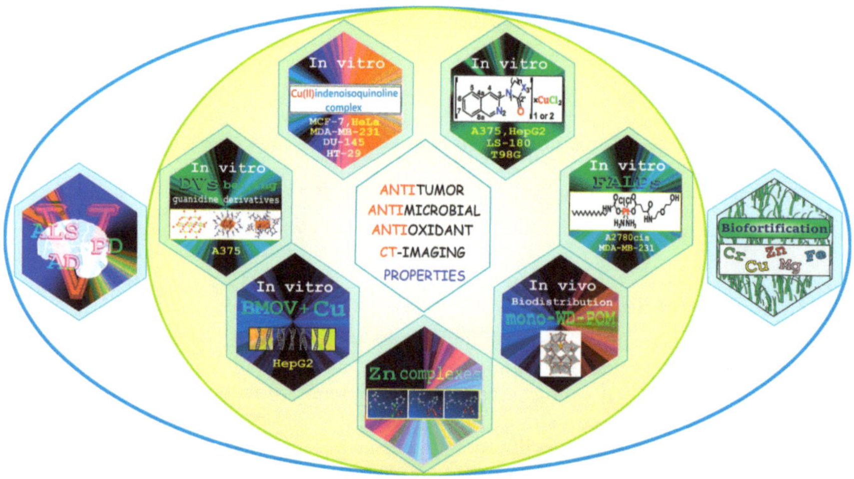


**Figure 1.** Graphical summary of metals and metal compounds/complexes overviewed in this Special Issue: neurodegenerative disorders, in vitro and in vivo studies, anticancer and antimicrobial activity, antioxidant and CT-imaging properties, biodistribution, and biofortification.

## 2. An Overview of Published Articles

The first paper published in this Special Issue entitled "CNS-Related Effects Caused by Vanadium at Realistic Exposure Levels in Humans: A Comprehensive Overview Supplemented with Selected Animal Studies" [contribution 1] is an attempt to provide thorough knowledge on the influence of vanadium on neurodegeneration in humans. Hence, it summarizes data on the neurological side effects and neurobehavioral alterations in subjects in relation to vanadium exposure and collects information about other adverse health outcomes of environmental exposure to this metal in humans and findings on the neurotoxic effects of vanadium after inhalation in laboratory animals and those naturally exposed to urban air. It also compiles data on the levels of vanadium in the biological fluids and brain of subjects with neurodegenerative disorders such as AD, PD, and ALS and briefly summarizes new therapies for these illnesses. Data overviewed in this review clearly indicate that more attention should be paid to vanadium, which is a well-known transition metal capable of disturbing the oxidation–reduction balance in the organism, as an environmental risk factor for the health of the general population and to its role in the etiopathogenesis of neurodegenerative diseases in which oxidative stress appears to be a part of the pathophysiological mechanism. They also suggest that vanadium could help in identifying people with a higher risk of detrimental effects on the central nervous system and serve as a predictor of some neurodegenerative diseases. However, more extensive epidemiological studies are needed to confirm this statement.

The original articles included in this SI focus on (1) the assessment of the in vitro effect of copper against the toxicity of bis(maltolato)oxovanadium (IV), BMOV [contribution 2], (2) the evaluation of the structure–activity relationship of fatty acid-like platinum [Pt(IV)] prodrug (FALP) derivatives (a class of Pt-based mitochondria-targeting metallodrugs) in terms of anticancer activity and cellular responses [contribution 3], and (3) the assessment of the

DNA interaction with a novel copper(II) indenoisoquinoline complex and its cytotoxic activity against five adenocarcinoma cell lines [contribution 4]. Other topics addressed in the articles include in vitro anticancer and antimicrobial activities along with the antioxidant potential of copper(II) complexes and 1-(isoquinolin-3-yl)heteroalkyl-2-ones [contribution 5], the antimicrobial and in vitro antitumor activity of decavanadate-bearing guanidine derivatives [contribution 6], and in vitro computed tomography (CT) imaging properties along with the in vivo potential toxic effects of monolacunary Wells–Dawson polyoxometalate (mono-DW POM) and the mono-WD POM tissue distribution [contribution 7]. Moreover, the antibacterial effect of Zn(II) carbosilane iminopyridine complexes with inorganic ligands was assessed against certain Gram-positive and Gram-negative bacteria [contribution 8]. In addition, a strategy for bioelement supplementations, which can help to solve current problems associated with the deficiency of some macro- and micro-elements in the human body and the related negative health effect, was assessed [contribution 9].

The study conducted in an in vitro model on a hepatic cell line (HepG2) by Rivas-García and co-workers [contribution 2] showed that HepG2 cells exposed to BMOV (3 mg V/L) for 32 h reduces cell viability, whereas the treatment of HepG2 cells with $CuCl_2$ (3 mg Cu/L) during incubation with BMOV has the opposite effect. In addition, under the 32 h treatment of the cells with $CuCl_2$ (3 mg Cu/L) in the presence of BMOV (3 mg V/L), the ND1/ND4 deletion of the mitochondrial DNA and the nuclear damage caused by the BMOV exposure was slightly limited and markedly reduced, respectively. Thus, these findings clearly show that the 32 h incubation of HepG2 cells with $CuCl_2$ in the presence of BMOV does not only enhance the cytotoxic response but also protects against BMOV toxicity, thereby indicating that Cu can effectively prevent the harmful effects caused by V. However, as stressed by the authors, there is a need for further studies to better determine the effects resulting from the interactions between V and Cu, as it may be helpful in limiting the toxic effects of V and enhancing the potential therapeutic applications of this element [contribution 2].

Another in vitro study carried out on two human cancer cell lines (A2780cis and MDA-MB-231) [contribution 3] demonstrated that the head group modifications of FALPs can markedly affect their cytotoxicity profiles. More precisely, by using two FALP model compounds, the authors revealed that one of them, i.e., the one with hydrophilic modification, was able to readily penetrate cancer cells and mitochondria, subsequently inducing a cascade of events leading to mitochondrial and DNA damage, which effectively eradicated cancer cells. The other model with hydrophobic modification exhibited significantly lower toxicity (notably lower uptake and weaker cellular responses). Thus, the results obtained from this study clearly showed that increased hydrophobicity, as highlighted by the authors, may not necessarily enhance the cellular uptake of therapeutic molecules [contribution 3]. These findings open a way for new drug development strategies.

The findings obtained from the in vitro study on human cell lines from cervix cancer, breast cancer, breast triple-negative cancer, colorectal cancer, and prostate cancer (i.e., HeLa, MCF-7, MDA-MB-231, HT-29, and DU-145) conducted by Molinaro and co-workers [contribution 4] showed that the new organometallic compound, i.e., Cu(II)-indenoisoquinoline complex (WN198), can inhibit topoisomerase I in a dose-dependent manner, starting at 1 μM and with optimum efficiency at 2 μM, and that this complex triggers autophagy evidenced by Beclin-1 accumulation and LC3-II formation. This study also demonstrated that WN198 is able to bind to DNA by intercalation, as shown by the melting curves and fluorescence measurements, where the main interaction took place via the aromatic ring. Based on these results, the authors suggest that Cu-derived indenoisoquinoline topoisomerase I inhibitor WN198 may be a promising antitumorigenic agent for the development of future DNA-damaging treatments and emphasize the need for determining the toxicity characteristic of this Cu complex with respect to membrane permeability and cellular uptake. In turn, the study conducted by Balewski and co-workers [contribution 5] on other Cu(II) complexes derived from 1-(isoquinolin-3-yl)heteroalkyl-2-one ligands (i.e., 1-(isoquinolin-3-yl)azetidin-2-one, 1-(isoquinolin-3-yl)imidazolidin-2-

one, 1-(isoquinolin-3-yl)-3-methylimidazolidin-2-one, and 1-ethyl-3-(isoquinolin-3-yl)-3-imidazolidin-2-one, L1-4, respectively) evaluated anticancer activity against human cancer A375 (melanoma), HepG2 (hepatoma), LS-180 (colon cancer), and T98C (glioblastoma) cell lines and demonstrated that Cu complexes such as dichloro{bis [1-(isoquinolin-3-yl)azetidin-2-one]}Cu(II) (C1), dichloro{bis [1-(isoquinolin-3-yl)imidazolidin-2-one]}Cu(II) (C2), dichloro [1-(isoquinolin-3-yl)-3-methylimidazolidin-2-one]}Cu(II) (C3), and dichloro [1-ethyl-3-(isoquinolin-3-yl)imidazolidin-2-one]}Cu(II) (C4) have greater potency against HepG2, LS-180, and T98C cells than the known antitumor agent *etoposide*, whereas free ligands L1-4 are inactive in all investigated cell lines. Moreover, the authors showed that dichloro{bis [1-(isoquinolin-3-yl)imidazolidin-2-one]}Cu(II) compound (C2) is more selective towards cancers cells, compared with the non-cancerous human normal skin fibroblasts (CCD-1059sK), than compounds C1, C3, and C4. In addition, they found that the treatment of HepG2 and T98G cells with the copper C2 compound at a concentration that did not inhibit cancer cell growth resulted in an increase in the cytotoxic effects of chemotherapeutics such as *etoposide, 5-fluorouracil*, and *temozolomide*. Furthermore, microbiological tests, in which the activity of L2-4 ligands and their Cu compounds (C2-4) were evaluated, showed that only L3 exhibited moderate anti-*Candida* activity, while antiradical tests revealed that only the copper C4 complex exhibited the strongest antioxidant potential. Further, the in vitro antitumor and antimicrobial activity was also examined by Dumitrescu's research team [contribution 6] with respect to decavanadate (DV)-bearing guanidine derivatives (DV-GDs). The isopolyoxovanadate decavanadate ($V_{10}$) and its derivatives are perhaps the most widely studied POV in biology, affecting key biochemical and cellular processes and showing several biomedical applications [4,5,7–12,14]. The authors showed that all investigated compounds, i.e., $(Hpbg)_4[H_2V_{10}O_{28}]\cdot 6H_2O$ (DV-GD1), $(Htbg)_4[H_2V_{10}O_{28}]\cdot 6H_2O$ (DV-GD2), $(Hgnd)_2(Hgnu)_4[V_{10}O_{28}]$ (DV-GD3), and $(Hgnu)_6[V_{10}O_{28}]\cdot 2H_2O$ (DV-GD4), inhibited the growth of some Gram-negative and Gram-positive bacteria, and DV-GD3 was the most active of these complexes. As far as the cytotoxicity assayed against A375 human melanoma cells and BJ human fibroblasts is concerned, the authors found that all compounds exhibited high cytotoxicity against both melanoma and fibroblast cells, but DV-GD1 was the most active in cancer cells.

Another study included in this SI presents the results of an experiment on monolacunary Wells–Dawson polyoxometalate (mono-WD POM), in which the authors evaluated the in vitro CT imaging properties, in vivo potential toxic effects, and tissue distribution of $\alpha_2$-$K_{10}P_2W_{17}O_{61}\cdot 20H_2O$ [contribution 7]. As outlined in the introduction, POMs have been described as compounds with emergent biomedical activities, as well as being well known to target several cellular key proteins and biological processes [7–10]. In this study, they found that the acute oral and intravenous administration of mono-WD POM did not induce either physical changes or mortality in rats. As suggested, the lack of mortality and physical alterations noted after the acute oral administration could be linked with the low absorption rate of mono-WD POM due to its large molecular weight. As emphasized, the low bioavailability of POM-based compounds and thus their low toxicity after *per os* treatment indicates that they could be considered as promising candidates for the development of a new oral CT contrast agent. In turn, a good survival rate was reached after intravenous administration, as stressed by the investigators, which indicates that mono-DW POM could also be developed as a new intravenous CT agent. However, the dose-dependent side effects confirmed by biochemical and histological analysis indicate that further studies are needed to develop safe CT contrast agents.

Another study refers to antibacterial activity of Schiff base Zn(II) complexes with three different inorganic ligands, i.e., chloride, nitrate, and acetate (G0[NCPh($o$-N)$ZnCl_2\cdot 2H_2O$], G0[NCPh($o$-N)$Zn(NO_3)_2\cdot 2H_2O$, and G0[NCPh($o$-N)$Zn(O_2CCH_3)_2$, respectively) [contribution 8]. The authors showed that these complexes exhibit moderate antibacterial activity against planktonic bacterial cells of *Staphylococcus aureus* and *Escherichia coli* strains and that this antibacterial effect resulted from the metal complexation to the Schiff base ligand. On the other hand, they found that the impact of the inorganic ligands was not significant for the antibacterial effect but was important for complex solubility. The obtained findings suggest that Zn(II) iminopyridine complexes could be promising candidates for antibacterial therapy.

Finally, the study on the UV-C seed surface sterilization and biofortification of wheat sprouts with iron (Fe), zinc (Zn), magnesium (Mg), and chromium (Cr) [contribution 9] showed that UV-C radiation is effective in preventing infections during seed germination and does not reduce either the growth or development of sprouts or nutrient bioassimilation. Thus, the results of this study clearly showed that irradiation by UV-C light can effectively stimulate plant growth.

## 3. Conclusions and Outlook

The studies described in this SI have provided valuable information about some pharmacologically interesting metals and metal-containing compounds/complexes that were examined for their potential therapeutic uses and as possible CT contrast agents for clinical application. The findings obtained from these studies may lay the groundwork for future research on new therapeutics based on metal-containing compounds/complexes that may ensure better treatment results and guarantee potential clinical success. Additionally, the results of the study focused on the UV-C seed surface sterilization and biofortification of wheat sprouts with certain elements as an effective strategy for bioelement supplementations revealed that irradiation by UV-C light could be used as an effective tool to obtain fortified food in a cheap and quick method.

We believe that the information provided in this SI will appeal to readers who are interested in metals and metal-containing compounds/complexes, as well as human health and nutrition in general, and to those interested in metal interactions and diseases such as cancer, AD, PD, and ALS, which constitute one of the single most important public health challenges.

**Author Contributions:** Conceptualization, A.Ś., M.A. and J.L.; writing—original draft preparation, A.Ś.; writing—review and editing, A.Ś., M.A. and J.L.; visualization, A.Ś.; supervision, A.Ś. and M.A.; project administration, A.Ś. and M.A.; funding acquisition, M.A. All authors have read and agreed to the published version of the manuscript.

**Funding:** This study received Portuguese national funds from the Foundation for Science and Technology (FCT) through projects UIDB/04326/2020, UIDP/04326/2020, and LA/P/0101/2020 (M.A.).

**Acknowledgments:** The authors would like to thank all the contributing authors and reviewers.

**Conflicts of Interest:** The authors declare no conflicts of interest.

**List of Contributions:**

1. Ścibior, A.; Llopis, J.; Dobrakowski, P.P.; Męcik-Kronenberg, T. CNS-Related Effects Caused by Vanadium at Realistic Exposure Levels in Humans: A Comprehensive Overview Supplemented with Selected Animal Studies. *Int. J. Mol. Sci.* **2023**, *24*, 9004. https://doi.org/10.3390/ijms24109004.
2. Rivas-García, L.; López-Varela, A.; Quiles, J.L.; Montes-Bayón, M.; Aranda, P.; Llopis, J., Sánchez-González, C. Elucidating the Therapeutic Potential of Bis(Maltolato)OxoVanadium(IV): The Protective Role of Copper in Cellular Metabolism. *Int. J. Mol. Sci.* **2023**, 24, 9367. https://doi.org/10.3390/ijms24119367.
3. Kshetri, M.; Jogadi, W.; Alqarni, S.; Datta, P.; Cheline, M.; Sharma, A.; Betters, T.; Deonya, B.; Zheng, Y.-R. Exploring the Impact of Head Group Modifications on the Anticancer Activities of Fatty Acid-Like Platinum(IV) Prodrugs: A Structure-Activity Relationship Study. *Int. J. Mol. Sci.* **2023**, *24*, 13301. https://doi.org/10.3390/ijms241713301.
4. Molinaro, C.; Wambang, N.; Pellegrini, S.; Henry, N.; Lensink, M.F.; Germain, E.; Bousquet, T.; De Ruyck, J.; Cailliau, K.; Pelinski, L.; Martoriati, A. Synthesis and Biological Activity of a New Indenoisoquinoline Copper Derivative as a Topoisomerase I Inhibitor. *Int. J. Mol. Sci.* **2023**, *24*, 14590. https://doi.org/10.3390/ijms241914590.

5. Balewski, Ł.; Plech, T.; Korona-Głowniak, I.; Hering, A.; Szczesio, M.; Olczak, A.; Bednarski, B.J.; Kokoszka, J.; Kornicka, A. Copper(II) Complexes with 1-(isoquinolin-3-yl)heteroalkyl-2-ones: Synthesis, Structure and Evaluation of Anticancer, Antimicrobial and Antioxidant Potential. *Int. J. Mol. Sci.* **2024**, *25*, 8. https://doi.org/10.3390/ijms25010008.
6. Dumitrescu, A.; Maxim, C.; Badea, M.; Rostas, A.M.; Ciorîtă, A.; Tirsoaga, A.; Olar, R. Decavanadate-Bearing Guanidine Derivatives Developed as Antimicrobial and Antitumor Species. *Int. J. Mol. Sci.* **2023**, *24*, 17137. https://doi.org/10.3390/ijms242417137.
7. Stojanović, M.; Čolović, M.B.; Lalatović, J.; Milosavljević, A.; Savić, N.D.; Declerck, K.; Radosavljević, B.; Ćetković, M.; Kravić-Stevović, T.; Parac-Vogt, T.N.; Krstić, D. Monolacunary Wells-Dawson Polyoxometalate as a Novel Contrast Agent for Computed Tomography: A Comprehensive Study on In Vivo Toxicity and Biodistribution. *Int. J. Mol. Sci.* **2024**, *25*, 2569. https://doi.org/10.3390/ijms25052569.
8. De la Mata Moratilla, S.; Casado Angulo, S.; Gómez-Casanova, N.; Copa-Patiño, J.L.; Heredero-Bermejo, I.; de la Mata, F.J.; García-Gallego, S. Zinc(II) Iminopyridine Complexes as Antibacterial Agents: A Structure-to-Activity Study. *Int. J. Mol. Sci.* **2024**, *25*, 4011. https://doi.org/10.3390/ijms25074011.
9. Czarnek, K.; Tatarczak-Michalewska, M.; Dreher, P.; Rajput, V.D.; Wójcik, G.; Gierut-Kot, A.; Szopa, A.; Blicharska, E. UV-C Seed Surface Sterilization and Fe, Zn, Mg, Cr Biofortification of Wheat Sprouts as an Effective Strategy of Bioelement Supplementation. *Int. J. Mol. Sci.* **2023**, *24*, 10367. https://doi.org/10.3390/ijms241210367.

## References


1. Ścibior, A.; Pietrzyk, Ł.; Plewa, Z.; Skiba, A. Vanadium: Risks and possible benefits in the light of a comprehensive overview of its pharmacotoxicological mechanisms and multi-applications with a summary of further research trends. *J. Trace Elem. Med. Biol.* **2020**, *61*, 126508. [CrossRef]
2. Ścibior, A. Overview of research on vanadium-quercetin complexes with a historical outline. *Antioxidants* **2022**, *11*, 790. [CrossRef] [PubMed]
3. Ścibior, A. Vanadium (V) and magnesium (Mg)—In vivo interactions: A review. *Chem. Biol. Int.* **2016**, *258*, 214–233. [CrossRef]
4. De Sousa-Coelho, A.L.; Fraqueza, G.; Aureliano, M. Repurposing therapeutic drugs complexed to vanadium in cancer. *Pharmaceuticals* **2024**, *17*, 12. [CrossRef]
5. De Sousa-Coelho, A.L.; Aureliano, M.; Fraqueza, G.; Serrão, G.; Gonçalves, J.; Sánchez-Lombardo, I.; Link, W.; Ferreira, B.I. Decavanadate and metformin-decavanadate effects in human melanoma cells. *J. Inorg. Biochem.* **2022**, *235*, 111915. [CrossRef]
6. Čolović, M.B.; Lacković, M.; Lalatović, J.; Mougharbel, A.S.; Kortz, U.; Krstić, D.Z. Polyoxometalates in Biomedicine: Update and Overview. *Curr. Med. Chem.* **2020**, *27*, 362–379. [CrossRef] [PubMed]
7. Sánchez-Lara, E.; Treviño, S.; Sánchez-Gaytán, B.L.; Sánchez-Mora, E.; Eugenia Castro, M.; Meléndez-Bustamante, F.J.; Méndez-Rojas, M.A.; González-Vergara, E. Decavanadate Salts of Cytosine and Metformin: A Combined Experimental-Theoretical Study of Potential Metallodrugs against Diabetes and Cancer. *Front. Chem.* **2018**, *6*, 402. [CrossRef] [PubMed]
8. Amante, C.; De Sousa-Coelho, A.L.; Aureliano, M. Vanadium and melanoma: A systematic review. *Metals* **2021**, *11*, 828. [CrossRef]
9. Aureliano, M.; Gumerova, N.I.; Sciortino, G.; Garribba, E.; Rompel, A.; Crans, D.C. Polyoxovanadates with emerging biomedical activities. *Coord. Chem. Rev.* **2021**, *447*, 214143. [CrossRef]
10. Aureliano, M.; Gumerova, N.I.; Sciortino, G.; Garribba, E.; McLauchlan, C.C.; Rompel, A.; Crans, D.C. Polyoxidovanadates' interactions with proteins: An overview. *Coord. Chem. Rev.* **2022**, *454*, 214344. [CrossRef]
11. Carvalho, F.; Aureliano, M. Polyoxometalates impact as anticancer agents. *Int. J. Mol. Sci.* **2023**, *24*, 5043. [CrossRef] [PubMed]
12. Aureliano, M.; De Sousa-Coelho, A.L.; Dolan, C.C.; Roess, D.A.; Crans, D.C. Biological Consequences of Vanadium Effects on Formation of Reactive Oxygen Species and Lipid Peroxidation. *Int. J. Mol. Sci.* **2023**, *24*, 5382. [CrossRef] [PubMed]
13. Ferretti, V.; León, I. An Overview of Vanadium and Cell Signaling in Potential Cancer Treatments. *Inorganics* **2022**, *10*, 47. [CrossRef]
14. Gonzalez-Cano, S.I.; Flores, G.; Guevara, J.; Morales-Medina, J.C.; Treviño, S.; Diaz, A. Polyoxidovanadates a new therapeutic alternative for neurodegenerative and aging diseases. *Neural Regen. Res.* **2024**, *19*, 571–577. [CrossRef] [PubMed]

International Journal of
*Molecular Sciences*

MDPI

*Review*

# CNS-Related Effects Caused by Vanadium at Realistic Exposure Levels in Humans: A Comprehensive Overview Supplemented with Selected Animal Studies

**Agnieszka Ścibior [1,*], Juan Llopis [2,3], Paweł Piotr Dobrakowski [4] and Tomasz Męcik-Kronenberg [5]**

1 Laboratory of Oxidative Stress, Department of Biomedicine and Environmental Research, Institute of Biological Sciences, Faculty of Medicine, The John Paul II Catholic University of Lublin, Konstantynów St. 1J, 20-708 Lublin, Poland
2 Department of Physiology, Institute of Nutrition and Food Technology "José Mataix", Biomedical Research Centre, University of Granada, 18100 Armilla, Spain; jllopis@ugr.es
3 Sport and Health Research Centre, University of Granada, 18016 Granada, Spain
4 Psychology Institute, Humanitas University in Sosnowiec, Jana Kilińskiego St. 43, 41-200 Sosnowiec, Poland; paweldobrakowski@interia.pl
5 Department of Pathomorphology, Faculty of Medical Sciences in Zabrze, Medical University of Silesia, 3 Maja St. 13, 41-800 Zabrze, Poland; patolog@interia.pl
* Correspondence: agnieszka.scibior@kul.pl; Tel.: +48-81-45-45-615

**Abstract:** Neurodegenerative disorders, which are currently incurable diseases of the nervous system, are a constantly growing social concern. They are progressive and lead to gradual degeneration and/or death of nerve cells, resulting in cognitive deterioration or impaired motor functions. New therapies that would ensure better treatment results and contribute to a significant slowdown in the progression of neurodegenerative syndromes are constantly being sought. Vanadium (V), which is an element with a wide range of impacts on the mammalian organism, is at the forefront among the different metals studied for their potential therapeutic use. On the other hand, it is a well-known environmental and occupational pollutant and can exert adverse effects on human health. As a strong pro-oxidant, it can generate oxidative stress involved in neurodegeneration. Although the detrimental effects of vanadium on the CNS are relatively well recognized, the role of this metal in the pathophysiology of various neurological disorders, at realistic exposure levels in humans, is not yet well characterized. Hence, the main goal of this review is to summarize data on the neurological side effects/neurobehavioral alterations in humans, in relation to vanadium exposure, with the focus on the levels of this metal in biological fluids/brain tissues of subjects with some neurodegenerative syndromes. Data collected in the present review indicate that vanadium cannot be excluded as a factor playing a pivotal role in the etiopathogenesis of neurodegenerative illnesses, and point to the need for additional extensive epidemiological studies that will provide more evidence supporting the relationship between vanadium exposure and neurodegeneration in humans. Simultaneously, the reviewed data, clearly showing the environmental impact of vanadium on health, suggest that more attention should be paid to chronic diseases related to vanadium and to the assessment of the dose–response relationship.

**Keywords:** vanadium; neurodegenerative diseases; neurological/neurobehavioral effects; environmental/occupational exposure; humans; body fluids; brain

## 1. Introduction

The present review is an attempt to provide thorough knowledge on the influence of vanadium (V) on neurodegeneration in humans. After the Introduction section, in which the main goals are formulated, the review comprises a few main sections and subsections displayed in Figure 1 for the reader's convenience. Section 2 provides information about the search strategy, i.e., it includes the sources and date of searching, the key terms used to identify records relevant to the topic, and two flowcharts of the literature review process. One of them refers to humans and the other one is focused on animals. Section 3 with a Polish accent provides a concise summary of basic information about certain neurogenerative diseases, i.e., Alzheimer disease (AD), Parkinson disease (PD), and Amyotrophic Lateral Sclerosis (ALS). Section 4, composed of three Sections 4.1–4.3 overviews, highlight selected issues on vanadium. The first subsection presents a brief background related to this metal, the second one summarizes data on the impact of some factors on the concentration of vanadium in tissues and body fluids, and the last one collects information about adverse health outcomes of environmental exposure to vanadium in humans with the exception of neurotoxic effects. Section 5 with eight Sections 5.1–5.8 compiles data on the content of vanadium in biological specimens from patients with AD, PD, and ALS, summarizes the neurotoxic effects of environmental exposure to this metal in humans, overviews findings from studies on the neurotoxic effects of vanadium after inhalation in animals (in both laboratory animals and those naturally exposed to urban air), and collects information about the neurological side effects and neurobehavioral changes in humans exposed to vanadium occupationally, subjects with acute vanadium poisoning, and the elderly. Some of these, Sections 5.1, 5.4 and 5.6, illustrate selected issues related to the topic on the timeline. Section 6 briefly summarizes new therapies of neurodegenerative illnesses and, finally, Section 7 provides a summary and conclusions and indicates further trends in research on neurodegenerative disorders.

**Figure 1.** Graphical summary of the overviewed issues.

One of the main goals of this review is to collect concise knowledge of the impact of vanadium on neurodegeneration in humans. It was important to provide objective information about a possible association between vanadium exposure and neurodegenerative disorders in human subjects. In other words, we tried to collect evidence on neurodegenerative alterations in relation to vanadium pollution in humans. Hence, a reliable analysis of the literature data has been carried out and selected literature findings from studies of the neurotoxic effects of this element in man have been overviewed and illustrated in an accessible form to anyone interested in neurodegenerative disorders and vanadium in general. Some results on the neurotoxic effects of this metal in animals have been presented, as well, to offer the reader a deeper insight into studies related to vanadium and neurotoxicity. Additionally, a brief historical framework has been provided to draw the reader's attention to the research on vanadium in the context of neurodegenerative processes in terms of time.

## 2. Methodology—Literature Search Strategy

### 2.1. Databases: General Outline

The literature search in English-language databases (i.e., PubMed, Scopus, and Web of Science) was conducted from November 2022 to January 2023 to collect relevant data. Only research articles written in English were reviewed and only abstracts in English from papers published in other languages were included. In the case of articles with an unavailable full text, correspondence contact was made, i.e., an email was sent to the corresponding author with a request for sending the full paper. In the absence of a reply, the information provided in the abstracts was included in the current report. To ensure comprehensive search of the literature, the reference lists of selected articles collected from the above-mentioned databases were manually reviewed to identify additional records (i.e., full-text papers or abstracts) that were potentially relevant to the topic.

### 2.2. Query Terms Used for the Literature Search on Neurotoxic Effects of Vanadium in Humans

The search was focused on the 'Title' and 'Abstract', and such keywords as 'vanadium or vanadate and Alzheimer disease', 'vanadium or vanadate and neurodegeneration', 'vanadium or vanadate and dementia', and 'vanadium or vanadate and neurodegenerative diseases' were used to obtain the records on the levels of vanadium in biological specimens, i.e., in the brain, cerebrospinal fluid (CSF), blood, urine, and hair of patients with neurodegenerative illness and on the content of this metal in the whole blood, serum, urine, and CSF of humans exposed to vanadium occupationally, subjects with acute vanadium poisoning and the elderly with neurological side effects and neurobehavioral changes. Additionally, such key terms as 'vanadium', 'particulate matter', 'Parkinson disease', 'Alzheimer disease', 'Amyotrophic lateral sclerosis', 'dementia', 'neurodegeneration', 'neurotoxicity', and 'humans' in various combinations were employed to obtain the data on the neurotoxic effects of vanadium-associated $PM_{2.5}/PM_{10}$ exposure in man.

### 2.3. Query Terms Used for the Literature Search on Neurotoxic Effects of Vanadium after Inhalation in Animals

The search terms, focused on the 'Title' and 'Abstract', such as 'vanadium', 'vanadate', 'pollution', 'exposure', 'accumulation', 'neurotoxicity', 'animals', 'inhalation studies', 'brain', 'mice', 'rat', 'animal model', 'environmental pollution', 'neurodegeneration', 'neuroinflammation', 'olfactory dysfunction' and those linked with 'AND', were used in the search strategy to obtain the records limited to neurotoxic effects of vanadium after inhalation in animals.

### 2.4. Search Results and Literature Review Flowchart on Neurotoxic Effects of Vanadium in Humans

The adopted strategy of searching by using specific keywords allowed the detection of records in PubMed (NCBI), Scopus, and WoS relevant to the topic of neurotoxic effects of vanadium in humans. The flow chart provided below (Figure 2) shows the process employed to identify records on neurotoxic effects of vanadium in humans.

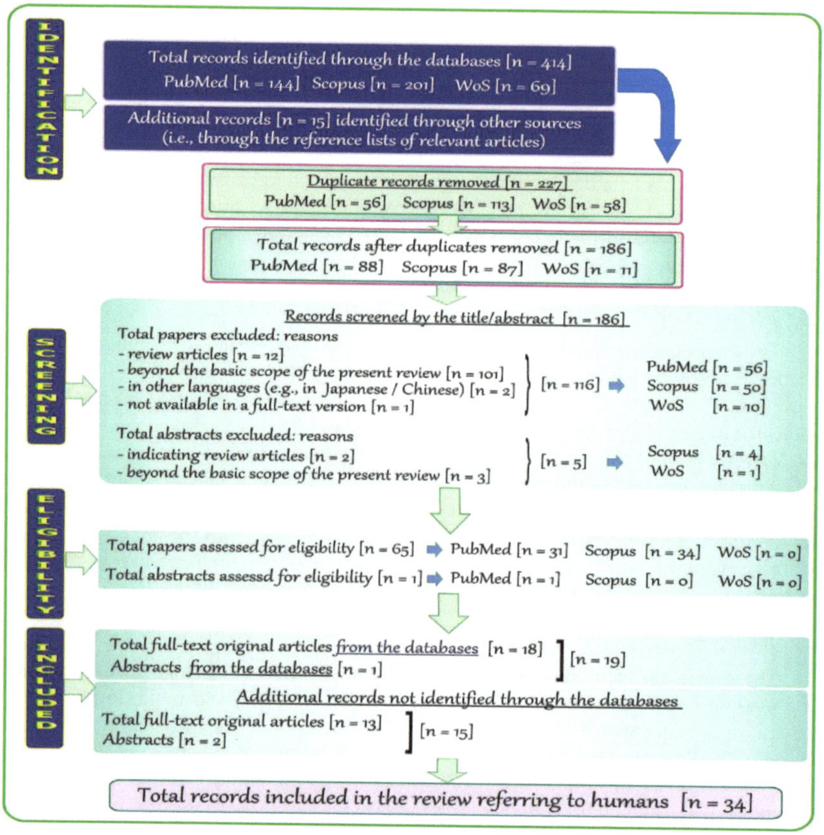


**Figure 2.** Flow chart of the systematic literature review on neurotoxic effects of vanadium in humans.

A total of 414 records published in English and other languages were identified through the databases listed above; i.e., 144 through PubMed, 201 through Scopus, and 69 through WoS. After removal of 227 duplicate items (PubMed: 56, Scopus: 113, and WoS: 58), the remaining records (n = 186) were initially screened by the titles and abstracts. Afterwards, 121 records published in, e.g., Japanese or Chinese and papers unavailable in a full-text version, as well as review articles and those that did not address the topic were excluded (PubMed: 56, Scopus: 50, and WoS: 10). Moreover, a total of 5 abstracts (Scopus: 4 and WoS: 1) that pointed to review papers, and were far beyond the scope of the present report, were excluded as well. Next, a total of 65 potentially relevant full-text articles (PubMed: 31 and Scopus: 34) and 1 abstract (PubMed) were further examined. Finally, a total of 18 full-text original papers and 1 abstract (n = 19) were included in the current review. Moreover, 15 additional records (i.e., 13 full-text original articles and 2 abstracts) relevant to the topic were included by a manual review of bibliographies.

### 2.5. Search Results and Literature Review Flowchart on Neurotoxic Effects of Vanadium after Inhalation in Animals

The adopted strategy of searching by using specific key terms also allowed detection of records in PubMed (NCBI), Scopus, and WoS relevant to the topic of neurotoxic effects of vanadium after inhalation in animals. The flow chart (Figure 3) shows the steps of the methodology described in this subsection.

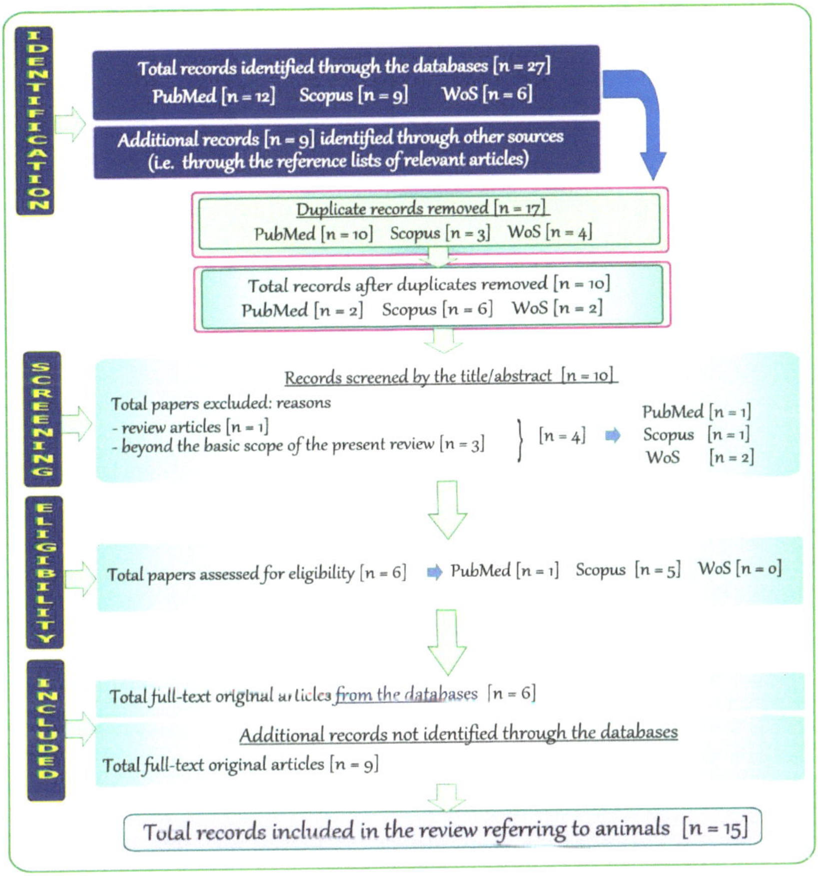


**Figure 3.** Flow chart of the systematic literature review on neurotoxic effects of vanadium after inhalation in animals.

At the beginning, a total of 27 records published in English were revealed through the databases, i.e., 12 through PubMed, 9 through Scopus, and 6 through WoS. Next, a total of 17 duplicate records (PubMed: 10, Scopus: 3, and WoS: 4) were removed and the remaining items (n = 10; PubMed: 2, Scopus: 6, and WoS: 2) were initially screened by the titles and abstracts. Afterwards, records that were far beyond the topic addressed in the present review were excluded (n = 4; PubMed: 1, Scopus: 1, and WoS: 2), and six potentially relevant items (i.e., only full-text original articles) were further examined. Finally, a total of six full-text original papers in English and nine additional records (i.e., the relevant full-text original articles) were included (n = 15).

## 3. Neurodegenerative Disorders—A Brief Outline with a Polish Accent

Neurodegenerative disorders are currently incurable diseases of the nervous system, which are a constantly growing social concern. They are progressive and lead to gradual degeneration and/or death of nerve cells, which in turn results in cognitive deterioration or impaired motor functions. Due to the lack of satisfactory methods of treatment, neurodegenerative diseases have been arousing interest of many research centers. New therapies that would ensure better treatment results and contribute to a significant slowdown in the progression of the disease are constantly being sought.

### 3.1. Alzheimer Disease (AD)

Alzheimer's disease (AD) is the most common form of dementia, accounting for 50 to 75% of cases. In Poland, about 500 thousand people suffer from various dementia syndromes, including over half (about 300 thousand) that have been diagnosed with AD. In 30 years this number may triple. The incidence of AD in the age group 65–69 years is 2–3/1000 people/year, but in the range of 85–89 years it is already 37–40/1000 people/year [1].

The etiology of AD is multifactorial and still not fully understood. Only a few forms have a genetic basis (the polymorphism of the gene encoding apolipoprotein E—APOE or mutations in the amyloid precursor protein or presenilin 1 and 2 genes). Most cases are probably related to the deposition of pathological proteins: β-amyloid protein and hyperphosphorylated tau protein. In addition, vascular changes often coexist, which directly damage neurons and stimulate Alzheimer's degeneration processes [2]. New research draws attention to other mechanisms. Inflammatory processes caused by active microglia, damage to cholinergic neurons, oxidative stress (OS) or glucose metabolism disorders observed in the course of dementia require treating as a systemic disease and paying more attention to the correlation between the brain and other organs [3,4].

So far, approved drugs, including cholinesterase inhibitors, N-methyl-D-aspartate (NMDA) receptor antagonist or their combination, usually provide temporary and incomplete symptomatic relief that may be accompanied by serious side effects. Only in 2021 did a new class of substances join, monoclonal antibodies (aducanumab and two years later lecanumab). Both substances have been approved by the Food and Drug Administration (FDA). Due to controversies regarding their effectiveness and side effects, the European Medicines Agency (EMA) has not yet approved these drugs. However, these substances also only slow down the development of the disease. In addition, their effectiveness is greatest when they are included at the beginning of the disease. Until recently, amyloid and tau protein were the main targets for most drugs in development for AD. It is now believed that it is unlikely that antibodies against amyloid alone will be sufficient to halt or reverse the course of the disease [2].

### 3.2. Parkinson Disease (PD)

Parkinson's disease (PD) is the second most common neurodegenerative illness. In Poland, about 80,000 people suffer from PD at various stages of advancement. It is estimated that among people over 65 years of age, 1 person in 100 suffers from the disease, while in the population over 85 years of age, 5 out of 100 [5].

Most patients are treated with orally administered levodopa. In addition, carbidopa preparations, dopamine agonists, catechol-O-methyltransferase inhibitors (COMT), monoamine oxidase B inhibitors (MAO-B), anticholinergic drugs and amantadine are used. Although these drugs can bring significant relief in some motor symptoms of PD, additional therapies were needed to solve other serious obstacles regarding quality of life as PD progresses. In terms of new strategies that would allow for better treatment outcomes and significantly slow down the progression of the disease, there are various approaches and therapies, such as gene therapy, cell therapy or recombinant protein therapy. However, most of them are still in the clinical trial phase and are not yet available to patients [6].

### 3.3. Amyotrophic Lateral Sclerosis (ALS)

Amyotrophic lateral sclerosis (ALS) is a rare disease, but it is the most common motor neuron disease among adults [7]. As a result of neurodegenerative processes, there is damage to the corticospinal and corticobulbar tracts, as well as to the motor nuclei of the cranial nerves and the anterior horns of the spinal cord.

The incidence of ALS is small with 1–2 cases per 100,000 people. The number of new cases per year is 4–5 per 100,000 people. In the Polish population, the number of patients can therefore be estimated at around two–three thousand [8].

There are currently two registered drugs (by FDA) for ALS, i.e., riluzole and edaravone. Riluzole slightly prolongs survival time, while edaravone slows down the progression of the disease to a small extent. However, owing to a better understanding of the molecular mechanisms of the disease in recent years, new potential targets for new drugs have been discovered. Details in this field are provided in another section of the present review.

## 4. Vanadium—Selected Issues in a Nutshell

### 4.1. Background

Vanadium (symbol: V) is part of Group 5 (5B) of the periodic table. It is a transition metal existing in a wide range of valence states, but the most common are +2, +3, +4, and +5. As a vanadate anion, i.e., $VO_3-$ (pentavalent form), it predominates extracellularly and as a vanadyl cation, i.e., $VO^{2+}$ (tetravalent form), it occurs mainly intracellularly [9,10].

Vanadium is widely dispersed in the environment. It occurs in the air, soil, and water in variable concentrations and can be found naturally in about 65 different minerals [11]. It plays a large role in the industrial sector and exhibits multidirectional biological effects. This metal is also essential to some species in trace amounts, but can be toxic in excess. Moreover, in some conditions, vanadium can exert antioxidant properties, as evidenced by animal studies [12–15]. More details on the issues mentioned above have previously been provided in two review papers [16,17].

It should be emphasized that vanadium is at the forefront among the different elements studied for their potential therapeutic use, due to its anti-viral, anti-bacterial, anti-parasitic, anti-fungal, anti-allergic, anti-cancer, anti-diabetic, anti-hypercholesterolemic, anti-inflammatory, anti-ulcer, cardio-protective, nephro-protective, and neuro-protective activity [18]. On the other hand, it can accumulate and act as a strong pro-oxidant [19], which limits the use of this metal in the treatment of some modern-age diseases in humans. Therefore, minimization of the adverse effects and pro-oxidative activity of vanadium may contribute to the development of effective therapeutic approaches in which this element could be used. It should also be highlighted that OS, i.e., a deleterious process generated by vanadium, has been reported to be implicated in neurogenerative mechanisms in the pathogenesis of some neurodegenerative disorders [20] that have been described in one of the previous sections of the current report. A brief summary of the background data on vanadium are graphically presented in Figure 4.

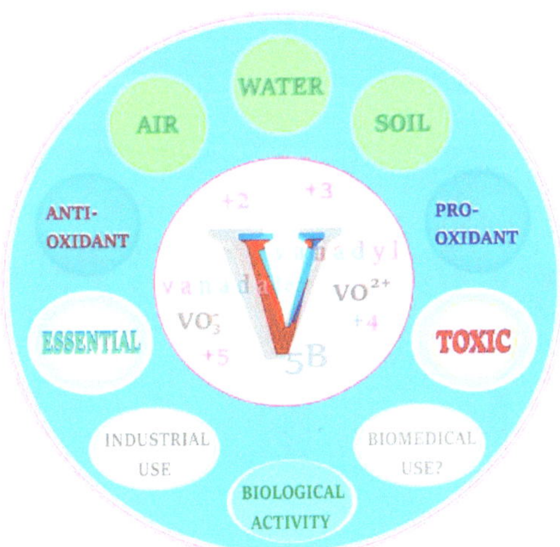


**Figure 4.** Summary of basic information about vanadium (V). $VO_3-$: vanadate anion, $VO^{2+}$: vanadyl cation.

### 4.2. The Impact of Certain Factors on the Concentration of Vanadium in Tissues and Body Fluids

The level of vanadium in body fluids and tissues is influenced by several factors, such as age, diet, health status, or occupational exposure [11,21,22]. As for the latter, elevated concentrations of this metal with a well-known industrial value [23] have repeatedly been noted in body fluids of vanadium-exposed workers [24]. The environmental vanadium sources and geographical location are important as well [25–27], and more details in this field are provided in further section of the present review. External contamination can also play a pivotal role. For example, Minoia et al. [28] observed the release of vanadium, commonly used for steel production [23], from stainless steel needles and syringes during blood collection. Moreover, the presence of vanadium in certain biomedical materials (e.g., in implants), reviewed previously [16], from which vanadium can be released into surrounding tissues may be another factor affecting the level of this metal in the body. Recently, Łanocha-Arendarczyk and co-workers [29] demonstrated a higher concentration of vanadium in spongy bone of the femoral head of patients with osteoarthritis inhabiting urban areas of Western Pomeranian (Poland), who had prosthetic implants and were occupationally exposed to various chemical substances, including heavy metals. Patients with implants also had a higher concentration of vanadium in cartilage [29]. As for the health status, Sampath et al. [30] have recently supported the hypothesis that an imbalance of vanadium (the concentration of which was significantly reduced in the serum following treatment) is a factor in the etiology of one of mental illnesses, i.e., bipolar disorder. Furthermore, dietary supplements frequently used by athletes and body builders to improve performance [31], in which vanadium may be present in the form of, e.g., vanadyl sulfate [11], are equally important. As reported, the prolonged consumption of such supplements, with daily doses up to 18 mg V/day/person, may pose a risk of adverse health effects [31]. Additionally, smoking should be taken into account as one of the determinants of the concentration of vanadium in the organism [32], as vanadium is contained in cigarettes at the mean level of 1.83 μg/g and approximately 30% of this metal is released in the smoke [33]. Therefore, both active and passive smokers may be exposed to higher levels of this element. However, the analysis of the available literature does not reveal any data on a rise in the concentration of vanadium in body fluids resulting from smoking. As reported in a study carried out by Toro-Román and co-workers [34], physical training is another factor that can influence the level of vanadium in the organism. The authors demonstrated that regular physical training, especially aerobic sport modalities, causes an increase in the concentration of this metal in the serum. All the above-mentioned factors are graphically summarized in Figure 5.

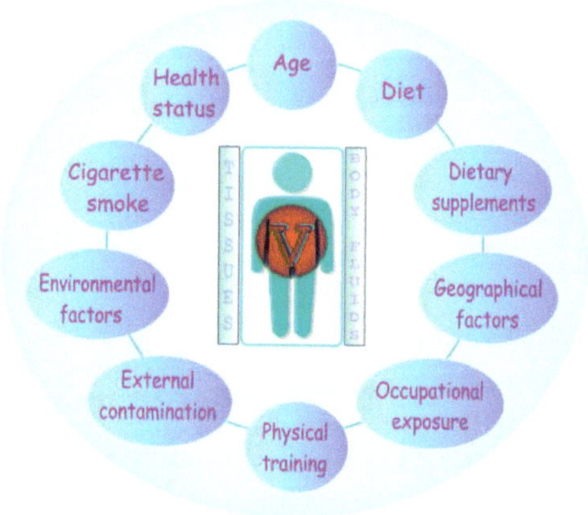


**Figure 5.** Factors influencing the content of vanadium in tissues and body fluids. Elaborated on the basis of available literature data cited in Section 4.2. V: vanadium.

*4.3. Environmental Vanadium Exposure in Humans—Adverse Health Outcomes Excluding Neurotoxic Effects*

The present section is focused on the environmental exposure of humans to vanadium, and offers the reader a concise summary of the potential harmful effects related to both non-anthropogenic and anthropogenic contamination of the environment with this metal (Figure 6). Simultaneously, it refers to the further parts of the current review, in which neurotoxic effects resulting from, e.g., the environmental exposure to vanadium in humans and animals are discussed and where selected issues related to vanadium and neurodegenerative disorders described in the previous sections of the current review are illustrated on the timeline for the reader's convenience.

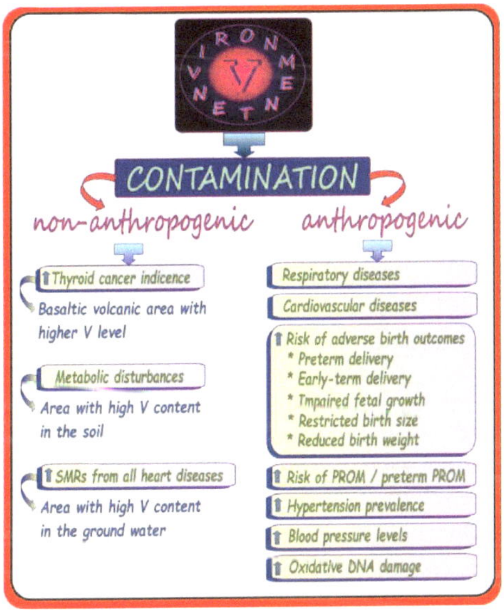


**Figure 6.** Summary of adverse health outcomes of environmental vanadium exposure in humans excluding neurotoxic effects. Elaborated on the basis of available literature data [25–27,35–45]. SMRs: standardized mortality rates, PROM: premature rupture of membranes, V: vanadium. ↑: increase.

As presented in Figure 6, not only anthropogenic human activity resulting in the release of vanadium into the environment but also non-anthropogenic vanadium sources may lead to serious health consequences. As far as the non-anthropogenic contamination with vanadium is concerned, Malandrino et al. [27] determined the concentrations of certain metals (also vanadium) in the tap water and lichens of the Mt. Etna volcanic area (Catania province), where an elevated thyroid cancer incidence has been reported to characterize the environmental pollution of that region, examined the urine of residents inhabiting the area to evaluate the level of biocontamination. The analyses demonstrated an eight-fold higher urinary vanadium concentration in the residents from the volcanic part of the country (0.16 μg/g creatinine) than in those living in the control area: 0.02 μg/g creatinine (i.e., in Palermo and Messina provinces). In turn, the concentration of vanadium in the tap water (19.9 μg/L) was almost 40 times higher than that found in the water from the control region (0.5 μg/L) [27] and often exceeded the maximum admissible concentration [42]. A significant correlation between the urinary vanadium concentration and the water vanadium level was also observed. These data suggest a possible relationship between the pollutants, including vanadium present in the volcanic regions and the increase in the thyroid cancer risk. A higher level of vanadium was also noted in the urine and some internal organs, e.g., kidneys (1.7-fold) and lungs (1.9-fold), in villagers inhabiting northeastern Thailand (where some metabolic problems such as, e.g., distal renal tubular acidosis occur), compared to residents in the central area. Additionally, examinations of the level of vanadium in the soil showed that the surface soil in the northeastern areas exhibited a three-fold higher vanadium concentration than the central part of the country [44]. As suggested by the authors, the high vanadium intake following the prolonged consumption of natural agricultural products and the iron deficiency noted in northeastern Thailand may have led to the elevated tissue and urine vanadium contents. At this point, the findings of the interaction of dietary iron levels on the toxicity of vanadium in chicks conducted by Blalock and Hill [46] are worth mentioning. The authors reported a rise in the degree of toxicity in iron-deficient animals accompanied by an increase in vanadium. More precisely, they showed that the iron-deficient animals retained more vanadium in the blood and liver, but the absorption of this metal was not influenced by the concentration of iron in the diet. In the case of villagers from northeastern Thailand, the concentration of vanadium in the liver was about 1.9-fold higher than in dwellers from the central part of the country [44]. Based on these data, it may be supposed that the accumulation of vanadium in tissues accompanied by iron deficit may have led to the development of renal acidosis in the villagers inhabiting northeastern Thailand. In turn, the work of Seko et al. [43] revealed the highest standardized mortality rates (SMRs) from all heart diseases in one of the eight secondary medical districts in Yamanashi prefecture (i.e., in Fuji-hokuroku, a high-vanadium area) in Japan, where the groundwater used as a source of tap water by habitants is relatively rich in vanadium (its concentration is about 50 μg/L around the mountain). These data allow suggesting an adverse effect of the vanadium-rich groundwater from Mt. Fuji on the health of habitants living around the mountain in that region.

As for the relationship between adverse health outcomes and vanadium contained in particulate matter with a diameter of 2.5 ($PM_{2.5}$), Cakmak et al. [37] reported an association between vanadium in $PM_{2.5}$ and an increased risk of respiratory and cardiovascular diseases. Dai et al. [47], who analyzed the differential effects of $PM_{2.5}$ species and sources on blood markers of inflammation and endothelial dysfunction in a large longitudinal cohort, found that among the species examined (K, S, Se, Al, Si, Fe, Ni, V, Cu, Zn, and Na) vanadium was associated with increased levels of both intercellular adhesion molecule-1 (ICAM-1) and vascular cell adhesion molecule-1 (VCAM-1). Additionally, they noted an association between oil combustion rich in vanadium and ICAM-1 and VCAM-1. Thus, these findings clearly suggest that particles containing vanadium derived from oil combustion may be related to endothelial dysfunction. Moreover, they are important in the context of the studies conducted by Blankenberg et al. [48], who showed that higher levels of ICAM-1 and VCAM-1 are related to future death from cardiovascular causes, and the studies conducted by Pradhan et al. [49], who noted that elevated levels of ICAM-1 are independently associated with the development of accelerated atherosclerosis among otherwise healthy men, even in the absence of acute coronary occlusion. Another study conducted by the Columbia Center for Children's Environmental Health provided evidence that the exposure to $PM_{2.5}$ associated vanadium and elemental carbon from heating oil and/or traffic at the levels characteristic of urban areas may be linked to asthma morbidity in very young urban children, i.e., at the age of two [41]. Another epidemiological study carried out in the United States, in which the relative risk of hospitalization due to cardiovascular and respiratory diseases in patients at the age of 65 or older was estimated, provided evidence that communities with a higher vanadium content in $PM_{2.5}$ have a higher risk of hospitalizations associated with $PM_{2.5}$ exposure, and the relationship between PM and health varies seasonally and regionally, likewise with the $PM_{2.5}$ chemical composition [36]. Moreover, Chen et al. [50], who examined the association between long-term exposure to $PM_{2.5}$ elemental components and mortality in a large pooled European cohort, found that the long-term exposure, especially to vanadium in $PM_{2.5}$, was associated with an increased mortality risk. Wang et al. [51] also observed an increased risk of mortality upon long-term exposure to $PM_{2.5}$ with higher vanadium content among older adults in the southeastern US. In turn, in a study including 11 European cohorts, Wolf et al. [52] found a positive association between long-term exposure to vanadium-containing PM and the incidence of coronary events. A positive association between long-term exposure to vanadium-containing $PM_{2.5}$ and the lung cancer incidence, as well as a negative association between vanadium-containing $PM_{2.5}$ and lung function, were reported as well [53,54]. As suggested by Alfaro-Moreno et al. [55], "*changes in the concentration of PM components help explain the changes in biological effects and support the hypothesis that particle composition plays an important role in particle-induced toxicity*". Studies conducted by these authors aimed to determine the relationships between specific components present in $PM_{10}$ and cell viability, cytokine secretion, DNA damage, and E-selectin expression to test the hypothesis that the composition plays an important role in PM effects. They identified a strong negative association of vanadium with cellular viability and a strong positive association of this metal with cytokine production, i.e., interleukin 6 (IL-6) and tumor necrosis factor (TNF-$\alpha$) as well as a discrete positive association with the comet longitude. The study conducted by Campen et al. [56], who investigated the effects of inhaled PM-associated transition metals (V and Ni), also supported the hypothesis of the adverse health effects exerted by metals contained in PM air pollution and provided insight into potential interactions regarding the toxicity of PM-bound elements. Interesting findings were obtained by Sørensen et al. [45] who examined the relationship between the personal exposure to some transition metals such as V, Cr, Fe, Ni, Cu, and Pt in $PM_{2.5}$ and oxidative DNA damage in human volunteers (i.e., in 49 students from central Copenhagen). The authors indicated that, among the six examined metals mentioned above, only vanadium and Cr were associated with the 7-hydro-8-oxo-2′-deoxyguanosine (8-oxodG) level in lymphocytes independently of the $PM_{2.5}$ mass. Thus, these results clearly suggest that vanadium and Cr contained in $PM_{2.5}$ can induce oxidative DNA damage. As stressed by the authors, the effect was independent of particle mass and/or other toxic compounds present in the particulate mixture. In turn, Basu et al. [35], who studied the influence of fine particulate matter and its constituents on low birth weight among full-term infants in California, demonstrated that the exposure to specific constituents of $PM_{2.5}$, especially such metals as vanadium, is associated with the largest reductions in birth weight. Moreover, a study conducted by Jin et al. [40] among 7290 pregnant women from China focused on evaluation of the relationship between urinary metal concentrations (including vanadium) and the risk of the premature rupture of membranes (PROM), revealing a positive association of the urinary vanadium level with PROM and preterm PROM. These results suggest that pregnant women who are exposed to higher levels of vanadium (in urban areas of China) have an increased risk of PROM and preterm PROM. Additionally, Jiang et al. [38] conducted a nested case–control study in the province of Hubei (China) to investigate the relationship between maternal urinary vanadium concentrations and the odds of delivering low birthweight (LBW) infants. The study revealed that higher levels of maternal urinary vanadium were associated with increased odds of delivering LBW infants. The authors pointed to the need for further studies on the urinary vanadium concentration during pregnancy to monitor changes in the level of vanadium during pregnancy and to assess the critical exposure-time window for the effect of vanadium exposure in fetal development. Further, the work conducted by Hu et al. [25] to estimate the association of prenatal exposure to vanadium with the risk of adverse birth outcomes in babies born to women in China, demonstrated a relationship between the prenatal exposure to vanadium and the increased risk of preterm delivery and early term delivery. Another study conducted by Hu et al. [26] to assess trimester-specific associations of vanadium exposure with ultrasound measures of fetal growth and birth size in a Chinese longitudinal cohort, indicated differences in the associations between prenatal exposure to vanadium and fetal growth depending on the time of exposure, and suggested that the first, early second, and late third trimester can be critical windows of vanadium exposure for fetal growth. Finally, Jiang et al. [39], who studied the association of vanadium exposure with hypertension prevalence and blood pressure levels in a general Chinese population, found that urinary vanadium concentrations were significantly associated with increased hypertension prevalence and blood pressure levels. In view of these results, the authors point to the need to pay more attention to vanadium, which is well known to enter the environment from various types of natural and anthropogenic sources. As vanadium has a role in contamination-related health risks, the necessity of monitoring areas with elevated vanadium concentrations has been repeatedly emphasized. Some researchers also propose the use of certain organisms as natural bioindicators of environmental vanadium pollution, and potential human health problems associated with vanadium exposure may act as an early warning system for hazards. This issue has been concisely summarized previously [17]. To sum up, the data described above clearly indicate that the geographical location and elevated amounts of vanadium in the environmental media (i.e., the air, water, or soil) may have a serious impact on public health.

## 5. Vanadium Neurotoxicity—Overview of Selected Findings: Humans and Animals

### *5.1. Vanadium in Biological Samples of Humans with AD, PD, and ALS—A Summary of Data on the Timeline*

A summary of studies on the levels of vanadium in biological specimens from humans with such neurodegenerative diseases as AD, PD, and ALS is presented in Figure 7. As illustrated, brain (BR), cerebrospinal fluid (CSF), blood (BL), urine (U), and hair (H) were tested to determine the vanadium content. Among the above-mentioned biological samples, BL was used most often for determination of the concentration of vanadium in patients with neurodegenerative illnesses, BR and CSF were used at a comparable frequency, and U and H were the least often tested in terms of accumulation of vanadium in subjects suffering from neurodegenerative disorders.

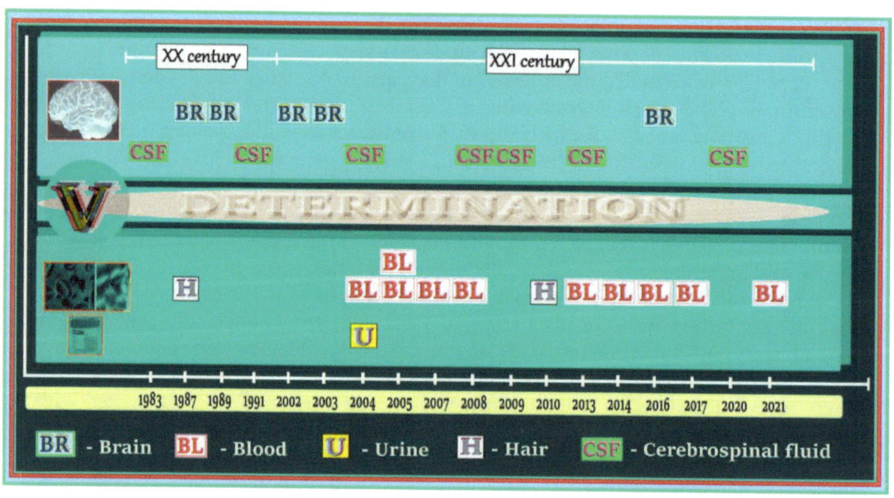


**Figure 7.** Summary of studies on the levels of vanadium in biological specimens in humans with certain neurodegenerative disorders on the timeline. Elaborated on the basis of data cited in Sections 5.2 and 5.3.

The first reference on the detection of vanadium in biological samples (i.e., CSF) obtained from patients with one of the neurodegenerative diseases, i.e., AD, appeared at the end of the 20th century, in 1983. Generally, in the 1980s and 1990s, there were not many investigations based on measurements of vanadium in biological specimens from subjects suffering from neurodegenerative illnesses. More papers on vanadium accumulation in this research area have been published since the beginning of the 21st century, which may reflect the increasing interest in the effects of vanadium on the development of neurodegenerative processes. To date, a total of 21 articles on the level of vanadium in AD, PD, or ALS patients have been issued. Except for five papers published in the 20th century, i.e., in 1983 (CSF), 1987 (BR), 1987 (H), 1989 (BR), and 1991 (CSF), sixteen articles were issued in the 21st century: in 2002 (BR), 2003 (BR), 2004 (CSF/BL/U), 2005 (BL), 2005 (BL), 2007 (BL), 2008 (CSF/BL), 2009 (CSF), 2010 (H), 2013 (CSF/BL), 2014 (BL), 2016 (BR), 2016 (BL), 2017 (BL), 2020 (CSF), and 2021 (BL).

### 5.2. Vanadium Content in Brain and Cerebrospinal Fluid of Patients with AD, PD, and ALS

Summarized values of the concentrations of vanadium in the different parts of the brain and in CSF are listed in Table 1.

**Table 1.** Summary of the results of the levels of vanadium in brain and cerebrospinal fluid of patients with neurodegenerative diseases.

| Patients | Number | Age (Mean/*Median*/*Ranges*) | Concentration (Mean/*Median*) * | Biological Fluids/ Tissues | Unit | Ref. |
|---|---|---|---|---|---|---|
| AD cases | | | | | | |
| Control<br>AD | n = 10<br>n = 14 | 88<br>78 | 3.0<br>5.0 | FC<br>FC | ng/g<br>ng/g | [57] |
| Control<br>AD | n = 10<br>n = 14 | 88<br>78 | <LOQ<br><LOQ | VF<br>VF | ng/mL<br>ng/mL | [57] |
| Control<br>AD | n = 54<br>n = 173 | *73 (60–94)*<br>*75 (52–86)* | *3.2*<br>*2.9* ↓ | CSF<br>CSF | µg/L<br>µg/L | [58] |
| Control<br>AD | n = 36<br>n = 33 | 38.5/70.4<br>~76 | ND<br>ND | CSF<br>CSF | n/a<br>n/a | [59] |
| Control †<br>AD †<br>Control ‡<br>AD ‡ | n = 30<br>n = 28<br>n = 30<br>n = 30 | 61 (52–69)<br>63.5 (57–72)<br>~63 (55–72)<br>64.5 (61–75) | 0.0352<br>0.0978<br>0.0274<br>0.0162 ↓ | Hippocampus<br>Hippocampus<br>Hippocampus<br>Hippocampus | µg/g dw<br>µg/g dw<br>µg/g dw<br>µg/g dw | [60] |
| Control †<br>AD †<br>Control ‡<br>AD ‡<br>Control †,‡<br>AD †,‡ | n = 30<br>n = 28<br>n = 30<br>n = 30<br>n = 60<br>n = 58 | <br>61 (52–69)<br>63.5 (57–72)<br>~63 (55–72)<br>64.5 (61–75) | 0.0308<br>0.0141 ↓<br>0.0272<br>0.0133 ↓<br>(0.0117–0.0506)<br>(0.0086–0.0209) | CC<br>CC<br>CC<br>CC<br>Hippocampus and CC<br>Hippocampus and CC | µg/g dw<br>µg/g dw<br>µg/g dw<br>µg/g dw<br>µg/g dw<br>µg/g dw | |
| Control<br>AD | n = 54<br>n = 174 | 73<br>74 | 3.2 (2.3–5.1)<br>*2.9 (<2.2–5.2)* ↓ | CSF<br>CSF | µg/L<br>µg/L | [61] |
| Control<br>AD | n = 4 (nos = 8)<br>n = 4 (nos = 8) | age-matched<br>78–88 | n/a [f]<br>n/a [f] | C and PC<br>C and PC | ppm<br>ppm | [62] |
| ALS cases | | | | | | |
| Control<br>ALS | n = 10<br>n = 17 | 45 *(26–77)*<br>64 *(47–85)* | *0.09 (0.03–0.19)*<br>*0.14* ↑ *(0.06–0.77)* | CSF<br>CSF | µg/L<br>µg/L | [63] |
| Control<br>ALS | n = 5<br>n = 8 | ~*56 (43–66)*<br>~*53 (34–66)* | 0.20<br>0.14 | GM<br>GM | µg/g dw<br>µg/g dw | [64] |
| Control<br>ALS | n = 5<br>n = 8 | ~56 (43–66)<br>~53 (34–66) | 0.14<br>0.08 | WM<br>WM | µg/g dw<br>µg/g dw | [64] |
| Control<br>ALS | nos = 3–8 | -<br>- | 0.16<br>0.15 | FL (GM)<br>FL (GM) | µg/g dw<br>µg/g dw | [64] |
| Control<br>ALS | nos = 3–8 | -<br>- | 0.23<br>0.13 | PL (GM)<br>PL (GM) | µg/g dw<br>µg/g dw | [64] |
| Control<br>ALS | nos = 3–8 | -<br>- | 0.21<br>0.09 | OL (GM)<br>OL (GM) | µg/g dw<br>µg/g dw | [64] |
| Control<br>ALS | nos = 3–8 | -<br>- | 0.10<br>0.18 | TL (GM)<br>TL (GM) | µg/g dw<br>µg/g dw | [64] |

**Table 1.** *Cont.*

| Patients | Number | Age (Mean/*Median*/*Ranges*) | Concentration (Mean/*Median*) * | Biological Fluids/ Tissues | Unit | Ref. |
|---|---|---|---|---|---|---|
| Control<br>ALS | nos = 3–8 | -<br>- | 0.21<br>0.15 | C (GM)<br>C (GM) | µg/g dw<br>µg/g dw | [64] |
| Control<br>ALS | nos = 3–8 | -<br>- | 0.26<br>0.16 | BG (GM)<br>BG (GM) | µg/g dw<br>µg/g dw | [64] |
| Control<br>ALS | nos = 3–8 | -<br>- | 0.12<br>0.09 | FL (WM)<br>FL (WM) | µg/g dw<br>µg/g dw | [64] |
| Control<br>ALS | nos = 3–8 | -<br>- | 0.06<br>0.07 | PL (WM)<br>PL (WM) | µg/g dw<br>µg/g dw | [64] |
| Control<br>ALS | nos = 3–8 | -<br>- | 0.22<br>0.08 | OL (WM)<br>OL (WM) | µg/g dw<br>µg/g dw | [64] |
| Control<br>ALS | nos = 3–8 | -<br>- | 0.19<br>0.08 | TL (WM)<br>TL (WM) | µg/g dw<br>µg/g dw | [64] |
| Control<br>ALS | nos = 3–8 | -<br>- | 0.20<br>0.09 | C (WM)<br>C (WM) | µg/g dw<br>µg/g dw | [64] |
| ALS **<br>ALS ## | n = 31<br>n = 6 | *65 (59–70)*<br>*70 (64–82)* | *0.30*<br>*0.35* | CSF<br>CSF | µg/L<br>µg/L | [65] |
| PD cases | | | | | | |
| Control<br>PDC | n = 5<br>n = 4 | ~56 (43–66)<br>~58 (51–65) | 0.20<br>0.21 | GM<br>GM | µg/g dw<br>µg/g dw | [64] |
| Control<br>PDC | n = 5<br>n = 4 | ~56 (43–66)<br>~58 (51–65) | 0.14<br>0.10 | WM<br>WM | µg/g dw<br>µg/g dw | [64] |
| Control<br>PDC | nos = 3–8 | -<br>- | 0.16<br>0.04 | FL (GM)<br>FL (GM) | µg/g dw<br>µg/g dw | [64] |
| Control<br>PDC | nos = 3–8 | -<br>- | 0.23<br>0.33 | PL (GM)<br>PL (GM) | µg/g dw<br>µg/g dw | [64] |
| Control<br>PDC | nos = 3–8 | -<br>- | 0.21<br>0.14 | OL (GM)<br>OL (GM) | µg/g dw<br>µg/g dw | [64] |
| Control<br>PDC | nos = 3–8 | -<br>- | 0.10<br>0.25 | TL (GM)<br>TL (GM) | µg/g dw<br>µg/g dw | [64] |
| Control<br>PDC | nos = 3–8 | -<br>- | 0.21<br>0.15 | C (GM)<br>C (GM) | µg/g dw<br>µg/g dw | [64] |
| Control<br>PDC | nos = 3–8 | -<br>- | 0.26<br>0.16 | BG (GM)<br>BG (GM) | µg/g dw<br>µg/g dw | [64] |
| Control<br>PDC | nos = 3–8 | -<br>- | 0.12<br>0.08 | FL (WM)<br>FL (WM) | µg/g dw<br>µg/g dw | [64] |
| Control<br>PDC | nos = 3–8 | -<br>- | 0.06<br>0.13 | PL (WM)<br>PL (WM) | µg/g dw<br>µg/g dw | [64] |
| Control<br>PDC | nos = 3–8 | -<br>- | 0.22<br>0.10 | OL (WM)<br>OL (WM) | µg/g dw<br>µg/g dw | [64] |
| Control<br>PDC | nos = 3–8 | -<br>- | 0.19<br>0.10 | TL (WM)<br>TL (WM) | µg/g dw<br>µg/g dw | [64] |
| Control<br>PDC | nos = 3–8 | -<br>- | 0.20<br>0.09 | C (WM)<br>C (WM) | µg/g dw<br>µg/g dw | [64] |

**Table 1.** *Cont.*

| Patients | Number | Age (Mean/*Median*/*Ranges*) | Concentration (Mean/*Median*) * | Biological Fluids/ Tissues | Unit | Ref. |
|---|---|---|---|---|---|---|
| PD | n = 13 | n/a | *0.0372*<br>1.0541 ↑ (1 wk)<br>0.2061 ↑ (2 wk)<br>0.4691 ↑ (4 wk)<br>0.2944 ↑ (6 wk)<br>0.1276 ↑ (8 wk) | CSF<br>CSF<br>CSF<br>CSF<br>CSF<br>CSF | µmol/L<br>µmol/L<br>µmol/L<br>µmol/L<br>µmol/L<br>µmol/L | [66] |
| Control<br>PD | n = 13<br>n = 26 | ~64<br>~65 | 0.12<br>0.07 ↓ | CSF<br>CSF | ng/mL<br>ng/mL | [67] |
| Control<br>PD | n = 12<br>n = 9 | 70<br>73 | n/a<br>n/a # | CN<br>CN | µg/g dw<br>µg/g dw | [68] |

* All values are given as reported in the cited reports. n: number of patients; n/a: not available, wk: week. AD: Alzheimer disease, ALS: amyotrophic lateral sclerosis, PDC: parkinsonism-dementia complex, PD: Parkinson disease, FC: frontal cortex, VF: ventricular fluid, CSF: cerebrospinal fluid, LOQ: level of quantitation, GM: gray matter, WM: white matter, FL: frontal lobe, PL: parietal lobe, OL: occipital lobe, TL: temporal lobe, C: cerebellum, CC: cerebral cortex, PC: parietal cortex, BG: basal ganglia, CN: caudate nucleus, DW: dry weight, nos: number of samples. ND: not detectable. ↑ Statistically significant increase. ↓ Statistically significant decrease. †, ‡ Eastern Canada and United Kingdom, respectively. # Detected only in one PD brain (the value for the vanadium level is not provided in the cited report). ʃ Too low value for the level of vanadium that is difficult to read from the graph provided in the cited report. **, ## ALS patients with a spinal and bulbar onset, respectively.

As for the level of vanadium in CSF of patients with AD, PD, or ALS, Gerhardsson et al. [58,61] found a significantly lowered concentration of this metal in CSF in subjects with AD, compared with the control. As emphasized by the authors, the observed effect could be related to the binding of vanadium to chelating proteins like Aβ and apolipoprotein E (APOE) in the CSF and brain, thereby detoxifying its pro-oxidative activity. In turn, Hershey et al. [59] did not detect vanadium in CSF of demented Alzheimer patients. In the case of PD, significantly reduced vanadium content was noted by Bocca et al. [67] in CSF of PD subjects, in comparison with the control. However, another study conducted by Shi et al. [66] who determined multiple chemical elements, including vanadium in CSF of subjects suffering from Parkinson disease before and after intracerebral autotransplantation of the adrenal medulla, showed a significantly higher level of vanadium in the 1st, 2nd, 4th, 6th, and 8th week. As concluded by the authors, the observed changes could be induced by the type of operation used. As for ALS, Roos et al. [63] noted a significantly elevated vanadium concentration in CSF of patients with ALS, whereas Patti et al. [65], who compared metal levels in CSF of ALS subtypes (i.e., spinal vs. bulbar clinical onset), did not find any significant alterations in the concentrations of vanadium in CSF between spinal and bulbar onset patients.

As far as the level of vanadium in the brain of AD, PD, or ALS patients is concerned, studies conducted by Szabo and co-workers [57] to examine the relationship between different metal concentrations, including vanadium in the brain and ventricular fluid (VF) of AD subjects and nondemended elderly controls, did not reveal any significant changes in the content of this metal in the frontal cortex; vanadium was not detected in VF. In turn, Ward and Mason [60] conducted studies to investigate many metals, including vanadium, in the brain tissue, i.e., in the hippocampus and cerebral cortex (CC) of Alzheimer's disease patients and age-matched controls from Eastern Canada and the United Kingdom. They demonstrated a significantly lowered vanadium concentration in both study areas of the brain of the AD subjects, compared to the control. Given the imbalance of other metals (i.e., Br, Ca, S, Si, Se, and Zn) found by the authors in the brain of the AD individuals, it can be assumed that the changes in the level of vanadium and the above-mentioned elements may have resulted from the altered brain metabolism in the response to pathogenic conditions. Another example was reported by Srivastava and Jain [62], who focused on elemental analysis in two AD brain regions, i.e., the parietal cortex and the cerebellum. In this case, however, the concentrations of vanadium are difficult to comment, as they were presented together with other metals in the graph, whose scale does not allow reading the concentration of this element correctly due to its too low values. Further, other researchers analyzing the regional metal concentration, including vanadium in the control and PD brain, detected vanadium in the caudate nucleus (CN) in only one PD brain but the exact levels of this metal in the CN of the PD and control brains were not provided [68]. Finally, Gellein and co-workers [64] carried out studies on the concentrations of selected metals, including vanadium, in formalin-fixed brain tissue collected in 1979–1983 from eight Guamanian patients with ALS, four subjects with Parkinsonism-dementia complex (PDC) and five with no known neurological disorders. The authors did not find any significant differences in the level of vanadium in the different brain regions between both patients groups and the control individuals. As stressed by the researchers, the limited number of subjects, inter alia, may have been responsible for the absence of statistically significant differences (Table 1).

### 5.3. *Vanadium Content in Blood, Urine, and Hair of Patients with AD, PD, and ALS*

Summarized values of the concentrations of vanadium in blood (including serum (S), plasma (P), and whole blood (WB)) and in urine (U) and hair (H) are listed in Table 2.

**Table 2.** Summary of the results of the levels of vanadium in blood, urine, and hair of patients with neurodegenerative diseases.

| Patients | Number | Age (Mean/*Median*/Ranges) | Concentration (Mean/*Median*) * | Biological Fluids/Other | Unit | Ref. |
|---|---|---|---|---|---|---|
| AD cases | | | | | | |
| Control | n = 54 | *73 (60–94)* | <2.2 | plasma | μg/L | [58] |
| AD | n = 173 | *75 (52–86)* | <2.2 | plasma | μg/L | |
| Control | n = 30 | 74 | 0.0613 | serum | μg/L | [69] |
| AD | n = 30 | ~80 | 0.0673 | serum | μg/L | |
| Control | n = 40 | 65 (57–87) | 0.04 (0.03–0.09) | serum | μg/L | [70] |
| AD | n = 34 | 72 (54–84) | 0.08 ↑ (0.03–0.53) | serum | μg/L | |
| Control | n = 17 | ~74 | 0.17 | hair | μg/g | [71] |
| AD | n = 9 | ~83 | 0.06 | hair | μg/g | |
| Control | n = 18 | age-matching | 39.9 | serum | ng/g | [72] |
| AD | n = 30 | 60 (45–84) | 76.3 ↑ | serum | ng/g | |
| Control | n = 161 | >60 | n/a # | plasma | n/a | [73] |
| AD | n = 92 | >60 | n/a # ↑ | plasma | n/a | |
| Control | n = 124 | 44.8 (20–84) | *0.06 (0.05–0.08)* | serum | μg/L | [74] |
| AD | n = 53 | 74.5 (58–86) | *0.04 (0.03–0.07)* | serum | μg/L | |
| Control | n = 44 | >45 | *0.06* | serum | ng/mL | [75] |
| AD | n = 60 | 74.6 | 0.05 | serum | ng/mL | |
| Control | n = 44 | >45 | 0.09 | WB | ng/mL | [75] |
| AD | n = 60 | 74.6 | 0.13 | WB | ng/mL | |
| PD cases | | | | | | |
| Control | n = 124 | 44.8 (20–84) | *0.06 (0.05–0.08)* | serum | μg/L | [74] |
| PD | n = 71 | 65.5 (41–81) | *0.11 (0.06–0.16)* | serum | μg/L | |
| Control | n = 44 | ~52 | 0.06 | serum | ng/mL | [76] |
| PD | n = 71 | 65.5 | 0.11 ↑ | serum | ng/mL | |
| Control | n = 13 | ~64 | 0.07 | WB | ng/mL | [67] |
| PD | n = 26 | ~65 | 0.22 ↑ | WB | ng/mL | |
| Control | n = 13 | ~64 | 0.15 | serum | ng/mL | [67] |
| PD | n = 26 | ~65 | 0.17 | serum | ng/mL | |

**Table 2.** *Cont.*

| Patients | Number | Age (Mean/*Median*/Ranges) | Concentration (Mean/*Median*) * | Biological Fluids/Other | Unit | Ref. |
|---|---|---|---|---|---|---|
| Control | n = 13 | ~64 | 0.10 | urine | ng/mL | [67] |
| PD | n = 26 | ~65 | 0.20 ↑ | urine | ng/mL | |
| ALS cases | | | | | | |
| Control | n = 9 | 45 *(26–77)* | *0.07* | plasma | μg/L | [63] |
| ALS | n = 15 | 64 (47–85) | *0.07* | plasma | μg/L | |
| Control | n = 100 | 51 | 0.08 | hair | μg/g | [77] |
| ALS | n = 100 | 55 | 0.05 | hair | μg/g | |

* All values are provided as reported in the cited reports. n: number of patients; n/a: not available. AD: Alzheimer disease, PD: Parkinson disease, ALS: amyotrophic lateral sclerosis, WB: whole blood, ↑ Statistically significant increase. # The values for the level of vanadium have not been provided by the authors. ↑ Higher level but not statistically significant, compared to the control.

As shown in Table 2, the studies conducted by Bocca et al. [67] did not indicate any significant differences in the level of vanadium in the serum of AD patients, compared to the control. In turn, in the case of WB, the concentration of vanadium in AD subjects was noticeably higher (by 1.4-fold), in comparison with the control, but this increase did not reach a statistically significant level. A clear trend toward an increase in the serum vanadium level in AD patients, compared to healthy control, was also observed by González-Dominguez et al. [69]; in this case, the concentration of vanadium in the serum of AD subjects turned out to be about 1-fold higher than that noted in the control individuals. Moreover, the same researchers showed that the level of vanadium in the serum of subjects with mild cognitive impairment was clearly higher (1.1-fold) than in the control, but slightly lower than that in the AD patients. Based on these findings, it cannot be excluded that the concentration of vanadium in the serum of AD patients may reach a significant level with the progress of the disease. Guan et al. [73] also found a higher, but not statistically significant, level of vanadium in the plasma of the AD group, compared to the control. In turn, a statistically significant increase (by 2-fold) in the concentration of vanadium in the serum of AD patients, compared to healthy subjects, was recorded by Paglia et al. [70]. These authors also indicated that the level of vanadium in the serum was significantly elevated in subjects with mild cognitive impairment. Additionally, they revealed that vanadium had the best diagnostic power in the discrimination between AD and healthy individuals. Lavanya et al. [72] also noted a statistically significant increase (by 1.9-fold) in the concentration of vanadium in the serum of AD patients, compared to the control. However, the concentrations of this metal in the serum reported by these researchers are slightly surprising. To sum up, on the basis of the findings described above, it can be suggested that vanadium may play a role in the pathogenesis of AD disease. However, the studies conducted by Gerhardsson et al. [58], Akanle et al. [71], and Alimonti et al. [74] analyzing the level of vanadium in the plasma, hair, and serum of subjects with AD, respectively, did not reveal any significant alterations in the concentration of this metal.

As for the Parkinson disease, the studies conducted by Bocca et al. [67] indicated a significant increase in the concentrations of vanadium in WB and U in PD patients and a clear trend toward an increase in the level of this metal in the serum of these subjects, compared to the control. The levels of vanadium in the PD cases were about 3-fold, 2-fold, and 1-fold higher, respectively, in comparison with those noted in the control individuals. Moreover, the serum vanadium level was found to be positively correlated with the serum antioxidant status, which turned out to be lower in the PD patients than in the control. Further, a statistically significant increase (1.8-fold) in the level of vanadium in the serum in PD patients, compared to healthy individuals, was found by Forte et al. [76]. However, the authors demonstrated that antioxidant capacity and oxidative damage were reduced and elevated, respectively, in PD subjects. An increased level of vanadium in the serum (1.8-fold) in PD patients, compared to control subjects, and reduced antioxidant capacity and enhanced oxidative status in the serum of these patients, were also reported by Alimonti et al. [74]. Thus, the results described above clearly point to alterations in the level of vanadium in body fluids of PD patients and do not rule out the involvement of this metal, which is able to generate OS in the pathogenesis of Parkinson disease. Simultaneously, these findings indicate that vanadium may be considered as a potential biomarker of this neurodegenerative syndrome and suggest that it can serve as a predictor of PD evolution and treatment of this illness.

As far as ALS is concerned, to date, there are only two papers presenting the levels of vanadium in ALS subjects and healthy controls. For example, Royce–Nagel et al. [77] who determined the level of vanadium in hair, did not find elevated levels of this metal in hair samples from individuals with ALS. Therefore, based on these findings, they concluded that hair vanadium levels cannot serve as a biomarker of primary pathogenic events in ALS. Similarly, no significant changes in the level of vanadium in the serum of ALS subjects, compared to the control, were noted by Roos and co-workers [63]. However, in this case, the authors recorded disturbances in the concentration of this metal in CSF of ALS patients, which has already been described in the previous section of the present report.

### 5.4. *Vanadium in Biological Samples of Humans Exposed to Vanadium Occupationally, Subjects with Acute Vanadium Poisoning and the Elderly with Neurological Side Effects/Neurobehavioral Changes—A Summary of Data on the Timeline*

A summary of studies on the levels of vanadium in biological specimens from humans exposed to vanadium occupationally, subjects with acute vanadium poisoning, and the elderly with neurological side effects and certain neurobehavioral alterations is presented in Figure 8. As illustrated, CSF, WB, S, and U were tested to determine the vanadium content.

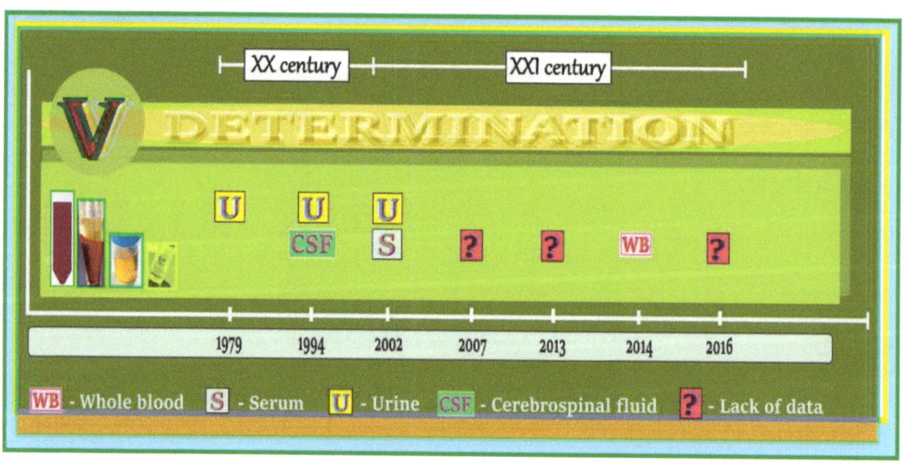


**Figure 8.** Summary of studies on the levels of vanadium in some biological specimens in humans exposed to vanadium occupationally, subjects with acute vanadium poisoning and the elderly with noted neurological side effects/neurobehavioral alterations on the timeline. Elaborated on the basis of data cited in Section 5.5.

To our knowledge, the first report on the level of vanadium in a biological sample, i.e., in the urine (U) of patients with some neurological disorders, was published in the late 20th century, i.e., in 1979, and referred to workers occupationally exposed to this metal. Several years later, in 1994, another paper appeared in which the concentration of vanadium was recorded in the CSF and U of female patients with acute vanadium poisoning and certain neurological alterations. In the 21st century, there are not many investigations in this research field. More precisely, only one paper on the levels of vanadium in the S and U of occupationally exposed subjects with neurological changes was published in 2002. In subsequent years, i.e., in 2007, 2013, and 2016, three articles described some neurological side effects in occupationally exposed workers, but no information about the concentration of vanadium in physiological fluids of these persons was provided. Finally, in 2014, one paper was published on the level of vanadium in the WB of elderly subjects with some neurological alterations. More details on the type of neurological side effects and neurobehavioral changes along with summarized values of the concentration of vanadium in some biological samples from the subjects mentioned above are provided below.

### *5.5. Neurological Side Effects/Neurobehavioral Changes and Vanadium Content in Body Fluids of Humans Exposed to Vanadium Occupationally, Subjects with Acute Vanadium Poisoning and the Elderly*

The present section provides a brief outline of the neurological side effects/neurobehavioral alterations of vanadium in humans. Details on this topic are concisely summarized in Table 3 and described below.

**Table 3.** Summary of the neurological side effects and neurobehavioral changes as well as the levels of vanadium in body fluids of humans exposed to vanadium occupationally, subjects with acute vanadium poisoning and the elderly.

| Item | Number | Age (Mean/Ranges) | Concentration (Mean/Range) * | Body Fluids | Unit | Findings | Ref. |
|---|---|---|---|---|---|---|---|
| Occupationally exposed | | | | | | | |
| WEV | n = 49 | 41.4 (20.1–57.1) | 7.5 (2.18–46.4)<br>C: 0.8 (0.3–3.1) | serum | µg/L | ↓ attention<br>↓ visuospatial abilities | [78] |
| | n = 49 | 41.4 (20.1–57.1) | 14.4 (2.12–95.4)<br>C: 0.4 (0.07–1.4) | urine | µg/L | | |
| WEV | n = 19<br>(2) | ~34 | 15.5 (3.0–35.2)<br>C: 6.0 | urine | µg/L | tremor of finger | [79] |
| WEV | n = 463 | 39.5 (20–60) | n/a | n/a | n/a | ↑ anger-hostility<br>↑ depression-dejection<br>↑ fatigue-inertia<br>↓ vigor-activity<br>↓ auditory/visual memory<br>↓ perception/motion speed<br>↓ coordination | [80] |
| WEV | n = 106 | n/a | n/a | n/a | n/a | ↙ vision-memory<br>↙ motor speed and accuracy<br>mood disorders (↗ negative moods) | [81] |
| WEV | n = 128 | n/a | n/a | n/a | n/a | ↙ hearing and visual memory<br>↙ movement, velocity, accuracy<br>↙ coordination | [82] |
| Acute poisoning cases | | | | | | | |
| Woman | n = 1 | 22 | 61.5 †/12.3 ††<br><20 | urine<br>CSF | µg/L<br>µg/L | right-sided brachiofacial paresis<br>right hemihypesthesia<br>right sensomotor hemiparesis<br>mixed amnestic and sensorimotor aphasia | [83] |
| Elderly subjects | | | | | | | |
| Male/female | n = 50 | ≥60 | 24.44 (6.0–63.0)<br>Y: 22.8 (6.0–49.0) | WB<br>WB | µg/L<br>µg/L | ↓ general cognitive function | [22] |

* All values are given as reported in the cited reports. n: number of people; n/a: not available. WEV: workers exposed to vanadium, C: control group, Y: young, CSF: cerebrospinal fluid, WB: whole blood. ↓: statistically significant reduction, ↑: statistically significant increase, ↙: decrease. ↗: increase. †, †† after 3 and 9 months, respectively. Numbers in squared brackets indicate the number of people with symptoms.

As shown in Table 3, one of the studies was conducted by Barth et al. [78] who used a modified Wisconsin Card Sorting Test, block design test, visual recognition test, simple reaction time, choice reaction, digit symbol substitution, and digit span to examine the effects of vanadium on attention, visuospatial and visuomotor functioning, reaction time, short-term memory, and prefrontal functioning in occupationally exposed subjects. The results of this study showed that vanadium, at the concentration of 14.2 μg/L in the urine, reduced neurobehavioral abilities, particularly visuospatial abilities and attention. A significant correlation between urinary and serum vanadium levels and cognitive deficit was also noted [78]. The concentrations of vanadium in the serum and urine of the occupationally vanadium-exposed people participating in these studies were about 9.4-fold and 36-fold higher, respectively, compared to those noted in the nonexposed subjects. Moreover, the serum and urinary vanadium levels in the same exposed workers were about 50-fold and 48-fold higher, compared to the most probable mean baseline serum (~0.15 μg/L) [84] and urinary (0.2–0.4 ng/mL) [21] vanadium concentrations, respectively, reported for humans not exposed to this metal. Another example is the results reported by Li et al. [80], who tested the hypothesis that occupational exposure to vanadium in a low-dose long-term exposure condition may lead to an early onset of neurobehavioral changes in vanadium-exposed Chinese workers. The neurobehavioral tests used by the authors, such as simple reaction time, digit symbol, Santa Ana dexterity, digital span, Benton visual retention, and pursuit aiming, included in the World Health Organization recommended Neurobehavioral Core Test Battery (WHO-NCTB), which is designed to identify adverse behavioral effects of chemical intoxication in the human population [85], pointed to poorer performance among exposed workers, compared to unexposed control subjects. The exposed workers exhibited increased anger-hostility, depression-dejection, and fatigue-interia and decreased vigor-activity. Longer mean reaction times and more counting errors were also observed in the exposed group. Moreover, vanadium exposure has been found to be associated with decreased coordination, auditory memory, and perception/motion speed. Unfortunately, no information about the levels of vanadium in the blood or urine of workers participating in these studies was provided. Further, with the use of the Santa Ana test, Benton visual retention test, and pursuit aiming II test, Zhou and co-workers [81] also investigated the effect of vanadium on neurobehavioral functions in workers exposed to this metal. The study revealed that the vanadium-exposed subjects showed poorer performance in the above-mentioned tests. The results obtained allow a conclusion that workers exposed to vanadium may show mood disorders, decreased vision-memory, and lower motor-speed and accuracy. Zhu et al. [82], who examined the effect of vanadium exposure on neurobehavioral function in workers with the use of WHO-NCTB tests as well (i.e., digit span, digit symbol, Santa Ana, Benton visual retention, and pursuit aiming II), found that the group exposed to this metal had lower test scores than the control. Their results indicate that the exposure to vanadium can manifest itself in decreased hearing/visual memory, movement, velocity, accuracy, and coordination. However, similarly in the studies conducted by Li et al. [80], no information about the levels of vanadium in the blood or urine of workers was provided. Compiled, the findings described by the above-mentioned authors clearly indicate that occupational exposure to vanadium can have adverse effects on neurobehavioral function.

In turn, studies conducted by Baierle et al. [22] to examine whether age-related cognitive deficit is associated with oxidative damage (especially with delta-aminolevulinate dehydratase inhibition, ALA-D) and to verify the influence of some metals, including vanadium on ALA-D activity and cognitive performance, showed that an increased level of vanadium was associated with reduced cognitive ability in Mini-Mental State Examination (MMSE) in an elderly group. Moreover, vanadium was noted to be negatively associated with ALA-D reactivation. Thus, the results of this study indicated that vanadium may contribute to a cognitive decline. It should also be mentioned that the concentration of vanadium in the whole blood of elderly subjects (i.e., 24.44 μg/L) was above that noted in serum of vanadium non-exposed humans (~0.15 μg/L; 0.016–0.939 μg/L) [84]. Another study conducted by Usutani et al. [79] on special physical examination of vanadium handling workers demonstrated tremor of fingers in the occupationally subjects with elevated levels of vanadium. More precisely, the concentration of this metal in the urine of these workers reached the value of 15.5 μg/L (3.0–35.2 μg/L). Finally, right-sided brachiofacial paresis, right hemihypesthesia, right sensorimotor hemiparesis, and mixed amnestic and sensorimotor aphasia were reported in a 22-year old woman who attempted suicide by oral ingestion of ~10–15 g ammonium metavanadate [83]. The concentration of vanadium in the urine of this woman 3 and 9 months after detoxification therapy (i.e., 61.5 μg/L and 12.3 μg/L) was about 205-fold and 41-fold higher, respectively [83], compared to the most probable mean baseline urinary vanadium levels, i.e., 0.2–0.4 ng/mL [21].

### 5.6. Vanadium and Neurotoxic Outcomes in Humans and Animals—A Summary of Data on the Timeline

A summary of studies on the neurotoxic effects of vanadium in humans exposed to vanadium-bound $PM_{2.5}/PM_{10}$, and in animals after inhalation, is illustrated in Figure 9. As shown, there are only a few literature reports describing the effects of neurotoxicity related to the exposure of humans to vanadium-bound $PM_{2.5}/PM_{10}$. They were published in 2013, 2015, and 2016 by Calderón–Garcidueña's research group [86–88] and in 2021 by Yu and co-workers [89]. More articles in the analyzed research field, i.e., 15 in total, refer to animals exposed to vanadium through inhalation. Three of these papers (published in 2003, 2008, and 2017) refer to animals naturally exposed to urban air and the other twelve reports (published in 2003, 2004, 2005, 2006, 2008, 2014, 2015, 2016, 2017, 2018, and 2021) provide results of the studies on laboratory animals (rats and mice). More details on the neurotoxic effects of vanadium in humans exposed to $PM_{2.5}/PM_{10}$ rich in this metal, as well as in animals, are concisely summarized in further sections of this review.

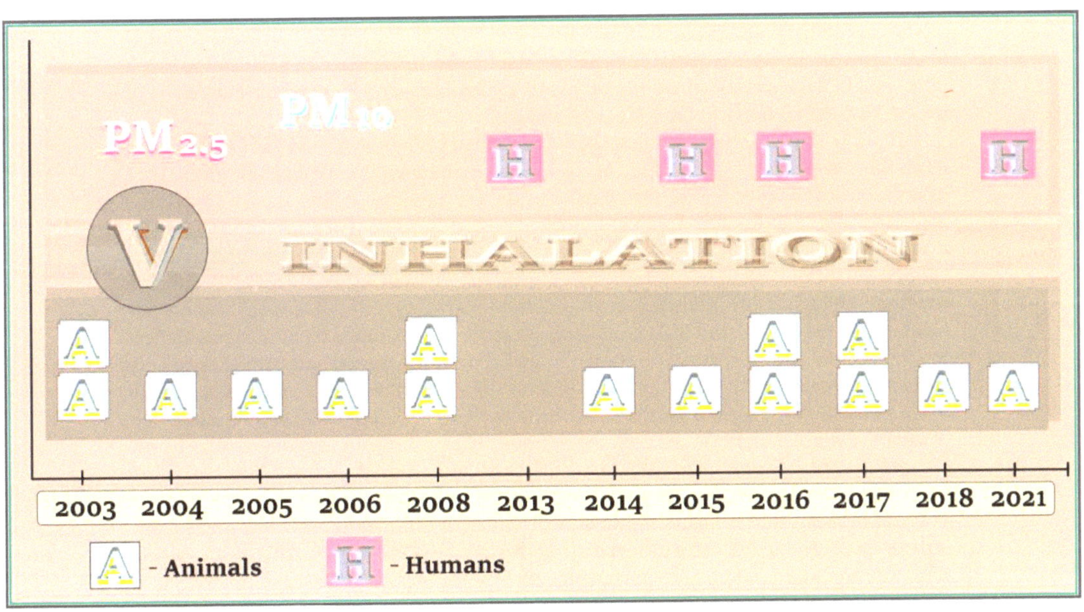


**Figure 9.** Summary of studies on neurotoxic effects of vanadium in humans and animals after inhalation on the timeline. Elaborated on the basis of data cited in Sections 5.7 and 5.8.

### 5.7. Neurotoxic Effects of Environmental Vanadium Exposure in Humans—A Summarizing Note

The potential negative impact of specific air pollution components on neurodegenerative disorders, including AD, PD, and ALS, has received more attention. Some reports indicate that certain environmental factors, including metals, are considered to increase the risk of these illnesses [90–92].

Vanadium, the environmental exposure to which occurs, inter alia, via inhalation [93,94], is a component of fine-particulate air pollution [37] posing a serious public health concern. This metal is present in both the $PM_{2.5}$ and $PM_{10}$ fraction and is mainly associated with mixed industrial/fuel oil combustion [95]. It should be emphasized that the combustion of heavy fuels, especially in oil-fired power plants, refineries, and industrial boilers, and coal are the major anthropogenic sources of atmospheric emission of vanadium [96,97], which is three times higher compared to vanadium releases from natural sources [98]. The findings obtained by Dye et al. [99] in cultured airway epithelial cells exposed to residual oil fly ash (ROFA), a component of ambient PM, revealed that vanadium was largely responsible for ROFA toxicity and that this metal induced its cellular effects, at least in part, through the generation of OS implicated in the progression of neurodegenerative diseases, as mentioned previously.

Since vanadium is a well-known environmental and occupational pollutant, the recognition of possible side effects resulting from the exposure to this metal in the context of neurodegenerative syndromes is crucial. Therefore, a possible association of vanadium with AD, PD, and ALS has been discussed in the present section, as it may provide valuable information on the potential pathophysiology of neurodegenerative illnesses. In other words, the main aim of this part of the review, which supplements the findings presented in the previous part of the current report on the association between vanadium in PM and an increased risk of adverse health outcomes excluding neurodegenerative changes in humans, is to summarize the data on a possible role of this metal in unbeneficial effects on the central nervous system (CNS) in humans. We believe that the findings presented in this chapter, that were gathered based on the literature search, will be useful to those interested in the effects of vanadium on neurodegeneration and increase the awareness of the environmental impact of this element on health.

Although the number of studies on environmental vanadium-bound PM pollution and its possible deleterious neurocognitive consequences in humans is limited (Figure 9), they deserve our comments. Figure 10 summarizes data on neurotoxic outcomes of vanadium-associated $PM_{2.5}/PM_{10}$ exposure in humans.

For example, Calderón –Garcidueñas et al. [86] examined the content of metals (including vanadium) resulting from anthropogenic activity in the frontal cortex of subjects residing in high and low pollution areas, i.e., in Mexico City (one of the world's most polluted city in the world) and in Tlaxcala and Veracruz (two control cities). The researchers found a higher (by 32%) concentration of vanadium in the frontal cortex in Mexico City subjects chronically exposed to environmental $PM_{2.5}$ containing industrial metals than in the subjects from control localities. The Mexico City residents also had higher levels of inflammatory mediators, i.e., cyclooxygenase-2 (COX2) mRNA and interleukin 1β (IL-1β) in the frontal cortex and COX2 in the olfactory bulb. Based on these findings, it can be suggested that humans exposed to a polluted urban environment rich in industrial $PM_{2.5}$-associated metals, including vanadium, which in some conditions may act as a pro-inflammatory agent [100] and induce the expression of COX2 mRNA [101], face a higher risk of neuroinflammation and detrimental effects on CNS. Another study conducted by Calderón–Garcidueña's research group [87] was focused on the examination of serum and CSF antibodies to neural and tight junction proteins and environmental pollutants in 139 children aged ~12 year living in Mexico City. They showed that the CSF-vanadium level in the Mexico City children significantly correlated with the CSF neural antibodies (S100 IgA/IgG). Moreover, the authors observed a correlation between the concentration of V in the serum (Vs) and myelin basic protein (MBP) IgG in control children with an elevated serum vanadium level which, as they stressed, could be related to emissions from the Tula refinery [87]. At this point, the results of studies conducted by Todorich et al. [102] and Soazo et al. [103] in a rodent model are worth mentioning. The researchers found that exposure to vanadium during early brain development produces hypomyelination [102], and postnatal vanadium intoxication leads to CNS myelin deficiency [103]. Moreover, Azeez and co-workers [104], who examined regional myelin and axon damage and neuroinflammation in adult mouse brains after long-term postnatal vanadium exposure, observed myelin damage which involved the midline corpus callosum and fibers in cortical gray matter, hippocampus, and diencephalon associated with axonal damage. Significant induction of TNF and IL-1β in the brain was noted as well. Similarly, Usende et al. [105] also observed intense destruction of myelin sheaths in vanadium exposed rats. As known, *"myelination is important in establishing connectivity in the growing brain by facilitating rapid and synchronized information transfer across the nervous system, which is essential to higher-order cognitive functions"* [106]. Therefore, due to the results obtained by Calderón–Garcidueñas and co-workers mentioned above and the findings obtained by Todorich et al. [102], Soazo et al. [103], Azeez and et al. [104], and Usende et al. [105], from a rodent model focused on the influence of vanadium on CNS, it should not be excluded that the exposure of industrial $PM_{2.5}$-associated V in humans can increase the risk of neurological/cognitive disorders. The studies carried out by Calderón–Garcidueñas and co-workers [88] also demonstrated significant derangements in CSF proteins, crucial for the development of AD and PD, in highly $PM_{2.5}$-exposed young Mexico City residents, compared to controls. The authors found that the concentration of amyloid-$\beta_{1\text{-}42}$ and brain-derived neurotrophic factor (BDNF) were significantly lowered. BDNF plays an important role in the proper functioning of the nervous system [107]. It is involved in plastic changes related to learning and memory and its reduced level has been reported to be associated with pathological conditions, including AD and PD [108]. At this point, the results obtained by Wang et al. [109], who determined the effects of exercise on the motor coordination in lactational V-exposed rats should be quoted. The researchers recorded impaired motor coordination and reduced plasma and cerebellum BDNF in the vanadium-intoxicated animals. Further, Calderón–Garcidueñas and co-workers [88] also observed an increase in total-α-synuclein (T-α-Syn) in the first childhood years related to cumulated $PM_{2.5}$ and a decrease after the age of 12 years, whereas d-α-synuclein exhibited a trend toward an increase with cumulated $PM_{2.5}$ [88]. Additionally, in Mexico City children, amyloid-$\beta_{1\text{-}42}$ correlated with T-α-Syn, and d-α-synuclein (d-α-Syn) correlated with tumor necrosis factor alpha (TNFα) and such interleukins as IL-6 and IL-10. Taken together, it can be suggested that young subjects living in highly industrialized areas and exposed to higher concentration of $PM_{2.5}$ rich in vanadium may face a higher risk of neurodegenerative syndromes. González–Maciel et al. [110] also identified combustion-derived nanoparticles (CDNPs) in neurons, glia, choroid plexus, and neurovascular units in young Mexico City residents; they were associated with pathology in mitochondria, endoplasmic reticulum (ER), mitochondria-ER contacts, axons, and dendrites. Finally, Yu et al. [89], who examined long-term exposure to ultrafine particles and PM constituents and the risk of ALS, observed increased odds ratios (ORs) for ALS in association with most air pollutants with the strongest associations for $PM_{2.5}$. They also found that, in the case of particle elements, road traffic non-tailpipe emissions of vanadium as well as Cu, Fe, Ni, S, and Si were associated with significantly higher ORs for ALS in both the $PM_{2.5}$ and $PM_{10}$ fraction. In turn, Cole-Hunter et al. [111], who examined the association between long-term exposure to ambient air pollution and PD mortality in seven European cohorts, provided evidence that long-term exposure to $PM_{2.5}$ at levels well below current EU air pollution limit values may contribute to PD mortality. However, no positive association with the vanadium component of $PM_{2.5}$ was detected.

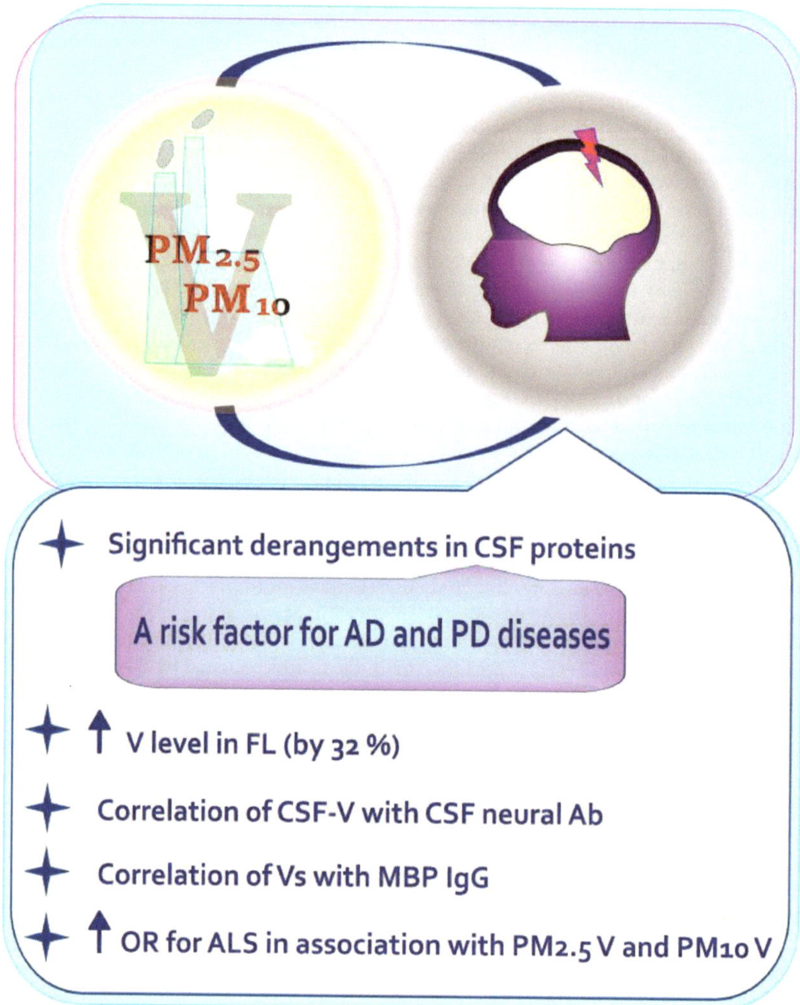


**Figure 10.** Summary of data on the neurotoxic effects of vanadium-associated $PM_{2.5}/PM_{10}$ exposure in humans. Elaborated on the basis of available literature data [86–89]. $PM_{2.5}$: particles with aerodynamic diameters less than 2.5 μm, $PM_{10}$: particles with aerodynamic diameters less than 10 μm, $PM_{2.5}V$: $PM_{2.5}$ vanadium fraction, $PM_{10}V$: $PM_{10}$ vanadium fraction, CSF: cerebrospinal fluid, CSF-V: CSF-V concentration, $V_S$: vanadium in the serum, MBP: myelin basic protein, IgG: immunoglobulin G, FL: frontal lobe, Ab: antibody, AD: Alzheimer disease, PD: Parkinson disease, OR: odd ratio; ALS: amyotrophic lateral sclerosis, V: vanadium. ↑: increase.

### 5.8. Neurotoxic Effects of Vanadium after Inhalation in Animals—A Summarizing Note

The present section is focused on the neurotoxic effects of vanadium both in laboratory animals and in those naturally exposed to urban air. Summarized data on this topic, whose main aim is to provide the reader with greater insight into the issues related to the influence of vanadium on CNS, are listed in Figure 11.

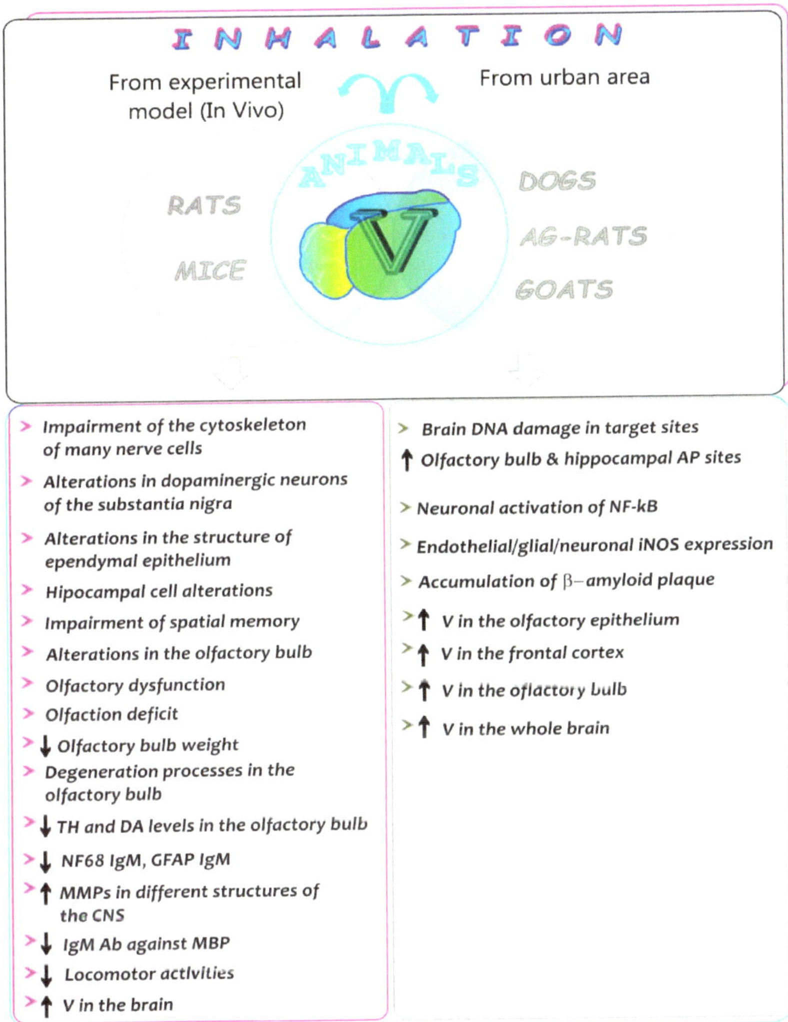


**Figure 11.** Summary of neurotoxic effects of vanadium after inhalation: studies in laboratory animals and those naturally exposed to urban air. Elaborated on the basis of the available literature data [112–126]. Ab: antibody, AG-rats: African giant rats, AP site: apurinic/apyrimidinic site, CNS: central nervous system, DA: dopamine, GFAP: glial fibrillary acidic protein; IgM; immunoglobulin M, iNOS: inducible nitric oxide synthase, MBP: myelin basic protein, MMPs: matrix metalloproteinases, NF-kB: nuclear factor kappa B, NF68: neurofilament 68 protein; TH: tyrosine hydroxylase, V: vanadium. ↑: increase, ↓: decrease.

As for the influence of environmental pollution with metals on animals living in urban area, Usende et al. [125], who investigated heavy metal pollutants (including vanadium) in selected organs of African giant rats from three agro-ecological zones of Nigeria (i.e., rainforest, woodland/grass savanna, and mangrove/freshwater swamp zones), showed that the concentrations of vanadium in the whole brain of rats from the mangrove/freshwater swamp zone (characterized by increased exploitation of minerals and pipeline vandalization by militants) were higher (by 73% and 71.5%) than those from the rainforest and woodland/grass savanna zones, respectively. These data do not seem to be surprising because vanadium is present in many minerals [11], and the increased exploitation, thereof, can lead to environmental pollution and accumulation in different tissues of animals living in these regions. An elevated concentration of vanadium was also noted by Igado et al. [119] in the olfactory bulb of goats from a relatively unindustrialized area in Nigeria. The results obtained by these authors may in turn suggest that even low scale industrial activities contribute to the accumulation of vanadium in the brain. In turn, studies conducted by Calderón–Garcidueñas et al. [113] showed a higher vanadium concentration in dogs exposed to urban pollution in South West Metropolitan Mexico City. This metal has been detected both in the frontal cortex and in the olfactory bulb, and its level turned out to be higher than the Ni level. Moreover, the hippocampal apurinic/apyrimidinic (AP) sites in the olfactory bulb in the exposed dogs were found to be significantly higher, compared with controls. Additionally, the dogs from the exposed region exhibited neuronal activation of NF-kB, endothelial/glial/neuronal inducible nitric oxide synthase (iNOS) expression, and β-amyloid plaque accumulation. As suggested by the authors, acceleration of AD-type pathology occurred in the dogs chronically exposed to the air pollutants.

An example of findings obtained in studies on laboratory animals is the results reported by Colin-Barenque et al. [118], who analyzed the possible role of matrix metalloproteinases 2 and 9 (MMP-2 and MMP-9) in changes observed in the brain of mice after chronic vanadium inhalation. The authors demonstrated that vanadium increased MMPs in different structures of the CNS and suggested that OS might be involved in the MMP activation. Another study conducted by Colin-Barenque and co-workers [118] that aimed to investigate the relationship between a simple olfactory function test, some enzyme activities, and morphological changes in granule cells from the olfactory bulb as a consequence of vanadium inhalation, demonstrated olfactory impairment at vanadium exposure in a mouse model. The authors observed loss of dendritic spines and necrotic neuronal death in the granule cells of the olfactory bulb [126]. As emphasized, olfactory disorders might be associated with OS generated by this metal. Other researchers examined the neurotoxic properties of vanadium in a mouse model, focusing on the effects of this element on the olfactory bulb to determine whether vanadium subchronic nasal exposure impairs neurobehavioral and neurochemical processes associated with olfactory function. The results revealed a significant decrease in the weight of the olfactory bulb as well as a reduced tyrosine hydroxylase (TH) level and depletion in the dopamine (DA) level in the olfactory bulb. The induction of degeneration processes in the olfactory bulb and in locomotor activities in the vanadium-exposed mice was also observed [120]. In turn, decreased numbers of TH-positive neurons in the substantia nigra and a substantial loss of dendritic spines of the medium-size spiny neurons were noted by Avila–Costa et al. [115], who examined nigrostriatal modifications after vanadium inhalation in a rodent model. Avila–Costa and co-workers [116] also found alterations in the structure of ependymal epithelium characterized by cilia loss, cell sloughing, and ependymal cell layer detachment from the basal membrane. Their studies also showed that vanadium inhaled by mice produced a time dependent loss of dendritic spines, necrotic-like cell death, and notorious alterations of the hippocampus CA1 neuropile, which correlated with spatial memory impairment [117]. Further, Montiel-Flores et al. [112] found a prominent impairment of the cytoskeleton of many nerve cells in rats inhaling vanadium and suggested that inhalation with this metal induces Alzheimer-like cell death. Keil et al. [124], who characterized particulate matter size, metal chemistry, and health effects of dust (including vanadium) collected from the Nellis Dunes Recreation Area (NDRA) located near Las Vegas (NV), found reduced IgM antibody production against neurofilament 68 protein (NF68) and glial fibrillary protein (GEAP) in dust-exposed mice, suggesting that geogenic dust from easily erodible sand dune surfaces at NDRA may have a potential health risk. In turn, a reduced concentration of IgM antibodies against myelin basic protein (MBP) was found by DeWitt et al. [122] in mice exposed to geogenic dust from arsenic-rich sediment (including vanadium) at NDRA. IgM antibodies against MBP were also reduced in mice exposed to geogenic dust (including vanadium) collected from active drainage surfaces (NDRA) [123]. Finally, studies of the levels of vanadium in some internal organs of rodents (in an inhalation model) conducted by Sánchez et al. [114] revealed elevated concentrations of this metal in the brain of mice.

## 6. New Therapies of Neurodegenerative Illnesses—A Summarizing Note

Currently, over 100 substances are being studied to modify the course of AD. More than amyloid and tau protein, it is focused on inflammatory processes. Inflammation is a major sign of aging, and chronic systemic inflammation is associated with reduced brain volume and impaired cognitive function. Although broad-spectrum anti-inflammatory drugs have not improved cognitive outcomes in patients with AD, recent efforts have focused on targeting specific aspects of inflammation that are harmful to the brain while preserving normal immune function. Therapies targeting synaptic plasticity and neuroprotection are also being studied. In addition to these pharmacological approaches, non-pharmacological interventions such as repetitive transcranial magnetic stimulation (rTMS) and transcutaneous direct current stimulation (tDCS), have the potential for clinical application [127]. It is estimated that a global delay in the onset of dementia by 5 years would reduce its incidence by half [128].

As for PD, new therapies focus mainly on disease modification and dopamine resistance symptoms. Some new disease-modifying therapies target α-Syn and its pathways, while others target genes and proteins involved in the pathogenesis of PD, including leucine-rich repeat kinase 2 (LRRK2), parkin and glucocerebrosidase. Great hopes are associated with prasinezumab: an antibody against α-Syn [129]. Disease-modifying pharmacotherapies (such as nilotinib, inosine and isradipine) are being re-examined for the treatment of PD. Antibody therapies, vaccines and immunotherapies aimed at removing abnormal proteins have emerged as promising approaches in preclinical models.

Investigated cellular therapies can be divided into rescue and regenerative therapies; rescue therapy aims to save neurons and slow down disease progression, while regenerative therapy focuses on replacing neurons. Despite these often expressed hopes, these therapies are not intended to cure PD. They are intended to rebuild lost dopaminergic pathway cells without side effects [6]. An alternative symptomatic therapy is adaptive deep brain stimulation (DBS), which involves implanting thin stimulating electrodes into deep parts of the brain. Although it is an invasive procedure, it is characterized by high safety and significant effectiveness in reducing symptoms [130]. Trials are also being conducted using MRI Guided Focused Ultrasound (MRgFUS). Damage of the subthalamic nucleus using MRgFUS can improve quality of life and probably has a beneficial effect on daily activities and the outcome of motor disorder assessment. It is noteworthy that MRgFUS can disrupt the blood–brain barrier, which could find application in the future in pharmacotherapy of Parkinson's disease, growth factors therapy, nanoparticle therapy or gene therapy [131]. Current efforts to overcome challenges in therapeutic development focus on individualizing therapy and precision in treatment; these principles are particularly important due to the heterogeneity of clinical PD subtypes.

As far as ALS is concerned, the first group of therapies currently being studied is the pharmacological approach. Among the substances that limit inflammation, mastinib stands out. Mastinib is a tyrosine kinase inhibitor on mast cells and macrophages in the central and peripheral nervous system. It had a protective effect on motor neurons through immunomodulation and reducing the inflammatory process in microglia. In patients with ALS receiving mastinib, there was a significantly slower progression of disability. Pegcetacoplan and ibudilast are also undergoing clinical trials. Pegcetacoplan is a synthetic cyclic peptide that binds to components C3 and C3b of the complement system. By blocking these components, it reduces the severity of the inflammatory process. In turn, ibudilast not only inhibits pro-inflammatory cytokines (TNF-α, IL-6) but also matrix metalloproteinase 9 (MMP9), which can potentially accelerate the progression of ALS. It also reduces the activation of astroglial cells, which is associated with the degree of upper motor neuron damage, and has a protective effect on nerve cells. Additionally, it accelerates the removal of TAR DNA-binding protein 43 (TDP-43) and superoxide dismutase 1 (SOD1) protein deposits in cell models with mutations typical for ALS.

There are also some hopes associated with tauroursodeoxycholic acid, which in preclinical studies, reduced OS, endoplasmic reticulum (ER) stress and apoptosis. Currently, its effectiveness is being studied in combination with sodium phenylbutyrate, a histone deacetylase inhibitor that improves transcription and post-transcriptional mechanisms.

Another therapeutic approach that has recently gained renewed interest is the targeting of muscle abnormalities in ALS. The first rationale is neuroprotective, as changes in the muscle and neuromuscular junction may play a role in retrograde degeneration. A second approach is symptomatic by increasing muscle contractility with two troponin activators in development: tirasemtiv and reldesemtiv.

Tofersen is an antisense oligonucleotide whose safety and efficacy have been studied in patients with ALS associated with a pathogenic mutation in the SOD1 gene. The SOD1 protein synthesized on the matrix of the mutated gene acquires additional toxic properties. Inhibiting its synthesis could therefore slow down the progression of the disease.

Many studies have evaluated the usefulness of stem cells in the treatment of ALS. However, no significant effect of treatment on respiratory function or survival time has been reported so far [132].

## 7. Summary, Conclusions, and Further Research Trends

Data summarized in the present review indicate that vanadium, which is able to produce an oxidation-reduction imbalance in the organism, cannot be excluded as a factor playing a role in neurological disorders, in which OS appears to be a part of the pathophysiological mechanism. They also suggest that the level of vanadium in body fluids could be considered as a potential biomarker and predictor of some neurodegenerative syndromes.

At this stage of knowledge, it should not be excluded that vanadium may help in identification of subjects with a higher risk of detrimental effects on CNS. However, more extensive epidemiological studies are needed to provide a conclusive link between pollution with vanadium and neurodegenerative processes in humans.

Finally, the data compiled in the current review clearly indicate that more attention should be paid to the chronic diseases related to vanadium and to the assessment of the environmental vanadium exposure. Additionally, the findings overviewed in this paper may lay the groundwork for further research on the interaction of vanadium with other elements in the context of neurodegenerative processes, especially those with antioxidant potential such as $Mg^{II}$, which has been used in the prevention and therapy of many diseases [133]. $Mg^{II}$ has been reported to be able to limit the pro-oxidative properties of vanadium in the rat liver [134], reduce the degeneration of brain cells, and improve cognitive function [135]. Therefore, a combination of $Mg^{II}$ with vanadium should be considered to be studied in the context of neurological disorders to check if $Mg^{II}$ is able to protect against the harmful effects of ROS and OS in the conditions of vanadium exposure. It is well known that ROS are implicated in vanadium deleterious effects and that OS is an underlying mechanism of vanadium-induced toxicity, as evidenced by the vivo studies [19,136]. Therefore, the examination of the consequences, character, and mechanisms of interactions of vanadium with $Mg^{II}$, as a potential 'antidote' limiting vanadium adverse effects and minimizing its strong pro-oxidant properties, is important, and even more crucial in the context of the occupational vanadium exposure and the growing environmental pollution with this metal. Through these kinds of studies, we can find out how important the role of $Mg^{II}$ plays in vanadium intoxication (Figure 12). Moreover, studies on the $V^{V}xMg^{II}$ combined effects allow us to better understand the modes of vanadium action and clarify the mechanisms underlying the potential protective role of $Mg^{II}$ in vanadium exposure in mammalian organisms.

As illustrated in Figure 12, OS is implicated in the progression of neurological illnesses, such as AD, PD, and ALS. ROS contributes to the oxidative modification and accumulation of misfolded proteins, which triggers a cascade of events resulting in neuronal damage and neuron death [137,138]. Therefore, detailed investigations targeting evaluations of the consequences and mechanisms of interactions of vanadium with an element having antioxidant potential, including $Mg^{II}$, or with some natural antioxidants during their combined administration, not only addresses a topic of high importance in toxicology, i.e., it refers to the search of approaches limiting vanadium toxicity, but they are also important in the context of possible antioxidant-based therapies for alleviating the severity of neurological diseases.

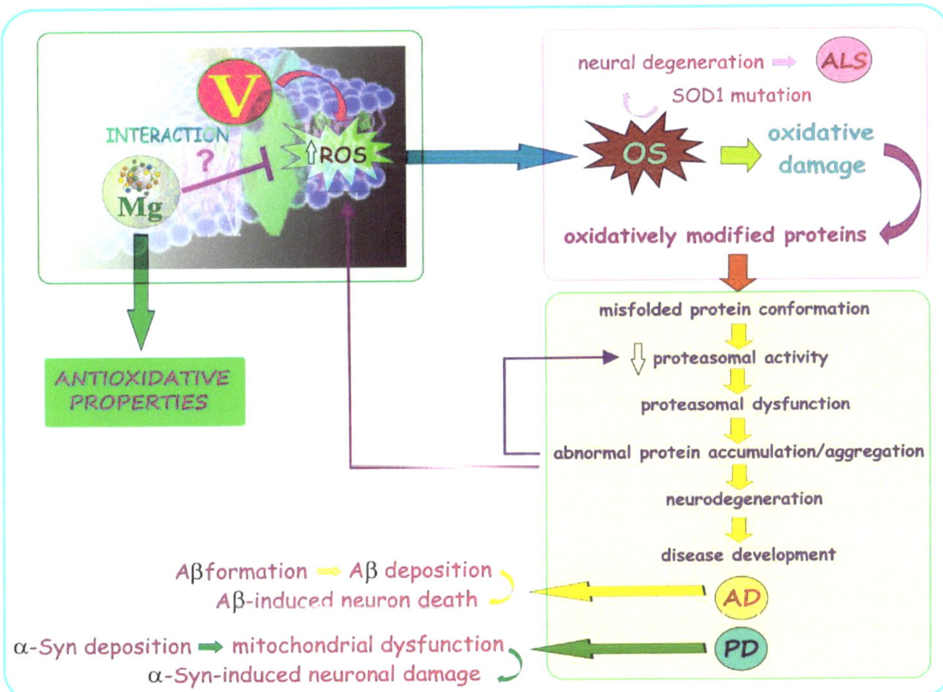


**Figure 12.** Vanadium in the pathological mechanisms of AD, PD, and ALS. V: vanadium, Mg: magnesium, ROS: reactive oxygen species, OS: oxidative stress, SOD1: superoxide dismutase 1, AD: Alzheimer disease, PD: Parkinson disease, ALS: amyotrophic lateral sclerosis, Aβ: amyloid beta, α-Syn: alpha synuclein. ↓: decrease, ↑: increase.

**Author Contributions:** Conceptualization, A.Ś.; methodology, A.Ś.; writing—original draft preparation, A.Ś., P.P.D. and T.M.-K.; writing—review and editing, A.Ś.; visualization, A.Ś.; supervision, A.Ś. and J.L. All authors have read and agreed to the published version of the manuscript.

**Funding:** This research received no external funding.

**Institutional Review Board Statement:** Not applicable.

**Informed Consent Statement:** Not applicable.

**Data Availability Statement:** Data sharing not applicable.

**Conflicts of Interest:** The authors declare no conflict of interest, financial or otherwise.

## References


1. NIK o Opiece Nad Chorymi Na Alzheimera. 16 May 2017. Available online: https://www.nik.gov.pl/aktualnosci/nik-o-opiece-nad-chorymi-na-alzheimera.html (accessed on 14 April 2023).
2. Du, X.; Wang, X.; Geng, M. Alzheimer's disease hypothesis and related therapies. *Transl. Neurodegener.* **2018**, *7*, 2. [CrossRef] [PubMed]
3. Kivipelto, M.; Mangialasche, F.; Snyder, H.M.; Allegri, R.; Andrieu, S.; Arai, H.; Baker, L.; Belleville, S.; Brodaty, H.; Brucki, S.M.; et al. World-Wide FINGERS Network: A global approach to risk reduction and prevention of dementia. *Alzheimer's Dement.* **2020**, *16*, 1078–1094. [CrossRef] [PubMed]
4. Livingston, G.; Huntley, J.; Sommerlad, A.; Ames, D.; Ballard, C.; Banerjee, S.; Brayne, C.; Burns, A.; Cohen-Mansfield, J.; Cooper, C.; et al. Dementia prevention, intervention, and care: 2020 report of the Lancet Commission. *Lancet* **2020**, *396*, 413–446. [CrossRef]
5. Budrewicz, S. *Choroba Parkinsona*; Elsevier Essentials Parkinson, Edra Urban & Partner: Wroclaw, Poland, 2018.
6. Elkouzi, A.; Vedam-Mai, V.; Eisinger, R.S.; Okun, M.S. Emerging therapies in Parkinson disease—Repurposed drugs and new approaches. *Nat. Rev. Neurol.* **2019**, *15*, 204–223. [CrossRef] [PubMed]
7. Banach, M.; Boguski, A.; Liberski, P. *Choroby Rdzenia Kręgowego*; Medycyna Praktyczna: Gdańsk, Poland, 2007.
8. Kiernan, M.C.; Vucic, S.; Cheah, B.C.; Turner, M.R.; Eisen, A.; Hardiman, O.; Burrell, J.R.; Zoing, M.C. Amyotrophic lateral sclerosis. *Lancet* **2011**, *377*, 942–955. [CrossRef] [PubMed]
9. Kabata-Pendias, A.; Pendias, H. *Biogeochemia Pierwiastków Śladowych*; Wydanie 2; Wydawnictwo Naukowe PWN: Warsaw, Poland, 1999.
10. Rehder, D. *Bioinorganic Vanadium Chemistry*; John Wiley and Sons: Chichester, UK, 2008; pp. 1–23.
11. Agency for Toxic Substances and Disease Registry (ATSDR). *Public Health Statement for Vanadium*; U.S. Department of Health and Human Services, Public Health Service: Atlanta, GA, USA, 2012.
12. Genet, S.; Kale, R.K.; Baquer, N.Z. Alterations in antioxidant enzymes and oxidative damage in experimental diabetic rat tissues: Effect of vanadate and fenugreek (*Trigonella foenum graecum*). *Mol. Cell. Biochem.* **2002**, *236*, 7–12. [CrossRef]
13. Ramachandran, B.; Ravi, K.; Narayanan, V.; Kandaswamy, M.; Subramanian, S. Protective effect of macrocyclic binuclear oxovanadium complex on oxidative stress in pancreas of streptozotocin induced diabetic rats. *Chem. Biol. Interact.* **2004**, *149*, 9–21. [CrossRef]
14. Tunali, S.; Yanardag, R. Protective effect of vanadyl sulfate on skin injury in streptozotocin-induced diabetic rats. *Hum. Exp. Toxicol.* **2013**, *32*, 1206–1212. [CrossRef]
15. Xie, M.; Chen, D.; Zhang, F.; Willsky, G.R.; Crans, D.C.; Ding, W. Effects of vanadium (III, IV, V)-chlorodipicolinate on glycolysis and antioxidant status in the liver of STZ-induced diabetic rats. *J. Inorg. Biochem.* **2014**, *136*, 47–56. [CrossRef]
16. Ścibior, A.; Pietrzyk, Ł.; Plewa, Z.; Skiba, A. Vanadium: Risks and possible benefits in the light of a comprehensive overview of its pharmacotoxicological mechanisms and multi-applications with a summary of further research trends. *J. Trace Elem. Med. Biol.* **2020**, *61*, 126508. [CrossRef]
17. Ścibior, A.; Wnuk, E.; Gołębiowska, D. Wild animals in studies on vanadium bioaccumulation—Potential animal models of environmental vanadium contamination: A comprehensive overview with a Polish accent. *Sci. Total Environ.* **2021**, *785*, 147205. [CrossRef]
18. Ścibior, A. Overview of research on vanadium-quercetin complexes with a historical outline. *Antioxidants* **2022**, *11*, 790. [CrossRef] [PubMed]
19. Ścibior, A.; Kurus, J. Vanadium and oxidative stress markers-in vivo model: A review. *Curr. Med. Chem.* **2019**, *26*, 5456–5500. [CrossRef] [PubMed]
20. Kim, G.H.; Kim, J.E.; Rhie, S.J.; Yoon, S. The role of oxidative stress in neurodegenerative diseases. *Exp. Neurobiol.* **2015**, *24*, 325–340. [CrossRef] [PubMed]
21. Kučera, J.; Sabbioni, E. Baseline vanadium levels in human blood, serum, and urine. In *Vanadium in the Environment. Part 2: Health Effects*; Nriagu, J.O., Ed.; John Wiley and Sons: New York, NY, USA, 1998; Volume 31, pp. 75–91.
22. Baierle, M.; Charão, M.F.; Göethel, G.; Barth, A.; Fracasso, R.; Bubols, G.; Sauer, E.; Campanharo, S.C.; Rocha, R.C.C.; Saint'Pierre, T.D.; et al. Are delta-aminolevulinate dehydratase inhibition and metal concentrations additional factors for the age-related cognitive decline? *Int. J. Environ. Res. Public Health* **2014**, *11*, 10851–10867. [CrossRef]
23. Yang, B.; He, J.; Zhang, G.; Gua, J. *Vanadium. Extraction, Manufacturing and Applications*, 1st ed.; Elsevier: Amsterdam, The Netherlands, 2021.
24. Nriagu, J.O. *Vanadium in the Environment, Part 2: Health Effects*; John Wiley and Sons: New York, NY, USA, 1998; pp. 1–424.
25. Hu, J.; Xia, W.; Pan, X.; Zheng, T.; Zhang, B.; Zhou, A.; Buka, S.L.; Bassig, B.A.; Liu, W.; Wu, C.; et al. Association of adverse birth outcomes with prenatal exposure to vanadium: A population-based cohort study. *Lancet Planet Health* **2017**, *1*, e230–e241. [CrossRef]
26. Hu, J.; Peng, Y.; Zheng, T.; Zhang, B.; Liu, W.; Wu, C.; Jiang, M.; Braun, J.M.; Liu, S.; Buka, S.L.; et al. Effects of trimester-specific exposure to vanadium on ultrasound measures of fetal growth and birth size: A longitudinal prospective prenatal cohort study. *Lancet Planet Health* **2018**, *2*, e427–e437. [CrossRef]
27. Malandrino, P.; Russo, M.; Ronchi, A.; Minoia, C.; Cataldo, D.; Regalbuto, C.; Giordano, C.; Attard, M.; Squatrito, S.; Trimarchi, F.; et al. Increased thyroid cancer incidence in a basaltic volcanic area is associated with non-anthropogenic pollution and biocontamination. *Endocrine* **2016**, *53*, 471–479. [CrossRef]

28. Minoia, C.; Pietra, R.; Sabbioni, E.; Ronchi, A.; Gatti, A.; Cavalleri, A.; Manzo, L. Trace element reference values in tissues from inhabitants of the European Community. III. The control of preanalytical factors in the biomonitoring of trace elements in biological fluids. *Sci. Total Environ.* **1992**, *120*, 63–79. [CrossRef]
29. Lanocha-Arendarczyk, N.; Kosik-Bogacka, D.I.; Kalisinska, E.; Sokołowski, S.; Kolodziej, L.; Budis, H.; Safranow, K.; Kot, K.; Ciosek, Z.; Tomska, N.; et al. Influence of environmental factors and relationships between vanadium, chromium, and calcium in human bone. *BioMed. Res. Inter.* **2016**, *2016*, 8340425. [CrossRef]
30. Sampath, V.P.; Singh, S.V.; Pelov, I.; Horesh, N.; Zannadeh, H.; Tirosh, O.; Erel, Y.; Lichtstein, D. Vanadium in bipolar disorders—Reviving an old hypothesis. *Int. J. Mol. Sci.* **2022**, *23*, 13901. [CrossRef] [PubMed]
31. European Food Safety Authority (EFSA) Panel on Opinion of the Scientific Panel on Dietetic products, Nutrition and Allergies on a request from the Commission related to the Tolerable Upper Intake Level of Vanadium. *EFSA J.* **2004**, *33*, 1–22. [CrossRef]
32. Bernhard, D.; Rossmann, A.; Wick, G. Metals in cigarette smoke. *Life* **2005**, *57*, 805–809. [CrossRef] [PubMed]
33. Adachi, A.; Asai, K.; Koyama, Y.; Matsumoto, Y.; Okano, T. Determination of vanadium in cigarettes by atomic absorption spectrophotometry. *Anal. Lett.* **1998**, *3*, 1769–1776. [CrossRef]
34. Toro-Román, V.; Bartolomé, I.; Siquier-Coll, J.; Alves, J.; Grijota, F.J.; Muñoz, D.; Maynar-Mariño, M. Serum vanadium concentrations in different sports modalities. *J. Trace Elem. Med. Biol.* **2021**, *68*, 126808. [CrossRef] [PubMed]
35. Basu, R.; Harris, M.; Sie, L.; Malig, B.; Broadwin, R.; Green, R. Effects of fine particulate matter and its constituents on low birth weight among full-term infants in California. *Environ. Res.* **2014**, *128*, 42–51. [CrossRef]
36. Bell, M.L.; Ebisu, K.; Peng, R.D.; Samet, J.M.; Dominici, F. Hospital admissions and chemical composition of fine particle air pollution. *Am. J. Respir. Crit. Care Med.* **2009**, *179*, 1115–1120. [CrossRef]
37. Cakmak, S.; Dales, R.; Kauri, L.M.; Mahmud, M.; Van Ryswyk, K.; Vanos, J.; Liu, L.; Kumarathasan, P.; Thomson, E.; Vincent, R.; et al. Metal composition of fine particulate air pollution and acute changes in cardiorespiratory physiology. *Environ. Poll.* **2014**, *189*, 208–214. [CrossRef]
38. Jiang, M.; Li, Y.; Zhang, B.; Zhou, A.; Zheng, T.; Qian, Z.; Du, X.; Zhou, Y.; Pan, X.; Hu, J.; et al. A nested case-control study of prenatal vanadium exposure and low brithweight. *Hum. Reprod.* **2016**, *31*, 2135–2141. [CrossRef]
39. Jiang, S.; Zhou, S.; Liu, H.; Peng, C.; Zhang, X.; Zhou, H.; Wang, Z.; Lu, Q. Concentrations of vanadium in urine with hypertension prevalence and blood pressure levels. *Ecotoxicol. Environ. Saf.* **2021**, *213*, 112028. [CrossRef]
40. Jin, S.; Xia, W.; Jiang, Y.; Sun, X.; Huang, S.; Zhang, B.; Zhou, A.; Zheng, T.; Xu, S.; Li, Y. Urinary vanadium concentration in relation to premature rupture of membranes: A birth cohort study. *Chemosphere* **2018**, *210*, 1035–1041. [CrossRef] [PubMed]
41. Patel, M.M.; Hoepner, L.; Garfinkel, R.; Chillrud, S.; Reyes, A.; Quinn, J.W.; Perera, F.; Miller, R.L. Ambient metals, elemental carbon, and wheeze and cough in New York City children through 24 months of age. *Am. J. Respir. Crit. Care Med.* **2009**, *180*, 1107–1113. [CrossRef] [PubMed]
42. Pellegriti, G.; De Vathaire, F.; Scollo, C.; Attard, M.; Giordano, C.; Arena, S.; Dardanoni, G.; Frasca, F.; Malandrino, P.; Vermiglio, F.; et al. Papillary thyroid cancer indicence in the volcanic area of Sicily. *J. Natl. Cancer Inst.* **2009**, *101*, 1575–1583. [CrossRef]
43. Seko, Y.; Watanabe, K.; Hasegawa, T. Vanadium in ground water from Mt. Fuji: Does it have health effect on habitants around the mountain? *Chin. J. Geochem.* **2006**, *25* (Suppl. S1), 60–61. [CrossRef]
44. Sitprija, V.; Eiam-Ong, S. Vanadium and metabolic problems. In *Vanadium in the Environment. Part 2: Health Effects*; Nriagu, J.O., Ed.; John Wiley and Sons: New York, NY, USA, 1998; Volume 31, pp. 91–120.
45. Sørensen, M.; Schins, R.P.F.; Hertel, O.; Loft, S. Transition metals in personal samples of $PM_{2.5}$ and oxidative stress in human volunteers. *Cancer Epidemiol. Biomarkers Prev.* **2005**, *14*, 1340–1343. [CrossRef]
46. Blalock, T.L.; Hill, C.H. Studies on the role of iron in the reversal of vanadium toxicity in chicks. *Biol. Trace Elem. Res.* **1987**, *14*, 225–235. [CrossRef]
47. Dai, L.; Bind, M.A.; Koutrakis, P.; Coull, B.A.; Sparrow, D.; Vokonas, P.S.; Schwarts, J.D. Fine particles, genetic pathways, and markers of inflammation and endothelial dysfunction: Analysis on particulate species and sources. *J. Expo. Sci. Environ. Epidemiol.* **2016**, *26*, 415–421. [CrossRef]
48. Blankenberg, S.; Rupprecht, H.J.; Bickel, C.; Peetz, D.; Hafner, G.; Tiret, L.; Meyer, J. Circulating soluble adhesion molecules and death in patients with coronary artery disease. *Circulation* **2001**, *104*, 1336–1342. [CrossRef]
49. Pradhan, A.D.; Rifai, N.; Ridker, P.M. Solube intercellular adhesion molecule-1, soluble vascular adhesion molecule-1, and the development of symptomatic peripheral artirial disease in men. *Circulation* **2002**, *106*, 820–825. [CrossRef]
50. Chen, J.; Rodopoulou, S.; de Hoogh, K.; Strak, M.; Jovanovic Andersen, Z.; Atkinson, R.; Bauwelinck, M.; Bellander, T.; Brandt, J.; Cesaroni, G.; et al. Long-term exposure to fine particle elemental components and natural and cause-specific mortality—A pooled analysis of eight European cohorts within the ELAPSE project. *Environ. Health Perspect.* **2021**, *129*, 47009. [CrossRef]
51. Wang, Y.; Shi, L.; Lee, M.; Liu, P.; Di, Q.; Zanobetti, A.; Schwartz, J.D. Long-term exposure to $PM_{2.5}$ and mortality among older adults in the Southeastern US. *Epidemiology* **2017**, *28*, 207–2014. [CrossRef]
52. Wolf, K.; Stafoggia, M.; Cesaroni, G.; Jovanovic Andersen, Z.; Beelen, R.; Galassi, C.; Hennig, F.; Migliore, E.; Penell, J.; Ricceri, F.; et al. Long-term exposure to particulate matter constituents and the incidence of coronary events in 11 European cohorts. *Epidemiology* **2015**, *26*, 565–574. [CrossRef]
53. Hvidtfeldt, U.A.; Chen, J.; Jovanovic Andersen, Z.; Atkinson, R.; Bauwelinck, M.; Bellander, T.; Brandt, J.; Brunekreef, B.; Cesaroni, G.; Concin, H.; et al. Long-term exposure to fine particle elemental components and lung cancer incidence in the ELAPSE pooled cohort. *Environ. Res.* **2021**, *193*, 110568. [CrossRef]

54. Eeftens, M.; Hoek, G.; Gruzieva, O.; Mölter, A.; Agius, R.; Beelen, R.; Brunekreef, B.; Custovic, A.; Cyrys, J.; Fuerte, E.; et al. Elemental composition of particulate matter and the association with lung function. *Epidemiology* **2014**, *25*, 648–657. [CrossRef]
55. Alfaro-Moreno, E.; Ponce-de-León, S.; Osorino-Vargas, A.R.; García-Cuellar, C.; Martínez, L.; Rosas, I. Potential toxic effects associated to metals and endotoxin present in $PM_{10}$: An ancillary study using multivariate analysis. *Inhal. Toxicol.* **2007**, *19* (Suppl. S1), 49–53. [CrossRef]
56. Campen, M.J.; Nolan, J.P.; Schladweiler, M.C.J.; Kodavanti, U.P.; Evansky, P.A.; Costa, D.L.; Watkinson, W.P. Cardiovascular and thermoregulatory effects of inhaled PM-associated transition metals: A potential interaction between nickel and vanadium sulfate. *Toxicol. Sci.* **2001**, *64*, 243–252. [CrossRef]
57. Szabo, S.T.; Harry, G.J.; Hayden, K.M.; Szabo, D.T.; Birnbaum, L. Comparison of metal levels between postmortem brain and ventricular fluid in Alzheimer's disease and nondemented elderly controls. *Toxicol. Sci.* **2016**, *150*, 292–300. [CrossRef]
58. Gerhardsson, L.; Lundh, T.; Minthon, L.; Londos, E. Metal concentrations in plasma and cerebrospinal fluid in patients with Alzheimer's disease. *Dement. Geriatr. Cogn. Disord.* **2008**, *25*, 508–515. [CrossRef]
59. Hershey, C.O.; Hershey, L.A.; Varnes, A.; Vibhakar, S.D.; Lavin, P.; Strain, W. Cerebrospinal fluid trace element content in dementia: Clinical, radiologic, and pathologic correlations. *Neurology* **1983**, *33*, 1350–1353. [CrossRef]
60. Ward, N.I.; Mason, J.A. Neutron activation analysis techniques for identifying elemental status in Alzheimer's disease. *J. Radioanal. Nucl. Chem.* **1987**, *113*, 515–526. [CrossRef]
61. Gerhardsson, L.; Blennow, K.; Lundh, T.; Londos, E.; Minthon, L. Concentrations of metals, β-amyloid and tau-markers in cerebrospinal fluid in patients with Alzheimer's disease. *Dement. Geriatr. Cogn. Disor.* **2009**, *28*, 88–94. [CrossRef] [PubMed]
62. Srivastava, R.A.; Jain, J.C. Scavenger receptor class B type I expression and elemental analysis in cerebellum and parietal cortex regions of the Alzheimer's disease brain. *J. Neurol. Sci.* **2002**, *196*, 45–52. [CrossRef] [PubMed]
63. Roos, P.M.; Vesterberg, O.; Syversen, T.; Flaten, T.P.; Nordberg, M. Metal concentration in cerebrospinal fluid and blood plasma from patients with amyotrophic lateral sclerosis. *Biol. Trace Elem. Res.* **2013**, *151*, 159–170. [CrossRef] [PubMed]
64. Gellein, K.; Garruto, R.M.; Syversen, T.; Sjøbakk, T.E.; Flaten, P. Concentrations of Cd, Co, Cu, Fe, Mn, Rb, V, and Zn in formalin-fixed brain tissue in amyotrophic lateral sclerosis and parkinsonism-dementia complex of guam determined by high-resolution ICP-MS. *Biol. Trace Elem. Res.* **2003**, *96*, 39–60. [CrossRef] [PubMed]
65. Patti, F.; Fiore, M.; Chisari, C.G.; D'Amico, E.; Lo Fermo, S.; Toscano, S.; Copat, C.; Ferrante, M.; Zappia, M. CSF neurotoxic metals/metalloids levels in amyotrophic lateral sclerosis patients: Comparison between bulbar and spinal onset. *Environ. Res.* **2020**, *188*, 109820. [CrossRef]
66. Shi, M.T. Determination of multiple chemical elements in CSF in Parkinson disease after intracerebral autotransplantation of the adrenal medulla. *Zhonghua Wai Ke Za Zhi* **1991**, *29*, 129–132.
67. Bocca, B.; Alimonti, A.; Petrucci, F.; Violante, N.; Sancesario, G.; Forte, G.; Senofonte, O. Quantification of trace elements by sector field indutcively coupled plasma mass spectrometry in urine, serum, blood and cerebrospinal fluid of patients with Parkinson's disease. *Spectrochim. Acta Part B* **2004**, *59*, 559–566. [CrossRef]
68. Uitti, R.J.; Rajput, A.H.; Rozdilsky, B.; Bickis, M.; Wollin, T.; Yuen, W.K. Regional metal concentrations in Parkinson's disease, other chronić neurological diseases, and control brains. *Can. J. Neurol. Sci.* **1989**, *16*, 310–314. [CrossRef]
69. González-Dominguez, R.; Garcia-Barrera, T.; Gómez-Ariza, J.L. Characterization of metal profiles in serum during the progression of Alzheimer's disease. *Metallomics* **2014**, *6*, 292–300. [CrossRef]
70. Paglia, G.; Miedico, O.; Cristofano, A.; Vitale, M.; Angiolillo, A.; Chiaravalle, A.E.; Corso, G.; Di Costanzo, A. Distinctive pattern of serum elements during the progression of Alzheimer's disease. *Sci. Rep.* **2016**, *6*, 22769. [CrossRef]
71. Akanle, O.A.; Spyrou, N.M.; Damyanova, A.A.; Shaw, D.M.; Ali, L. Investigation of elemental models in senile dementia and depressives using neutron activation analysis. *J. Radioanal. Nucl. Chem.* **1987**, *113*, 405–416. [CrossRef]
72. Lavanya, R.D.; Reddy, B.S.; Sattar, S.A.; Rao, A.D.P. Trace element imbalances in blood serum of Alzheimer's disease patients. *Spectre. Lett.* **2021**, *54*, 458–471. [CrossRef]
73. Guan, C.; Dang, R.; Cui, Y.; Liu, L.; Chen, X.; Wang, X.; Zhu, J.; Li, D.; Li, J.; Wang, D. Characterization of plasma metal profiles in Alzheimer's disease using multivariate statistical analysis. *PLoS ONE* **2017**, *12*, e0178271. [CrossRef] [PubMed]
74. Alimonti, A.; Ristori, G.; Giubilei, F.; Stazi, M.A.; Pino, A.; Visconti, A.; Brescianini, S.; Monti, M.S.; Forte, G.; Stanzione, P.; et al. Serum chemical elements and oxidative status in Alzheimer's disease, Parkinson disease and multiple sclerosis. *Neurotoxicology* **2007**, *28*, 450–456. [CrossRef]
75. Bocca, B.; Forte, G.; Petrucci, F.; Pino, A.; Marchione, F.; Bombi, G.; Senofonte, O.; Giubilei, F.; Alimonti, A. Monitoring of chemical elements and oxidative damage in patients affectd by Alzheimer's disease. *Ann. Ist. Super Sanita* **2005**, *41*, 197–203.
76. Forte, G.; Alimonti, A.; Pino, A.; Stanzione, P.; Brescianini, S.; Brusa, L.; Sancesario, G.; Violante, N.; Bocca, B. Metals and oxidative stress in patients with Parkinson's disease. *Ann. Ist. Super Sanita* **2005**, *41*, 189–195.
77. Royce-Nagel, G.; Cudkowicz, M.; Myers, D.; Nicholson, K.; Shui, A.; Schoenfeld, D.; Huang, X.; Brown, R.H. Vanadium, aluminium, magnesium, and manganese are not elevated in hair samples in amyotrophic lateral sclerosis. *Amyotroph. Lateral Scler.* **2010**, *11*, 492–493. [CrossRef]
78. Barth, A.; Schaffer, A.W.; Konnaris, C.; Blauensteiner, R.; Winker, R.; Osterode, W.; Rüdiger, H.W. Neurobehavioral effects of vanadium. *J. Toxicol. Environ. Health Part A* **2002**, *65*, 677–683. [CrossRef]
79. Usutani, S.; Nishiyama, K.; Sato, I.; Matsuura, K.; Sawada, Y.; Kawabata, T.; Hosokawa, Y.; Izumi, S. Results of the special physical examination of workers in a vanadium plant. *Jpn. J. Ind. Health* **1979**, *21*, 21–28. [CrossRef]

80. Li, H.; Zhou, D.; Zhang, Q.; Feng, C.; Zheng, W.; He, K.; Lan, Y. Vanadium exposure-induced neurobehavioral alterations among Chinese workers. *Neurotoxicology* **2013**, *36*, 49–54. [CrossRef]
81. Zhou, D.; Feng, C.; Lan, Y.J.; Wang, Z.; Huang, S.; Wang, M.; Zhu, T. Paired-control study on the effect of vanadium on neurobehavioral functions. *Sichuan Da Xue Xue Bao Yi Xue Ban* **2007**, *38*, 468–470.
82. Zhu, C.W.; Liu, Y.X.; Huang, C.J.; Gao, W.; Hu, G.L.; Li, J.; Zhang, O.; Lan, Y.J. Effect of vanadium exposure on neurobehavioral function in workers. *Zhonghua Lao Dong Wei Sheng Zhi Ye Bing Za Zhi* **2016**, *34*, 103–106. [CrossRef]
83. Schlake, H.P.; Bertram, H.P.; Husstedt, I.W.; Schuierer, G. Acute systemic vanadate poisoning presenting as cerebrovascular ischemia with prolonged reversible neurological deficits (PRIND). *Clin. Neurol. Neurosurg.* **1994**, *96*, 92–95. [CrossRef]
84. World Health Organization (WHO). International Atomic Energy Agency & Food and Agriculture Organization of the United Nations, 1996. Trace Elements in Human Nutrition and Health. World Health Organization. Available online: https://apps.who.int/iris/handle/10665/37931 (accessed on 17 January 2023).
85. Anger, W.K. Reconsideration of the WHO NCTB strategy and test selection. *Neurotoxicology* **2014**, *45*, 224–231. [CrossRef]
86. Calderón-Garcidueñas, L.; Serrano-Sierra, A.; Torres-Jardón, R.; Zhu, H.; Yuan, Y.; Smith, D.; Delgado-Chávez, R.; Cross, J.V.; Medina-Cortina, H.; Kavanaugh, M.; et al. The impact of environmental metals in young urbanites' brains. *Exp. Toxicol. Pathol.* **2013**, *65*, 503–511. [CrossRef]
87. Calderón-Garcidueñas, L.; Vojdani, A.; Blaurock-Bush, E.; Busch, Y.; Friedle, A.; Franco-Lira, M.; Sarathi-Mukherjee, P.; Martínez-Aguirre, X.; Park, S.B.; Torres-Jardón, R.; et al. Air pollution and children: Neural and tight junction antibodies and combustion metals, the role of barier breakdown and brain immunity in neurodegeneration. *J. Alzheimer's Dis.* **2015**, *43*, 1039–1058. [CrossRef]
88. Calderón-Garcidueñas, L.; Avila-Ramírez, J.; Calderón-Garcidueñas, A.; González-Heredia, T.; Acuña-Ayala, H.; Chao, C.; Thompson, C.; Ruiz-Ramos, R.; Cortéz-González, V.; Martínez- Martínez, L.; et al. Cerebrospinal fluid biomarkers in highly exposed $PM_{2.5}$ urbanites: The risk of Alzheimer's and Parkinson's diseases in young Mexico City residents. *J. Alzheimer's Dis.* **2016**, *52*, 597–613. [CrossRef]
89. Yu, Z.; Peters, S.; van Boxmeer, L.; Downward, G.S.; Hoek, G.; Kioumourtzoglou, M.A.; Weisskopf, M.G.; Hansen, J.; van den Berg, L.H.; Vermeulen, R.C.H. Long-term exposure to ultrafine particles and particulate matter constituents and the risk of amyotrophic lateral sclerosis. *Environ. Health Perspect.* **2021**, *129*, 097702. [CrossRef]
90. Johnson, F.O.; Atchison, W. The role of environmental mercury, lead, and pesticide exposure in development of amyotrophic lateral sclerosis. *Neurotoxicology* **2009**, *30*, 761–765. [CrossRef]
91. Willis, A.W.; Evanoff, B.A.; Lian, M.; Galarza, A.; Wegrzyn, A.; Schootman, M.; Racette, B.A. Metal emissions and urban incydent Parkinson disease: A community health study of medicare beneficiaries by using geographic information systems. *Am. J. Epidemiol.* **2010**, *172*, 1357–1363. [CrossRef]
92. Elonheimo, H.M.; Andersen, H.R.; Katsonouri, A.; Tolonen, H. Environmental substances associated with Alzheimer's disease-a scoping review. *Int. J. Environ. Res. Public Health* **2021**, *18*, 11839. [CrossRef]
93. Barceloux, D.G. Vanadium. *J. Toxicol. Clin. Toxicol.* **1999**, *37*, 265–278. [CrossRef]
94. International Agency for Research on Cancer (IARC). Cobalt in hard metals and cobalt sulfate, gallium arsenide, indium phosphide and vanadium pentoxide. *IARC Monogr. Eval. Carcinog. Risks Hum.* **2006**, *86*, 119–237.
95. De Hoogh, K.; Wang, M.; Adam, M.; Badaloni, C.; Beelen, R.; Birk, M.; Cesaroni, G.; Cirach, M.; Declercq, C.; Dedele, A.; et al. Development of land use regression models for particle composition in twenty study areas in Europe. *Environ. Sci. Technol.* **2013**, *47*, 5778–5786. [CrossRef]
96. Bosco, M.L.; Varrica, D.; Dongarria, G. Case study: Inorganic pollutants associated with particulate matter from an area near a petrochemical plant. *Environ. Res.* **2005**, *99*, 18–30. [CrossRef]
97. Moreno, T.; Querol, X.; Alastuey, A.; de la Rosa, J.; Sánchez de la Campa, A.M.; Mingullón, M.; Pandolfi, M.; González-Castanedo, Y.; Monfort, E.; Gibbons, W. Variations in vanadium, nickel and lanthanoid element concentrations in urban air. *Sci. Total Environ.* **2010**, *408*, 4569–4579. [CrossRef]
98. U.S. Department of Health and Hunam Services Public Health Service Agency for Toxic Substances and Disease Registry. *Toxicological Profile for Vanadium (TPV)*; U.S. Department of Health and Hunam Services Public Health Service Agency for Toxic Substances and Disease Registry: Atlanta, GA, USA, 2012.
99. Dye, J.A.; Adler, K.B.; Richards, J.H.; Dreher, K.L. Role of soluble metals in oil fly ash-induced airway epithelial injury and cytokine gene expression. *Am. J. Physiol.* **1999**, *277*, L498–L510. [CrossRef]
100. U.S. Environmental Protection Agency. Toxicological review of vanadium pentoxide ($V_2O_5$) (CAS No. 1314-62-1). In *Support of Summary Information on the Integrated Risk Information System (IRIS)*; U.S. Environmental Protection Agency Publ.: Washington, DC, USA, 2011; 210p.
101. Korbecki, J.; Baranowska-Bosiacka, I.; Gutowska, I.; Chlubek, D. Vanadium compounds as pro-inflammatory agents: Effects on cyclooxygenases. *Int. J. Mol. Sci.* **2015**, *16*, 12648–12668. [CrossRef]
102. Todorich, B.; Olopade, J.O.; Surguladze, N.; Zhang, X.; Neely, E.; Connor, J.R. The mechanism of vanadium-mediated developmental hypomyelination is related to destruction of oligodendrocyte progenitors through a relationship with ferritin and iron. *Neurotox. Res.* **2011**, *19*, 361–373. [CrossRef]
103. Soazo, M.; Garcia, G.B. Vanadium exposure through lactation produces behavioral alterations and CNS myelin deficyt in neonatal rats. *Neurotoxicol. Teratol.* **2007**, *29*, 503–510. [CrossRef]

104. Azeez, I.A.; Olopade, F.; Laperchia, C.; Andriolo, A.; Scambi, I.; Onwuka, S.K.; Bentivoglio, M.; Olopade, J.O. Regional myelin and axon damage and neuroinflammation in the adult mouse brain after long-term postnatal vanadium exposure. *J. Neuropathol. Exp. Neurol.* **2016**, *75*, 843–854. [CrossRef]
105. Usende, I.L.; Olopade, J.O.; Azeez, I.A.; Andrioli, A.; Bankole, M.O.; Olopade, F.E.; Nafady, A.A.; Bentivoglio, M. Neuroecotoxicology: Effects of environmental heavy metal exposure on the brain of African giant rats and the contribution of vanadium to the neuropathology. *IBRO Neurosci. Rep.* **2022**, *13*, 215–234. [CrossRef]
106. Nickel, M.; Gu, C. Regulation of central nervous system myelination in higher brain functions. *Neural Plast.* **2018**, *2018*, 6436453. [CrossRef]
107. Kazana, W.; Zabłocka, A. Brain-derived neurotrophic factor as a potential therapeutic tool in the treatment of nervous system disorders. *Postępy Hig. Med. Doświadczalnej* **2020**, *74*, 517–531. [CrossRef]
108. Miranda, M.; Morici, J.F.; Zanoni, M.B.; Bekinschtein, P. Brain-derived neurotrophic factor: A key molecule for memory in the healthy and the pathological brain. *Front. Cell. Neurosci.* **2019**, *13*, 363. [CrossRef]
109. Wang, D.C.; Lin, Y.Y.; Lin, H.T. Recovery of motor coordination after exercise is correlated to enhancement of brain-derived neurotrophic factor in lactational vanadium-exposed rats. *Neurosci. Lett.* **2015**, *600*, 232–237. [CrossRef]
110. González-Maciel, A.; Reynoso-Robles, R.; Torres-Jardón, R.; Mukherjee, P.S.; Calderón-Garcidueñas, L.G. Combustion-derived nanoparticles in key brain target cells and organelles in young urbanites: Culprit hidden in plain sight in Alzheimer's disease development. *J. Alzheimers Dis.* **2017**, *59*, 189–208. [CrossRef]
111. Cole-Hunter, T.; Zhang, J.; So, R.; Samoli, E.; Liu, S.; Chen, J.; Strak, M.; Wolf, K.; Weinmayr, G.; Rodopolou, S.; et al. Long-term air pollution exposure and Parkinson's disease mortality in a large pooled European cohort: An ELAPSE study. *Environ. Int.* **2023**, *171*, 107667. [CrossRef]
112. Montiel-Flores, E.; Mejía-García, O.A.; Ordoñez-Librado, J.L.; Gutierrez-Valdez, A.L.; Espinosa-Villanueva, J.E.; Dorado-Martínez, C.; Reynoso-Erazo, L.; Tron-Alvarez, R.; Rodríguez-Lara, V.; Avila-Costa, M.R. Alzheimer-like cell death after vanadium pentoxide inhalation. *Heliyon* **2021**, *7*, e07856. [CrossRef]
113. Calderón-Garcidueñas, L.; Maronpot, R.R.; Torres-Jardon, R.; Henríquez-Roldán, C.; Schoonhoven, R.; Acuña-Ayala, H.; Villarreal-Calderón, A.; Nakamura, J.; Fernando, R.; Reed, W.; et al. DNA damage in nasal and brain tissues of canines exposed to air pollutants is asscociated with evidence of chronić brain inflammation and neurodegeneration. *Toxicol. Pathol.* **2003**, *31*, 524–538. [CrossRef]
114. Sánchez, G.I.; López, I.; Mussali, P.; Bizarro, N.P.; Niño, G.; Saldivar, L.; Espejel, G.; Avila, M.; Morales, D.; Colin, L.; et al. Vanadium concentrations in lung, liver, kidney, testes and brain after the inhalation of 0.02 M of $V_2O_5$. An experimental model in mice. In Proceedings of the 42nd Annual Meeting of the Society of Toxicology, Salt Lake City, UT, USA, 9–13 March 2003.
115. Avila-Costa, M.R.; Montiel Flores, E.; Colin-Barenque, L.; Ordoñez, J.L.; Gutiérrez, A.L.; Niño-Cabrera, H.G.; Mussali-Galante, P.; Fortoul, T.I. Nigrostriatal modifications after vanadium inhalation: An immunocytochemical and cytological approach. *Neurochem. Res.* **2004**, *29*, 1365–1369. [CrossRef]
116. Avila-Costa, M.R.; Colín-Barenque, L.; Zepeda-Rodríguez, A.; Antuna, S.B.; Saldivar, L.; Espejel-Maya, G.; Mussali-Galante, P.; del Carmen Avila-Casado, M.; Reyes-Olivera, A.; Anaya-Martinez, V.; et al. Ependymal epithelium disruption after vanadium pentoxide inhalation. A mice experimental model. *Neurosci. Lett.* **2005**, *381*, 21–25. [CrossRef]
117. Avila-Costa, M.R.; Fortoul, T.I.; Niño-Cabrera, G.; Colín-Barenque, L.; Bizarro-Nevares, P.; Gutiérrez-Valdez, A.L.; Ordóñez-Librado, J.L.; Rodríguez-Lara, V.; Mussali-Galante, P.; Díaz-Bech, P.; et al. Hippocampal cell alterations induced by the inhalation of vanadium pentoxide ($V_2O_5$) promote memory deterioration. *Neurotoxicology* **2006**, *27*, 1007–1012. [CrossRef]
118. Colín-Barenque, L.; Martínez-Hernández, M.G.; Baiza-Gutman, L.A.; Avila-Costa, M.R.; Ordóñez-Librado, J.L.; Bizarro-Nevares, P.; Rodriguez-Lara, V.; Piñón-Zarate, G.; Rojas-Lemus, M.; Mussali-Galante, P.; et al. Matrix metalloproteinases 2 and 9 in central nervous system and their modification after vanadium inhalation. *J. Appl. Toxicol.* **2008**, *28*, 718–723. [CrossRef]
119. Igado, O.O.; Olopade, J.O.; Onwuka, S.K.; Chukwudi, A.C.; Daramola, O.A.; Ajufo, U.E. Evidence of environmental pollution in caprine brains ontained from a relatively unindustrialized area in Nigeria. *Afr. J. Biomed. Res.* **2008**, *11*, 305–309. [CrossRef]
120. Ngwa, H.A.; Kanthasamy, A.; Jin, H.; Anantharam, V.; Kanthasamy, A.G. Vanadium exposure induces olfactory dysfunction in an animal model of metal neurotoxicity. *Neurotoxicology* **2014**, *43*, 73–81. [CrossRef]
121. Colín-Barenque, L.; Pedraza-Chaverri, J.; Medina-Campos, O.; Jimenez-Martínez, R.; Bizarro-Nevares, P.; González-Villalva, A.; Rojas-Lemus, M.; Fortoul, T.I. Functional and morphological olfactory bulb modifications in mice after vanadium inhalation. *Toxicol Pathol.* **2015**, *43*, 282–291. [CrossRef]
122. DeWitt, J.; Buck, B.; Goossens, D.; Hu, Q.; Chow, R.; David, W.; Young, S.; Teng, Y.; Leetham-Spencer, M.; Murphy, L.; et al. Health effects following subacute exposure to geogenic dusts from arsenic-rich sediment at the Nellis Dunes Recreation Area, Las vegas, NV. *Toxicol. Appl. Pharmacol.* **2016**, *304*, 79–89. [CrossRef]
123. DeWitt, J.C.; Buck, B.J.; Goossens, D.; Teng, Y.; Pollard, J.; McLaurin, B.T.; Gerads, R.; Keil, D.E. Health effects following subacute exposure to geogenic dust collected from active drainage surfaces (Nellis Dunes Recreation Area, Las Vegas, NV). *Toxicol. Rep.* **2017**, *4*, 19–31. [CrossRef]
124. Keil, D.; Buck, B.; Goossens, D.; Teng, Y.; Leetham, M.; Murphy, L.; Pollard, J.; Eggers, M.; McLaurin, B.; Gerads, R.; et al. Immunotoxicological and neurotoxicological profile of health effects following subacute exposure to geogenic dust from sand dunes at the Nellis Dune Recreation Area, Las Vegas, NV. *Toxicol. Appl. Pharmacol.* **2016**, *291*, 1–12. [CrossRef]

125. Usende, I.L.; Emikpe, B.O.; Olopade, J.O. Heavy metal pollutants in selected organs of African giant rats from three agro-ecological zones of Nigeria: Evidence for their role as an environmental specimen bank. *Environ. Sci. Pollut. Res.* **2017**, *24*, 22570–22578. [CrossRef]
126. Colín-Barenque, L.; Bizarro-Nevares, P.; González Villalva, A.; Pedraza-Chaverri, J.; Medina-Campos, O.N.; Jimenez-Martínez, R.; Rodríguez-Rangel, D.S.; Reséndiz, S.; Fortoul, T.I. Neuroprotective effect of carnosine in the olfactory bulb after vanadium inhalation in a mouse model. *Int. J. Exp. Path.* **2018**, *99*, 180–188. [CrossRef]
127. Hara, Y.; Fillit, H.M. The dawn of a new era of Alzheimer's research and drug development. *J. Prev. Alzheimers Dis.* **2022**, *9*, 385–386. [CrossRef]
128. Sherzai, D.; Sherzai, A. Preventing Alzheimer's: Our Most Urgent Health Care Priority. *Am. J. Lifestyle Med.* **2019**, *13*, 451–461. [CrossRef]
129. Pagano, G.; Taylor, K.; Anzures-Cabrera, J.; Marchesi, M.; Simuni, T.; Marek, K.; Postuma, B.R.; Pavese, N.; Stocchi, F.; Azulay, J.P.; et al. Trial of prasinezumab in early-stage Parkinson's Disease. *N. Engl. J. Med.* **2022**, *387*, 421–432. [CrossRef]
130. Choroba Parkinsona. 5 December 2021. Available online: https://www.mp.pl/neurologia/choroba-parkinsona/232939,choroba-parkinsona (accessed on 14 April 2023).
131. Dobrakowski, P.P.; Machowska-Majchrzak, A.K.; Labuz-Roszak, B.; Majchrzak, K.G.; Kluczewska, E.; Pierzchała, K.B. MR-guided focused ultrasound: A new generation treatment of Parkinson's disease, essential tremor and neuropathic pain. *Interv. Neuroradiol.* **2014**, *20*, 275–282. [CrossRef]
132. Khamaysa, M.; Pradat, P.F. Status of ALS treatment, insights into therapeutic challenges and dilemmas. *J. Pers. Med.* **2022**, *12*, 1601. [CrossRef]
133. Szymczyk, H. Magnez—Pierwiastek niezbędny do prawidłowego funkcjonowania organizmu. *Farm. Współ.* **2016**, *9*, 217–223.
134. Ścibior, A.; Gołębiowska, D.; Niedźwiecka, I. Magnesium can protect against vanadium-induced lipid peroxidation in the hepatic tissue. *Oxid. Med. Cell. Longev.* **2013**, *2013*, 802734. [CrossRef]
135. Chui, D.; Chen, Z.; Yu, J.; Zhang, H.; Wang, W.; Song, Y.; Yang, H.; Zhou, L. *Magnesium in the Central Nervous System*; Vink, R., Nechifor, M., Eds.; University of Adelaide Press: Adelaide, AU, USA, 2011. Available online: https://www.ncbi.nlm.nih.gov/books/NBK507256/ (accessed on 17 January 2023).
136. Ścibior, A.; Gołębiowska, D.; Adamczyk, A.; Niedźwiecka, I.; Fornal, E. The renal effects of vanadate exposure: Potential biomarkers and oxidative stress as a mechanism of functional renal disorders—Preliminary studies. *BioMed Res. Int.* **2014**, *2014*, 740105. [CrossRef]
137. Bi, M.; Du, X.; Jiao, Q.; Chen, X.; Jiang, H. Expanding the role of proteasome homeostasis in Parkinson's disease: Beyond protein breakdown. *Cell Death Dis.* **2021**, *12*, 154. [CrossRef]
138. Tomaru, U.; Ito, T.; Ohmura, Y.; Higashikawa, K.; Miyajima, S.; Tomatsu, R.; Higashi, T.; Ishizu, A.; Kuge, Y.; Yoshioka, M.; et al. Decreased proteasomal function induces neuronal loss and memory impairment. *Am. J. Pathol.* **2021**, *191*, 144–156. [CrossRef]

International Journal of
*Molecular Sciences*

MDPI

*Article*

# Elucidating the Therapeutic Potential of Bis(Maltolato)OxoVanadium(IV): The Protective Role of Copper in Cellular Metabolism

**Lorenzo Rivas-García [1,2], Alfonso López-Varela [1], José L. Quiles [1,3,4], María Montes-Bayón [5], Pilar Aranda [1,2], Juan Llopis [1,2,*] and Cristina Sánchez-González [1,2]**

1 Department of Physiology, Institute of Nutrition and Food Technology "José Mataix Verdú", Biomedical Research Centre, University of Granada, 18016 Armilla, Spain; lorenrivas@ugr.es (L.R.-G.); alvarela@ugr.es (A.L.-V.); jlquiles@ugr.es (J.L.Q.); paranda@ugr.es (P.A.); crissg@ugr.es (C.S.-G.)
2 Sport and Health Research Centre, University of Granada, C/Menéndez Pelayo 32, 18016 Granada, Spain
3 Research and Development Functional Food Centre (CIDAF), Health Science Technological Park, Avenida del Conocimiento 37, 18016 Granada, Spain
4 Research Group on Foods, Nutritional Biochemistry and Health, Universidad Europea del Atlántico, Isabel Torres 21, 39011 Santander, Spain
5 Instituto de Investigación Sanitaria del Principado de Asturias (ISPA), Department of Physical and Analytical Chemistry, Faculty of Chemistry, University of Oviedo, C/ Julián Clavería 8, 33006 Oviedo, Spain; montesmaria@uniovi.es
* Correspondence: jllopis@ugr.es

**Abstract:** Vanadium (V) is a trace mineral whose biological activity, role as a micronutrient, and pharmacotherapeutic applications remain unknown. Over the last years, interest in V has increased due to its potential use as an antidiabetic agent mediated by its ability to improve glycemic metabolism. However, some toxicological aspects limit its potential therapeutic application. The present study aims to evaluate the effect of the co-treatment with copper (Cu) and bis(maltolato)oxovanadium(IV) (BMOV) as a possible strategy to reduce the toxicity of BMOV. Treating hepatic cells with BMOV reduced cell viability under the present conditions, but cell viability was corrected when cells were co-incubated with BMOV and Cu. Additionally, the effect of these two minerals on nuclear and mitochondrial DNA was evaluated. Co-treatment with both metals reduced the nuclear damage caused by BMOV. Moreover, treatment with these two metals simultaneously tended to reduce the ND1/ND4 deletion of the mitochondrial DNA produced with the treatment using BMOV alone. In conclusion, these results showed that combining Cu and V could effectively reduce the toxicity associated with V and enhance its potential therapeutic applications.

**Keywords:** vanadium; copper; mineral; nutrition; mitochondria; DNA

## 1. Introduction

Vanadium (V) is a widely distributed element in nature, and it is essential for some life forms. However, several aspects of V metabolism, such as its absorption, distribution, and biological and pharmacological activities still need to be fully defined. Currently, the research on V has increased, motivated by its increasing environmental levels due to its wide use in industrial processes [1], its use as a supplement for athletes, and the growing interest in the pharmacological applications of some vanadium compounds as hypoglycemic agents. However, before its use in the clinic, many metabolic aspects must be well established, such as digestive and metabolic interactions with other elements involved in the antioxidant defense [2].

During the last few years, multiple chemical compounds of V have been evaluated. Finally, bis(maltolato)oxovanadium(IV) (BMOV) is the V compound that reports the highest activity (i.e., BMOV is 2 to 3 times more potent than inorganic V as a glucose-lowering agent than vanadium sulfate), and the chemical V compound reporting the best tolerance [3]. In addition, BMOV can be administered to experimental animals in drinking water with similar effectiveness as vanadium sulfate [4]. BMOV has shown the ability to decrease fasting glucose and hemoglobin A1C levels in diabetic rats [5]. Moreover, treatment with V decreased the exogenous insulin necessity of diabetic rats, improving insulin sensitivity [6]. The antidiabetic mechanism associated with V could be mediated mainly by two effects. On the one hand, the vanadyl ion could stimulate cytosolic protein kinases and subsequently improve cellular signaling. On the other hand, another mechanism proposed is mediated by the translocation of the GLUT-4 glucose transporter from the intracellular compartment to the plasma membrane [7]. However, several limitations reduce the use of V in the clinic as an antidiabetic agent. The main limitation is the ability of V to generate reactive oxygen species (ROS) during metabolism [8]. ROS are molecules that can cause cellular damage and inflammation, leading to other complications. V stimulates the production of ROS in certain tissues [9], e.g., exposure to vanadium pentoxide ($V^{5+}$) is correlated with an oxidative biomarker, and this exposure causes fibroblast senescence and pulmonary fibrosis [10]. Moreover, the exposure of hepatic cells to BMOV causes damage to cellular DNA and mitochondrial metabolism [11].

Thus, several strategies to reduce V toxicity are required. In that way, combining V with other metallic elements has emerged as an innovative solution. Magnesium could reduce the prooxidant activity of V [12], and the combination of BMVO with $MnCl_2$ has revealed that this agent can decrease the damage produced by V in the nucleus and mitochondria [11].

Copper (Cu) is a transition metal whose main function in biological systems is the regulation of enzymatic processes [13]. The role of this mineral in antioxidant defense is noteworthy because it is a cofactor of several enzymes involved in this process [14]. Cu is involved in the enzymes superoxide dismutase (SOD) located in the cytosol [15], lysyl oxidase [16] in the connective tissue, and cytochrome C oxidase [16], which catalyzes the reaction between oxygen and hydrogen peroxide during cellular respiration. Cu exerts its antioxidant action mainly due to its role in Cu/Zn SOD [15], a metalloenzyme that protects against oxidative damage. Cu/Zn-SOD is located in the cytosol of most tissues and is an integral part of an organism's defense mechanism against the consequences of oxygen metabolism.

The interactions between Cu and V have yet to be elucidated. Several authors found, in general, no significant differences between Cu absorption and V compound absorption [17]. However, other authors suggest that V treatment may affect Cu metabolism in different tissues. For instance, V treatment significantly decreased Cu in tissues such as the liver and kidney [12]. Nevertheless, treatment with V increased the amount of Cu in the femur [18]. This association could be because the absorption of these metals occurs through a common process mediated by divalent metal transporter (DMT1) [19,20].

The present study aims to evaluate the effect of adding Cu on cell viability and the mitochondrial and nucleus DNA stability of cells previously treated with BMOV. Thus, these findings will provide new strategies for implementing V in therapy.

## 2. Results

### 2.1. Cell Viability (MTT Assay)

Figure 1A shows the effect of V treatment (as BMOV) on hepatic cells for 32 h. The intermediate doses (0.75 mg/L V and 1.5 mg/L V) significantly decreased cell viability compared with the control cells. The highest dose (3 mg/L V) significantly decreased cell viability compared with all the other groups.

A

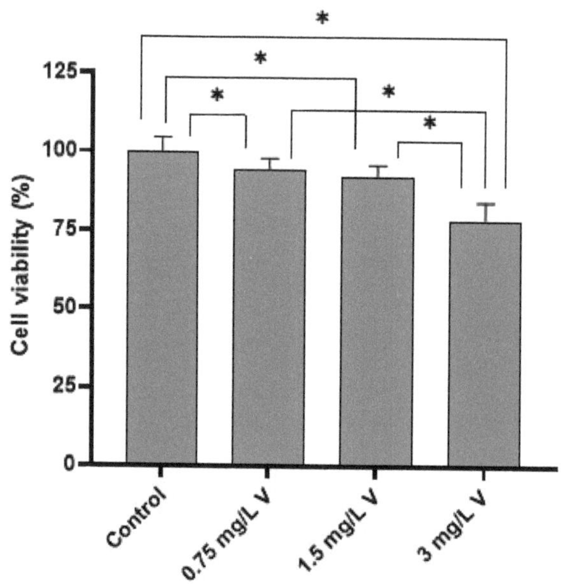


B

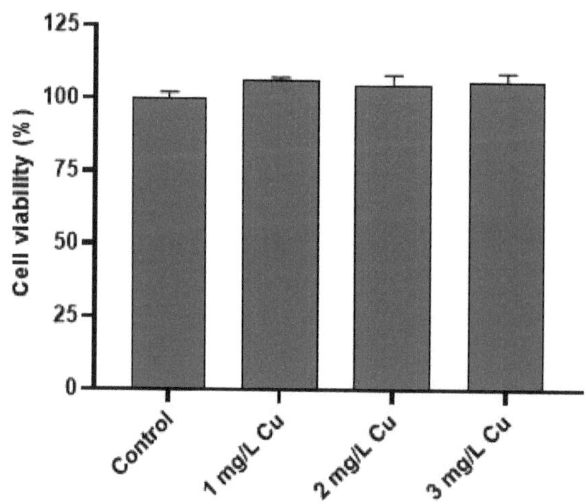


**Figure 1.** *Cont.*

C

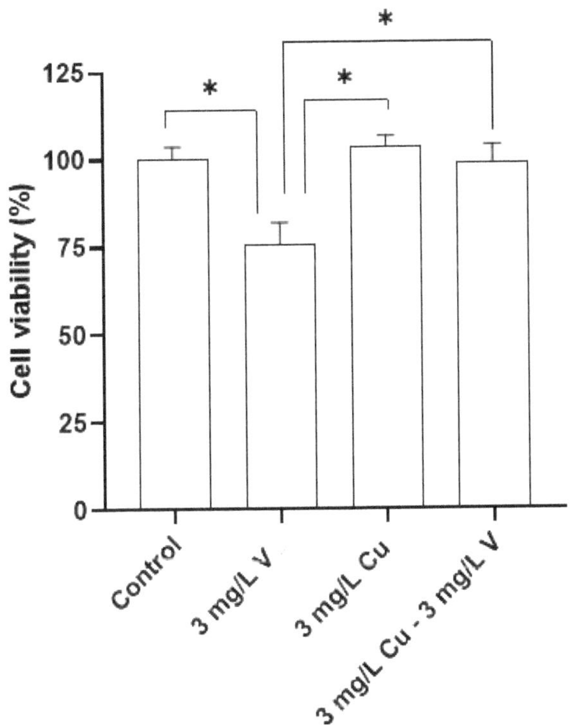


**Figure 1.** Cell viability of HepG2 cells. (**A**): Cells treated with different doses of BMOV; (**B**): cells treated with different doses of $CuCl_2$; (**C**): cells treated with the combination of both metals. * $p < 0.05$. ns.: nonsignificant. Results are presented as the mean ± SD

However, no changes were found in cell viability after treating with $CuCl_2$ for 32 h for all the doses studied compared with the control cells (Figure 1B). Moreover, combining both metals (Figure 1C) showed no modification in cell viability compared with the control cells.

*2.2. Metal Uptake*

The amounts of V and Cu were measured using the inductively coupled plasma mass spectrometry technique (ICP-MS). Treatment with 3 mg/L V (as BMOV) increased the cellular content of V and Cu compared with control cells (Table 1). Moreover, if cells were exposed to 3 mg/L Cu (as $CuCl_2$), it increased the uptake of Cu (Table 1). Furthermore, when cells were exposed to the combination of metals at the same time, the uptake of Cu increased, but less than after the exposure to Cu alone, and the uptake of V was similar to that found after treatment with BMOV alone.

**Table 1.** The cellular metal content of HepG2 treated with metallic solutions for 32 h. Data are expressed as mean ± SD. Superscript letters denote significant differences: [a]: vs. control cells; [b]: vs. 3 mg V/L; [c]: vs. 3 mg Cu/L.

| | V (ng/$10^6$ Cells) | Cu (ng/$10^6$ Cells) |
|---|---|---|
| Control | 4.8 ± 0.6 | 307 ± 110 |
| 3 mg/L V | 50 ± 0.4 [a] | 503 ± 110 [a] |
| 3 mg/L Cu | 4.84 ± 0.4 [b] | 5340 ± 1551 [a,b] |
| 3 mg/L Cu- 3 mg/L V | 42 ± 0.2 [a,c] | 2255 ± 707 [a,b,c] |

*2.3. ND1/ND4 mtDNA Deletion*

ND1/ND4 deletion was measured to evaluate the mtDNA status. Control cells and cells treated with 3 mg/L Cu for 32 h showed no significant differences in this parameter. In contrast, cells treated with 3 mg/L V and the combination (3 mg/L Cu–3 mg/L V) showed a significant increase in ND1/ND4 mtDNA deletion compared to control cells and cells treated with 3 mg/L Cu (Figure 2). No significant difference was found between cells treated with 3 mg/L V and cells treated with the combination (3 mg/L Cu–3 mg/L V).

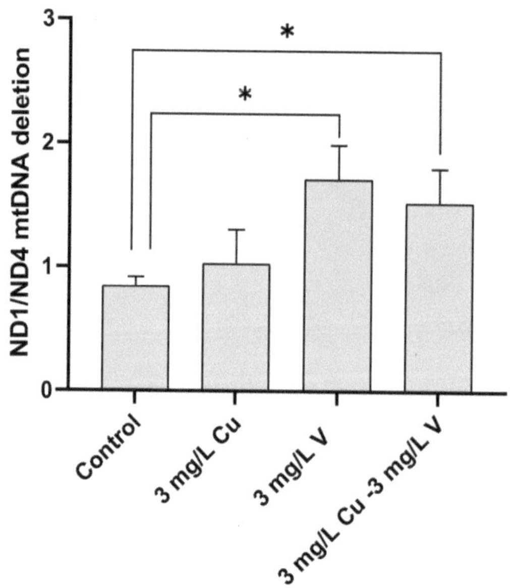


**Figure 2.** ND1/ND4 mtDNA deletion in HepG2 cells treated with metallic solutions for 32 h. * $p < 0.05$.

At our experimental conditions, the combined treatment of 3 mg/L Cu–3 mg/L V showed a tendency to reduce ND1/ND4 mtDNA deletion compared to the treatment with BMOV alone (3 mg/L V).

*2.4. Comet Assay*

The comet assay was conducted to evaluate the nuclear DNA damage (Table 2). The Olivé tail moment (OTM) was the parameter used to compare group differences. Table 2 shows no differences between the cells treated with 3 mg/L Cu and control cells. In contrast, those treated with 3 mg/L V had an OTM value almost 40 times higher than the control cells and the cells treated with 3 mg/L Cu. The cells treated with 3 mg/L V exhibited the highest OTM score. Additionally, the combination of metals (3 mg/L Cu–3 mg/L V) produced an OTM value of 10.3, which was statistically lower than the group treated with

3 mg/L V and statistically higher than both the control group and the group treated with 3 mg/L Cu. Figure 3 shows a representative image for each treatment from the comet assay.

**Table 2.** OTM for each group of HepG2 cells treated with the metallic solutions for 32 h. Superscript letters denote significant differences: [a]: vs. control; [b]: vs. 3 mg/L V. $p < 0.05$.

| | Tail Moment Value |
|---|---|
| Control | 0 |
| 3 mg/L V | 0 |
| 3 mg/L Cu | 35.6 [a] |
| 3 mg/L Cu–3 mg/L V | 10.3 [a,b] |

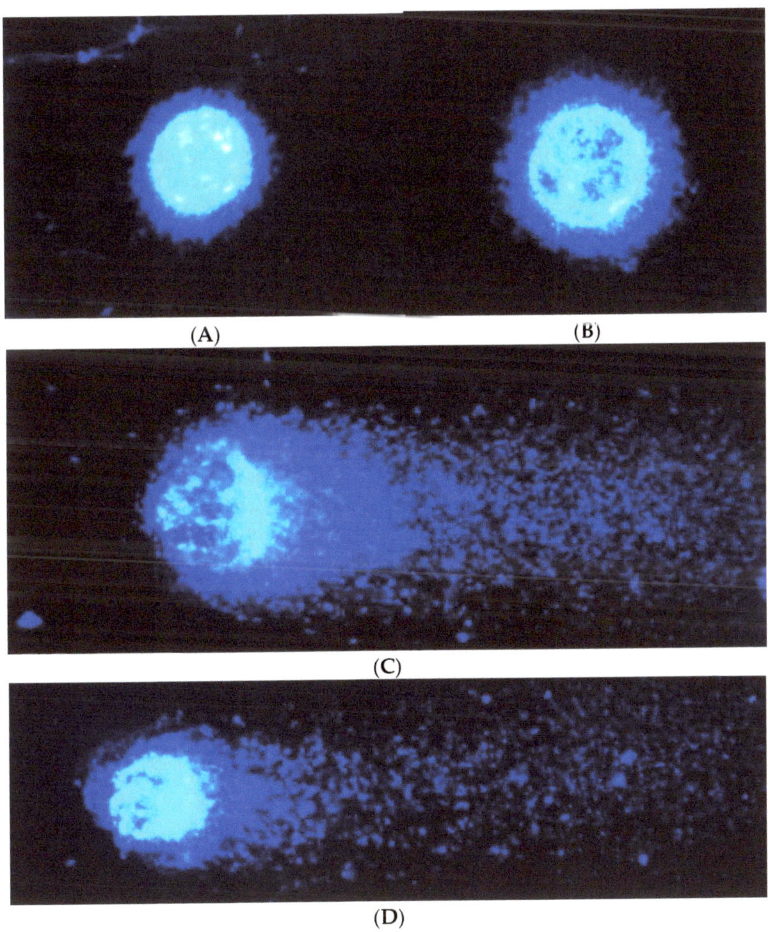

**Figure 3.** Comets from HepG2 cells (20x). (**A**) Control cells; (**B**) cells treated with 3 mg/L Cu; (**C**) cells treated with 3 mg/L V; and (**D**) cells treated with 3 mg/L Cu–3 mg/L V.

## 3. Discussion

Firstly, the effect of the addition of the metals studied was assessed using the MTT test, which is widely used to evaluate the impact of chemical compounds and drugs on cell metabolism. The forms of each metal used were the chemical compound with the highest therapeutic potential; BMOV was used for testing the V addition, and $CuCl_2$ was used for Cu. Moreover, a model of hepatic cells (HepG2) was used because the liver is the main organ responsible for metabolism in mammals. This cellular model has been previously used in other studies to evaluate the uptake, biological effect, and toxicity of metallic elements [11,21,22]. However, this cell line exhibits limitations compared with primary hepatocytes, such as low metabolic capacities, i.e., lower expression of some metabolic activities [23].

Under our experimental conditions, the incubation of hepatic cells with different doses of BMOV resulted in a significant decrease in cell viability compared with the control cells (Figure 1A). The significant decrease associated with the highest dose (3 mg/L V) is particularly relevant. These doses of V are similar to those obtained in the plasma of hyperglycemic rats treated with BMOV for 5 weeks [24] and are effective doses for lowering blood glucose levels. In fact, under these conditions, treatment with BMOV reduced hepcidin levels [25], increased intracellular erythropoietin signaling [25], and restored non-iron deficiency anemia [25]. However, the present study reported a toxic effect associated with the incubation with BMOV, which may be related to the prooxidant and pro-inflammatory effects reported in experimental animals [24,26,27]. It could also be related to the hepatotoxicity reported by other authors in experimental animals [28]. Furthermore, the effect of Cu addition was tested, and in this case, an increase in cell viability was observed for all studied doses (Figure 1B). This effect can be associated with the antioxidant capacity of Cu. Thus, the form of SOD located in the cytoplasm and nucleus requires the presence of Cu for its catalytic center to be active [15]. Therefore, an extra supply of Cu will promote the action of SOD and protect cells from potential free radicals, increasing their viability [15].

Subsequently, the effect of both metals on cellular metabolism was evaluated, and in this case, the toxic effect of BMOV was reduced by Cu (Figure 1C). The effect of decreasing BMOV toxicity thanks to supplementation with antioxidant metals has been previously reported. Rivas-García L et al. described that Mn addition (as $MnCl_2$) decreased BMOV toxicity by acting through the mitochondria. It has also been observed that Cu can protect other tissues from toxic effects mediated by other metals, such as aluminum [29].

The cellular uptake activity was monitored to further understand the cellular metabolism of both metals. Both metals could be transported through the cell membrane by the same transporter, DMT1 [19,20]. However, the information available on the capacity of V to modify the expression of the transporter is limited. Similarly, it is unknown whether overexposure to V could alter the functionality of the transporter [19]. Ścibior et al. described that the expression of DMT1 was not modified with V supplementation in the liver of rats, while in other tissues, such as the kidney, it was decreased [12]. Under the present conditions, individual exposure to hepatocytes increased the content of both metals (Table 1). Additionally, BMOV exposure increases the individual content of Cu. Thus, the presence of BMOV mediates the entry of Cu into cells. This same effect has been previously described because BMOV exposure increased the uptake of another antioxidant metal (Mn) [11]. These results are consistent with others in which only increases in Cu have been found in the kidneys of rats treated with V [30]. The co-exposure of hepatocytes to both metals led to repetition in the trends for individual treatments. Thus, when cells were exposed to BMOV and Cu in combination, the intracellular levels of both elements increased without reaching the values observed in individual exposures. In this case, the effect of Cu and BMOV administration was contrary to Mn and BMOV supplementation; in that case, the presence of Mn in the medium facilitated the exit of the other mineral elements [11]. However, in other in vivo studies, treatment with sodium metavanadate decreased the content of Cu in the liver, kidney, and spleen [12]. Furthermore, treatment with sodium metavanadate has also been described to potentially increase the content of Cu in a femur diaphysis [18].

Subsequently, ND1/ND4 deletion of mitochondrial DNA (mtDNA) was studied. ND1 and ND4 are subunits of mtDNA that encode complex I [31], one of the largest and most complicated protein complexes found in mitochondria. Mutations in these genes can lead to mitochondrial diseases, and studies have shown that certain mutations in ND1 and ND4 can cause complex I dysfunction, which can contribute to various health conditions [32]. Therefore, understanding the relationship between ND1/ND4 and complex I is crucial for understanding the complex biology of mitochondria and its role in cellular energy production [33].

The data reported are consistent with previously published findings [11]. Under the present conditions, the treatment with BMOV increased ND1/ND4 deletion in mtDNA (Figure 2), which could be associated with mitochondrial dysfunction, probably caused by the prooxidant damage induced by BMOV. However, treatment with Cu showed no influence on the ND1/ND4 deletion compared with control cells. These results are partially inconsistent with recent publications, which suggest that Cu plays a crucial role in mitochondrial function and signaling, affecting cell fate through metabolic reprogramming involving bioenergetics, dynamics, and mitophagy [14,34].

When the combined treatment was applied, the deletion of ND1/ND4 tended to increase compared with control cells, but to a lesser extent than with V treatment, indicating a corrective effect produced by Cu (Figure 2).

Finally, the impact of these metals on nuclear DNA was evaluated. Thus, the ROS generated by BMOV treatment interacted with nuclear DNA. In addition, the treatment with $CuCl_2$ showed no effect on the nuclear structure. When both metals were added together, the nuclear damage was reduced compared with BMOV treatment alone; this could be associated with the antioxidant activity of Cu.

## 4. Materials and Methods

### *4.1. Preparation of Metallic Solutions and Exposure to V and Cu*

BMOV was used as the complex of V and $CuCl_2$ as the source of Cu. The preparation of the BMOV complex was extemporaneous. $CuCl_2$ (Sigma Aldrich, St. Louis, MO, USA) was prepared in water to the concentration required. BMOV was synthesized according to the protocol described by Peters, K. et al. [35]. Briefly, the complex was synthesized by adding vanadyl sulfate and ligand to water and adjusting the pH to 9. The stability of the BMOV in the cell culture medium was previously determined [11].

Cells were incubated with BMOV at the following concentrations of V: 0.75, 1.5, and 3 mg/L V. For Cu exposure, cells were incubated with $CuCl_2$ at the following concentrations of Cu: 1, 2, and 3 mg/L Cu. The selection of doses in the present study was calculated based on the serum concentration found in vivo [24,27]. The Cu concentrations correspond to the physiological levels of Cu in the plasma of rats, twofold and threefold. MTT tests were performed at all the proposed concentrations of metals. However, only the highest concentration of each metal and the combination of both metals were used for the metallic uptake, mtDNA deletion, and comet assay. Cells were incubated with the metals for 32 h in all the experiments.

### *4.2. Cell Conditions*

The HepG2 cell line was supplied by the Cell Culture Resource Centre at the University of Granada (Spain). Cells were precultured in 25 $cm^2$ culture flasks (Thermo Fisher Scientific, Waltham, MA, USA) at 37 °C using RPMI 1640 medium (Sigma Aldrich, St. Louis, MO, USA) supplemented with 10% ($v/v$) fetal bovine serum and 2 mM glutamine in a humidified atmosphere of 5% $CO_2$. The culture medium for HepG2 was replaced once per day. After the cell density reached approximately $1 \times 10^6$ cell/mL, cells were detached with trypsinization using a 0.25% trypsin-EDTA solution (Sigma Aldrich, St. Louis, MO, USA) and collected using centrifugation at 1500 rpm for 5 min. Cell density was determined using a Neubauer chamber.

### 4.3. MTT Assay

Cell viability was evaluated by assessing the production of formazan by viable cells from a salt (3-(4,5-dimethylthiazol-2-yl)-2,5-diphenyltetrazolium bromide) (MTT). Briefly, cells were seeded at $1 \times 10^5$ cells per mL in 96-well plates (VWR, Radnor, PA, USA) and incubated for 24 h. Then, the media were refreshed, and the appropriate dilutions of each metallic solution were added. After incubation for 32 h, the medium was discarded and replaced with a 0.5 μg/mL solution of MTT (Sigma, St. Louis, MO, USA) prepared in PBS (Phosphate Buffered Saline, Sigma, St. Louis, MO, USA) for 1 h at 37 °C. Then, the MTT was removed, and DMSO was added to solubilize the formed crystals. Last, absorbance values were measured at 555 nm (555–690) using a microplate reader (Agilent Technologies, Winooski, VT, USA). All experiments were performed in triplicate.

### 4.4. Metal Uptake

HepG2 cells were seeded at $1 \times 10^5$ cells/mL in a 25 $cm^2$ cell culture flask. After 24 h, the medium was replaced, and cells were treated with the metallic solutions for 32 h. Then, hepatocytes were washed with PBS and diluted with a basic solution containing ammonium hydroxide, butanol, EDTA, and Triton X-100. The quantification of V and Cu was performed using an ICP-MS instrument (Agilent 7500, Agilent Technologies, Winooski, VT, USA) fitted with a Meinhard type nebulizer (Glass Expansion, Weilburg, Germany) and equipped with a He collision cell. A Milli-Q system (Millipore, Burlington, MA, USA) was used to obtain deionized water (18 MΩ). All reagents (ammonium hydroxide solution, butanol, EDTA, Triton X-100) used were of the highest available purity. A standard solution of 100 μg/L of Li, Mg, Sc, Co, Y, In, Ce, Ba, Pb, Bi, and U in 1% (*v*/*v*) HNO3 was prepared from a 1000 mg/L multi-element stock standard solution (Merck, Darmstadt, Germany) and used for daily optimization of the ICP-MS parameters. Single-element standard solutions for ICP-MS containing 1000 μg/mL of V and Cu were also provided by Merck (Merck, Darmstadt, Germany). Calibration curves were prepared using Ga as an internal standard and by diluting stock solutions of 1000 mg/L in 1% $HNO_3$. The accuracy of this method was evaluated by comparing it with the certified reference material SeronormTM Trace Elements Serum (Sero, Billingstad, Norway) and using recovery studies of spiked samples with multi-element standards. The calculated recoveries for each element were between 95% and 105% in all cases. For each element, a mean of five separate determinations of this reference material was used.

### 4.5. ND1/ND4 mtDNA Deletion

HepG2 cells were exposed to the metallic solutions as described in Section 2.4. Then, ND1/ND4 deletion was measured using the previously proposed methodology [36].

### 4.6. Comet Aassay

Hepatocytes were allowed to grow until 60% confluence was reached. Subsequently, cell cultures were exposed to the metallic solutions for 32 h. Then, cells were treated with trypsin–EDTA solution (Sigma, St Louis, MO, USA) and centrifuged for pellet collection. Cells were resuspended in 1 mL PBS. Microscope glass slides were pre-coated with 1% normal melting point (NMP) agarose on one side. A total of 30 μL of cell suspension were mixed with 65 μL of low melting point (LMP) agarose solution (final LMP agarose concentration 0.5%). Drops of each agarose-cell suspension were added on each pre-coated slide and placed for 1 h at 4 °C in the lysis solution (containing NaCl 2.225 M, $Na_2$EDTA 88.9 mM, Tris 8.8 mM, NaOH 0.22 M, 10% DMSO, and 1% Triton X-100). Electrophoresis (1 V/cm) was performed for 20 min at 4 °C using an electrophoresis solution ($Na_2$EDTA 1 mM and NaOH 300 mM, pH > 13). The solution on the slide was neutralized in triplicate (5 min every time) with Tris 0.4 M. Ethanol was used for fixation. Finally, the samples were stained with 2-(4-amidinophenyl)-1*H*-indole-6-carboxamidine (DAPI) (1 μg/mL) and analyzed using a fluorescence microscope (Eclipse Ni; Nikon Instruments Europe B.V., Badhoevedorp, The Netherlands). Over 150 nucleoids per sample (50 nucleoids per slide) stained with DAPI were scored using computerized image analysis (CaspLab, University of Wroclaw, Wroclaw, Poland). Cells containing damaged DNA appear like a comet with a bright head and a tail. In contrast, undamaged DNA appears as an intact nucleus with no tail. DNA damage was determined by comparing tail moments among groups.

### 4.7. Statistical Analysis

Descriptive statistical parameters (means and standard deviations of 8 samples for each experiment) were obtained for each studied variable. The means of independent variables were statistically compared among the groups. All analyses were performed using SPSS 26.0 (SPSS, Chicago, IL, USA). Differences were considered statistically significant at a probability level of <5%.

## 5. Conclusions

In conclusion, under the experimental conditions of this study, the exposure of HepG2 cells to 3 mg/L V administered as BMOV reduced cell viability, which was caused mainly by the effect of this metal in the damage of DNA on the cellular nucleus and mitochondria. These negative effects could be partially corrected by adding Cu (as $CuCl_2$), which could act mainly by decreasing the damage to nuclear DNA. This positive association between these two metals could mediate the decrease in the toxicity of V and enhance its potential applications in therapy. Nevertheless, further studies are needed to better determine the effects arising from these interactions to establish the role of V as a micronutrient and to reduce its toxic effects.

**Author Contributions:** Formal analysis, investigation, and writing—original draft: L.R.-G. and A.L.-V. Conceptualization, methodology, validation, resources, writing—review and editing: J.L.Q., M.M.-B., P.A., J.L. and C.S.-G. All authors have read and agreed to the published version of the manuscript.

**Funding:** We are grateful for the support and financing received from the Consejería de Innovación, Ciencia y Empresa, Andalusian Regional Government (Project P06-CTS-01435), and from the Spanish Ministry of Economy and Competitivity (SAF2011-29648).

**Institutional Review Board Statement:** Not applicable.

**Informed Consent Statement:** Not applicable.

**Data Availability Statement:** Not applicable.

**Acknowledgments:** This paper has been included in the Doctoral Program in Nutrition and Food Science of the University of Granada as part of the PhD dissertation of L.R.-G.

**Conflicts of Interest:** The authors declare no conflict of interest.

## References

1. Gummow, B. Vanadium: Environmental Pollution and Health Effects. In *Encyclopedia of Environmental Health*; Elsevier: Amsterdam, The Netherlands, 2011; pp. 628–636; ISBN 978-0-444-52272-6.
2. Treviño, S.; Diaz, A. Vanadium and Insulin: Partners in Metabolic Regulation. *J. Inorg. Biochem.* **2020**, *208*, 111094. [CrossRef]
3. Orvig, C.; Caravan, P.; Gelmini, L.; Glover, N.; Herring, F.G.; Li, H.; McNeill, J.H.; Rettig, S.J.; Setyawati, I.A. Reaction Chemistry of BMOV, Bis(Maltolato)Oxovanadium(IV), a Potent Insulin Mimetic Agent. *J. Am. Chem. Soc.* **1995**, *117*, 12759–12770. [CrossRef]
4. Sakurai, H.; Watanabe, H.; Tamura, H.; Yasui, H.; Matsushita, R.; Takada, J. Insulin-Mimetic Vanadyl—Dithiocarbamate Complexes. *Inorg. Chim. Acta* **1998**, *283*, 175–183. [CrossRef]
5. Thompson, K.H.; Lichter, J.; LeBel, C.; Scaife, M.C.; McNeill, J.H.; Orvig, C. Vanadium Treatment of Type 2 Diabetes: A View to the Future. *J. Inorg. Biochem.* **2009**, *103*, 554–558. [CrossRef]
6. Goldfine, A.B.; Patti, M.-E.; Zuberi, L.; Goldstein, B.J.; LeBlanc, R.; Landaker, E.J.; Jiang, Z.Y.; Willsky, G.R.; Kahn, C.R. Metabolic Effects of Vanadyl Sulfate in Humans with Non—Insulin-Dependent Diabetes Mellitus: In Vivo and in Vitro Studies. *Metabolism* **2000**, *49*, 400–410. [CrossRef]
7. Li, S.H.; McNeill, J.H. In Vivo Effects of Vanadium on GLUT4 Translocation in Cardiac Tissue of STZ-Diabetic Rats. *Mol. Cell. Biochem.* **2001**, *217*, 121–129. [CrossRef]
8. Ścibior, A.; Kurus, J. Vanadium and Oxidative Stress Markers—In Vivo Model: A Review. *Curr. Med. Chem.* **2019**, *26*, 5456–5500. [CrossRef]

9. Aureliano, M.; De Sousa-Coelho, A.L.; Dolan, C.C.; Roess, D.A.; Crans, D.C. Biological Consequences of Vanadium Effects on Formation of Reactive Oxygen Species and Lipid Peroxidation. *Int. J. Mol. Sci.* **2023**, *24*, 5382. [CrossRef]
10. He, X.; Jarrell, Z.R.; Liang, Y.; Ryan Smith, M.; Orr, M.L.; Marts, L.; Go, Y.-M.; Jones, D.P. Vanadium Pentoxide Induced Oxidative Stress and Cellular Senescence in Human Lung Fibroblasts. *Redox Biol.* **2022**, *55*, 102409. [CrossRef]
11. Rivas-García, L.; Quiles, J.L.; Varela-López, A.; Arredondo, M.; Lopez, P.; Diéguez, A.R.; Montes-Bayon, M.; Aranda, P.; Llopis, J.; Sánchez-González, C. In Vitro Study of the Protective Effect of Manganese against Vanadium-Mediated Nuclear and Mitochondrial DNA Damage. *Food Chem. Toxicol.* **2020**, *135*, 110900. [CrossRef]
12. Ścibior, A.; Adamczyk, A.; Gołębiowska, D.; Niedźwiecka, I.; Fornal, E. The Influence of Combined Magnesium and Vanadate Administration on the Level of Some Elements in Selected Rat Organs: V–Mg Interactions and the Role of Iron-Essential Protein (DMT-1) in the Mechanism Underlying Altered Tissues Iron Level. *Metallomics* **2014**, *6*, 907–920. [CrossRef]
13. Chen, L.; Min, J.; Wang, F. Copper Homeostasis and Cuproptosis in Health and Disease. *Signal Transduct. Target. Ther.* **2022**, *7*, 378. [CrossRef]
14. Ruiz, L.M.; Libedinsky, A.; Elorza, A.A. Role of Copper on Mitochondrial Function and Metabolism. *Front. Mol. Biosci.* **2021**, *8*, 711227. [CrossRef]
15. Plazas Guerrero, C.G.; Acosta Cota, S.D.J.; Castro Sánchez, F.H.; Vergara Jiménez, M.D.J.; Ríos Burgueño, E.R.; Sarmiento Sánchez, J.I.; Picos Corrales, L.A.; Osuna Martínez, U. Evaluation of Sucrose-Enriched Diet Consumption in the Development of Risk Factors Associated to Type 2 Diabetes, Atherosclerosis and Non-Alcoholic Fatty Liver Disease in a Murine Model. *Int. J. Environ. Health Res.* **2021**, *31*, 651–669. [CrossRef]
16. Tsang, T.; Davis, C.I.; Brady, D.C. Copper Biology. *Curr. Biol.* **2021**, *31*, R421–R427. [CrossRef]
17. Rucker, R.B.; Cui, C.T.; Tchaparian, E.H.; Mitchell, A.E.; Clegg, M.; Uriu-Hare, J.Y.; Keen, C.L. Dietary Vanadium, P-ATPase-7A Expression and the Influence on Lysyl Oxidase and Cu Accumulation in Rat Skin and Liver. In *Trace Elements in Man and Animals 10*; Roussel, A.M., Anderson, R.A., Favier, A.E., Eds.; Springer: New York, NY, USA, 2002; pp. 186–187; ISBN 978-0-306-46378-5.
18. Ścibior, A.; Gołębiowska, D.; Adamczyk, A.; Kurus, J.; Staniszewska, M.; Sadok, I. Evaluation of Lipid Peroxidation and Antioxidant Defense Mechanisms in the Bone of Rats in Conditions of Separate and Combined Administration of Vanadium (V) and Magnesium (Mg). *Chem.-Biol. Interact.* **2018**, *284*, 112–125. [CrossRef]
19. Treviño, S.; Díaz, A.; Sánchez-Lara, E.; Sanchez-Gaytan, B.L.; Perez-Aguilar, J.M.; González-Vergara, E. Vanadium in Biological Action: Chemical, Pharmacological Aspects, and Metabolic Implications in Diabetes Mellitus. *Biol. Trace Elem. Res.* **2019**, *188*, 68–98. [CrossRef]
20. Arredondo, M.; Muñoz, P.; Mura, C.V.; Núñez, M.T. DMT1, a Physiologically Relevant Apical Cu $^{1+}$ Transporter of Intestinal Cells. *Am. J. Physiol.-Cell Physiol.* **2003**, *284*, C1525–C1530. [CrossRef]
21. Cordier, W.; Yousaf, M.; Nell, M.J.; Steenkamp, V. Underlying Mechanisms of Cytotoxicity in HepG2 Hepatocarcinoma Cells Exposed to Arsenic, Cadmium and Mercury Individually and in Combination. *Toxicol. Vitr.* **2021**, *72*, 105101. [CrossRef]
22. Wang, P.; Wu, Q.; Wang, F.; Zhang, Y.; Tong, L.; Jiang, T.; Gu, C.; Huang, S.; Wang, H.; Bu, S.; et al. Evaluating Cellular Uptake of Gold Nanoparticles in HL-7702 and HepG2 Cells for Plasmonic Photothermal Therapy. *Nanomedicine* **2018**, *13*, 2245–2259. [CrossRef]
23. Kammerer, S.; Küpper, J.-H. Human Hepatocyte Systems for in Vitro Toxicology Analysis. *J. Cell. Biotechnol.* **2018**, *3*, 85–93. [CrossRef]
24. Sánchez-González, C.; Rivas-García, L.; López-Chaves, C.; Rodríguez-Nogales, A.; Algieri, F.; Gálvez, J.; Gómez-Aracena, J.; Vera-Ramírez, L.; Montes-Bayon, M.; Sanz-Medel, A.; et al. Exposure to Bis(Maltolato)Oxovanadium(IV) Increases Levels of Hepcidin MRNA and Impairs the Homeostasis of Iron but Not That of Manganese. *Food Chem. Toxicol.* **2014**, *73*, 113–118. [CrossRef] [PubMed]
25. Sánchez-González, C.; Rivas-García, L.; Rodríguez-Nogales, A.; Algieri, F.; Gálvez, J.; Aranda, P.; Montes-Bayón, M.; Llopis, J. Vanadium Decreases Hepcidin MRNA Gene Expression in STZ-Induced Diabetic Rats, Improving the Anemic State. *Nutrients* **2021**, *13*, 1256. [CrossRef]
26. Sanchez-Gonzalez, C.; Bermudez-Peña, C.; Guerrero-Romero, F.; Trenzado, C.E.; Montes-Bayon, M.; Sanz-Medel, A.; Llopis, J. Effect of Bis(Maltolato)Oxovanadium (IV) (BMOV) on Selenium Nutritional Status in Diabetic Streptozotocin Rats. *Br. J. Nutr.* **2012**, *108*, 893–899. [CrossRef]
27. Sanchez-Gonzalez, C.; Bermudez-Peña, C.; Trenzado, C.E.; Goenaga-Infante, H.; Montes-Bayon, M.; Sanz-Medel, A.; Llopis, J. Changes in the Antioxidant Defence and in Selenium Concentration in Tissues of Vanadium Exposed Rats. *Metallomics* **2012**, *4*, 814–819. [CrossRef]
28. Samira, M.; Mounira, T.; Kamel, K.; Yacoubi, M.T.; Ben Rhouma, K.; Sakly, M.; Tebourbi, O. Hepatotoxicity of Vanadyl Sulfate in Nondiabetic and Streptozotocin-Induced Diabetic Rats. *Can. J. Physiol. Pharmacol.* **2018**, *96*, 1076–1083. [CrossRef]
29. Sohrabi, M.; Gholami, A.; Azar, M.H.; Yaghoobi, M.; Shahi, M.M.; Shirmardi, S.; Nikkhah, M.; Kohi, Z.; Salehpour, D.; Khoonsari, M.R.; et al. Trace Element and Heavy Metal Levels in Colorectal Cancer: Comparison Between Cancerous and Non-Cancerous Tissues. *Biol. Trace Elem. Res.* **2018**, *183*, 1–8. [CrossRef]
30. Sánchez-González, C.; Moreno, L.; Aranda, P.; Montes-Bayón, M.; Llopis, J.; Rivas-García, L. Effect of Bis(Maltolato)Oxovanadium(IV) on Zinc, Copper, and Manganese Homeostasis and DMT1 MRNA Expression in Streptozotocin-Induced Hyperglycemic Rats. *Biology* **2022**, *11*, 814. [CrossRef]

31. Bonnet, C.; Augustin, S.; Ellouze, S.; Bénit, P.; Bouaita, A.; Rustin, P.; Sahel, J.-A.; Corral-Debrinski, M. The Optimized Allotopic Expression of ND1 or ND4 Genes Restores Respiratory Chain Complex I Activity in Fibroblasts Harboring Mutations in These Genes. *Biochim. Biophys. Acta (BBA)-Mol. Cell Res.* **2008**, *1783*, 1707–1717. [CrossRef]
32. Danhelovska, T.; Kolarova, H.; Zeman, J.; Hansikova, H.; Vaneckova, M.; Lambert, L.; Kucerova-Vidrova, V.; Berankova, K.; Honzik, T.; Tesarova, M. Multisystem Mitochondrial Diseases Due to Mutations in MtDNA-Encoded Subunits of Complex, I. *BMC Pediatr.* **2020**, *20*, 41. [CrossRef]
33. Bhatti, J.S.; Bhatti, G.K.; Reddy, P.H. Mitochondrial Dysfunction and Oxidative Stress in Metabolic Disorders—A Step towards Mitochondria Based Therapeutic Strategies. *Biochim. Biophys. Acta (BBA)-Mol. Basis Dis.* **2017**, *1863*, 1066–1077. [CrossRef]
34. Cobine, P.A.; Moore, S.A.; Leary, S.C. Getting out What You Put in: Copper in Mitochondria and Its Impacts on Human Disease. *Biochim. Biophys. Acta (BBA)-Mol. Cell Res.* **2021**, *1868*, 118867. [CrossRef] [PubMed]
35. Peters, K.G.; Davis, M.G.; Howard, B.W.; Pokross, M.; Rastogi, V.; Diven, C.; Greis, K.D.; Eby-Wilkens, E.; Maier, M.; Evdokimov, A.; et al. Mechanism of Insulin Sensitization by BMOV (Bis Maltolato Oxo Vanadium); Unliganded Vanadium (VO4) as the Active Component. *J. Inorg. Biochem.* **2003**, *96*, 321–330. [CrossRef] [PubMed]
36. Rivas-García, L.; Quiles, J.L.; Varela-López, A.; Giampieri, F.; Battino, M.; Bettmer, J.; Montes-Bayón, M.; Llopis, J.; Sánchez-González, C. Ultra-Small Iron Nanoparticles Target Mitochondria Inducing Autophagy, Acting on Mitochondrial DNA and Reducing Respiration. *Pharmaceutics* **2021**, *13*, 90. [CrossRef] [PubMed]

International Journal of
*Molecular Sciences*

MDPI

*Article*

# Exploring the Impact of Head Group Modifications on the Anticancer Activities of Fatty-Acid-like Platinum(IV) Prodrugs: A Structure–Activity Relationship Study

**Man Kshetri [1], Wjdan Jogadi [1], Suha Alqarni [1,2], Payel Datta [1,3], May Cheline [1], Arpit Sharma [1], Tyler Betters [1], Deonya Broyles [1] and Yao-Rong Zheng [1,*]**

1 Department of Chemistry and Biochemistry, Kent State University, 236 Integrated Sciences Building, Kent, OH 44242, USA; salqarn3@kent.edu (S.A.); pdatta1@kent.edu (P.D.); mcheline@kent.edu (M.C.)
2 Department of Chemistry, University of Bisha, Bisha 67714, Saudi Arabia
3 Department of Chemistry, Case Western Reserve University, Cleveland, OH 44106, USA
* Correspondence: yzheng7@kent.edu; Tel.: +1-330-672-2267

**Abstract:** We conducted the first comprehensive investigation on the impact of head group modifications on the anticancer activities of fatty-acid-like Pt(IV) prodrugs (FALPs), which are a class of platinum-based metallodrugs that target mitochondria. We created a small library of FALPs (**1–9**) with diverse head group modifications. The outcomes of our study demonstrate that hydrophilic modifications exclusively enhance the potency of these metallodrugs, whereas hydrophobic modifications significantly decrease their cytotoxicity. To further understand this interesting structure–activity relationship, we chose two representative FALPs (compounds **2** and **7**) as model compounds: one (**2**) with a hydrophilic polyethylene glycol (PEG) head group, and the other (**7**) with a hydrophobic hydrocarbon modification of the same molecular weight. Using these FALPs, we conducted a targeted investigation on the mechanism of action. Our study revealed that compound **2**, with hydrophilic modifications, exhibited remarkable penetration into cancer cells and mitochondria, leading to subsequent mitochondrial and DNA damage, and effectively eradicating cancer cells. In contrast, compound **7**, with hydrophobic modifications, displayed a significantly lower uptake and weaker cellular responses. The collective results present a different perspective, indicating that increased hydrophobicity may not necessarily enhance cellular uptake as is conventionally believed. These findings provide valuable new insights into the fundamental principles of developing metallodrugs.

**Keywords:** platinum(IV) prodrugs; structure–activity relationship; anticancer

## 1. Introduction

Platinum-based chemotherapy has been a cornerstone of cancer treatment for several decades [1,2]. Key agents such as cisplatin, carboplatin, and oxaliplatin have assumed critical roles in the management of diverse malignancies, encompassing testicular, ovarian, lung, head and neck, and colorectal cancers. These chemotherapeutic agents exert their anticancer effects by instigating the formation of DNA cross-links, which, in turn, impede cancer cell proliferation and elicit apoptosis [1,3,4]. Nonetheless, despite their widespread application, the clinical utilization of platinum-based drugs is encumbered by notable toxicity concerns, culminating in adverse effects such as nephrotoxicity, neurotoxicity, and ototoxicity [1,2]. Furthermore, the regrettably common development of drug resistance and the subsequent cancer relapse in patients underline an imperious necessity to explore novel approaches to platinum-based chemotherapy [5–7]. The pursuit of such innovative strategies is envisaged to surmount the prevailing limitations and offer improved therapeutic outcomes for cancer patients [8–35].

Fatty-acid-like Pt(IV) prodrugs (FALPs) have emerged as a promising new class of Pt-based anticancer agents that utilize innovative drug delivery strategies and cancer biology to overcome the challenges associated with conventional Pt(II) drugs [8,36–41]. Designed to mimic the structure of fatty acids, these prodrugs utilize non-covalent interactions with human serum albumin (HSA) for efficient drug delivery [36]. FALPs have demonstrated remarkable stability in whole human blood, reducing their rate of reduction via reducing agents. Furthermore, they possess a distinctive mechanism of action that involves accumulation in mitochondria, inducing mitochondrial damage with the release of Pt(II) payloads, and resulting in increased proapoptotic peroxidase activity and elevated reactive oxygen species (ROS) levels [39]. FALPs have shown potent in vitro activity against a broad range of cancer types and promising in vivo efficacy in various mouse models [42]. Importantly, FALPs can be readily chemically modified to alter their biological activities and chemical properties [37,41,43–46]. Recent studies have also demonstrated the potential of incorporating these novel Pt(IV) prodrugs into nanoparticles for drug delivery using either non-covalent encapsulation or covalent conjugation based on their amphiphilic structures [37]. Overall, FALPs represent a highly diverse and unique Pt scaffold with promising mechanisms of action that could serve as powerful tools in developing new approaches for cancer therapy. Although the modification of FALPs has predominantly centered around their carboxylic head group, there has not been a comprehensive exploration of the effects of modifying these groups on cellular responses.

This new study focuses on exploring the structure–activity relationship of FALP derivatives (**1–9** in Figure 1A), with a specific emphasis on how modifications of the carboxylic head group's hydrophobicity influence their anticancer activity and cellular responses (Figure 2A). Understanding the impact of hydrophobicity on the uptake of therapeutic molecules is widely recognized as a crucial factor in drug development [36,47–53]. Remarkably, contrary to the widely accepted notion that increased hydrophobicity enhances the cellular uptake of therapeutic molecules, the primary findings of this study demonstrate that FALPs with hydrophilic modifications exhibit exceptional penetration into cancer cells and mitochondria. This, in turn, triggers a cascade of events, leading to substantial mitochondrial and DNA damage, and effectively eradicating cancer cells. On the other hand, increased hydrophobicity in the modifications unexpectedly hinders cellular uptake and mitochondrial accumulation, resulting in weaker cellular responses and a lower in vitro therapeutic efficacy. These findings provide valuable new insights into the fundamental principles of developing metallodrugs.

A

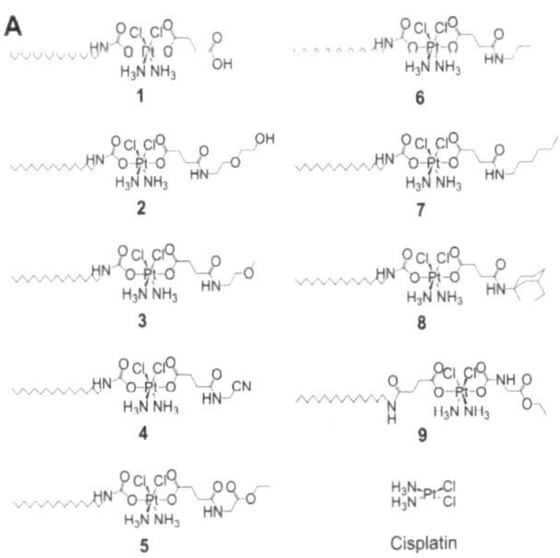

B

| $IC_{50}$ (μM) | A2780cis<br>Ovarian Cancer | MDA-MB-231<br>Breast Cancer | Cal. Log P of head groups |
|---|---|---|---|
| Cisplatin | 109.2 ± 13.5 | 92 ± 8.0 | - |
| **1** | 1.00 ± 0.06 | 6.3 ± 0.4 | - |
| **2** | 0.3 ± 0.06 | 0.76 ± 0.03 | −0.28 |
| **3** | 0.7 ± 0.04 | 1.9 ± 0.3 | 0.59 |
| **4** | 0.41 ± 0.05 | 1.63 ± 0.12 | −0.04 |
| **5** | 0.39 ± 0.08 | 5.22 ± 1.03 | 0.74 |
| **6** | 1.12 ± 0.07 | 1.5 ± 0.4 | 2.19 |
| **7** | 3.75 ± 0.47 | 7.05 ± 0.8 | 4.02 |
| **8** | 5.07 ± 0.71 | 6.09 ± 0.3 | 4.22 |
| **9** | 0.40 ± 0.01 | 1.42 ± 0.07 | - |

**Figure 1.** Cytotoxicity profiles of the fatty-acid-like Pt(IV) prodrugs. (**A**) Chemical structures of the Pt(IV) prodrugs (**1–9**) and cisplatin. (**B**) Table of $IC_{50}$ values of the Pt compounds against human cancer cells.

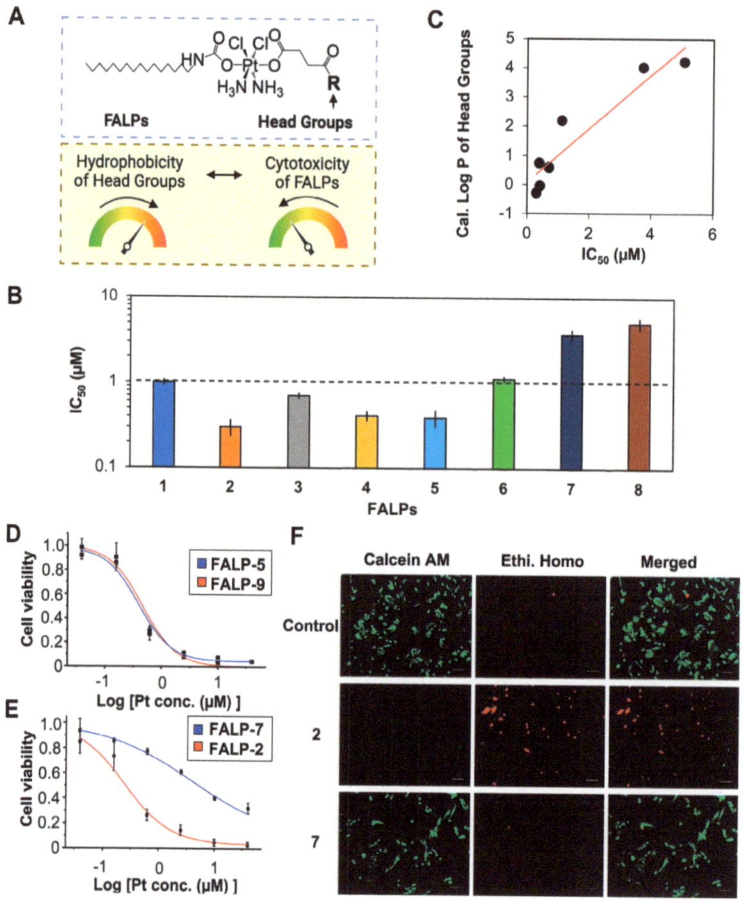


**Figure 2.** Structure–activity relationship of fatty-acid-like Pt(IV) prodrugs. (**A**) Graphical representation of the hydrophobicity of the head group tuning the cytotoxicity of FALPs. (**B**) A bar graph depicting the $IC_{50}$ values of FALPs (**2–8**) with varying levels of hydrophobicity in comparison to unmodified **1** against A2780cis ovarian cancer cells. (**C**) Correlation of the $IC_{50}$ values and the calculated Log P of the head groups of FALPs (**2–8**). (**D**) Killing curves of **5** and **9** against A2780cis cells for 24 h. (**E**) Killing curves of **2** and **7** against A2780cis cells for 24 h. (**F**) Live/dead cell assay images of A2780cis cells treated with **2** and **7** ([Pt] = 1 μM) for 24 h. Scale bar = 100 μm.

## 2. Results and Discussion

**Synthesis and characterization of FALP derivatives with various head groups**. The synthesis of the Pt(IV) prodrug (**2–8**) is depicted in Figure S1A in the Supplementary Materials. Briefly, amino moieties with different hydrophobicities (**10–16**) were conjugated to compound **1** via the HATU-catalyzed amide bond formation reaction. The final compounds were purified via flash column chromatography and/or recrystallization. The overall yields were 44–78%. The synthesis of the Pt(IV) prodrug (**9**) was accomplished in a similar manner, as shown in Figure S1B. The conjugated Pt(IV) compounds (**2–9**) were characterized via $^{1}H$ and $^{13}C$ NMR spectroscopy, electrospray ionization mass spectrometry (ESI-MS), and HPLC, and they can be found in the Supplementary Materials (Figures S2–S9). In the $^{1}H$ NMR spectra, the broad signal at ~6.6 ppm corresponds to the amine groups of the Pt(IV) center. The signal at ~2.8 ppm is the $CH_2$ group adjacent to the carbamate. The signals at 6.8–8.6 ppm are attributed to the amides in **2–8**. In ESI-MS, the isotopically resolved signals agree with the theoretical value of **2–9**. The HPLC analysis of the final product indicated that the purity of compounds **2–9** from the described synthetic method was >95%.

**Cytotoxicity profiles of FALP derivatives with various head groups**. The in vitro anticancer activity of the Pt(IV) prodrugs (**1–9**) was assessed using the 3-(4,5-dimethylthiazol-2-yl)-2,5-diphenyltetrazolium bromide (MTT) assay. This study utilized two human cancer cell lines, namely A2780cis and MDA-MB-231. A2780cis is an ovarian cancer cell line known for its resistance to conventional platinum chemotherapy, making it a formidable challenge to treat. On the other hand, MDA-MB-231 represents a triple-negative breast cancer cell line, which is currently recognized as one of the difficult-to-treat cancer types. The cells were treated with **1–9** or cisplatin for 24 h, and the cell viability was evaluated. The $IC_{50}$ values, which represent the concentration of the drug required to inhibit the growth of cells by 50%, are reported in the table in Figure 1B. The results show that **2–5** have lower $IC_{50}$ values compared to those of **1** and have lower cisplatin in general. For example, in the A2780cis ovarian cancer cell line, the $IC_{50}$ (**2**) = 0.30 ± 0.06 μM is 36 times lower than that of cisplatin ($IC_{50}$ = 109.2 ± 13.5 μM) and 3 times lower than that of **1** ($IC_{50}$ = 1.00 ± 0.06 μM). Notably, head group modifications do not always increase cytotoxicity more than **FALP-1**. For example, **7** and **8** exhibit much lower cytotoxicity than **1**. Overall, in the A2780cis ovarian cancer cell line, the $IC_{50}$ (**7**) = 3.75 ± 0.47 μM is three times higher than that of **1**, but it is still more potent than cisplatin. Notably, the $IC_{50}$ (**8**) = 5.07 ± 0.71 μM is 15 times higher than that of **2**. Overall, the head group modifications of FALPs result in an alternation of therapeutic effects, which is a promising way to fine-tune the anticancer activity of this class of metallodrugs.

**Structure–activity relationship of FALP derivatives with head groups of different hydrophobicity**. Our goal was to study the relationship between the structure and activity to gain a better understanding of how modifications to the head group affect the anticancer properties of FALP derivatives. We hypothesized that the hydrophobicity of these modifications is a key factor in determining the cytotoxicity of the compounds (Figure 2A). To test this hypothesis, we calculated the Log P values for all the modifications with the ALOGPS 2.1 program, which ranged from −0.28 to 4.22, as shown in Figure 1B. Compounds **2–5** had head group modifications with low hydrophobicity (or high hydrophilicity), while compounds **7** and **8** had head groups with high hydrophobicity (or low hydrophilicity). Our results, presented in Figure 2B, indicate that hydrophilic modifications lead to lower $IC_{50}$ values and a higher potency of FALPs, while hydrophobic modifications result in increased $IC_{50}$ values and reduced anticancer activity. To better illustrate the correlation between the hydrophobicity of the head group modifications, we plotted the calculated Log P values against the corresponding $IC_{50}$ values in Figure 2C, which clearly demonstrates the inverse impact of hydrophobicity on the anticancer activities of FALPs in general. Additionally, we sought to determine if this observation was solely based on hydrophobicity, so we engineered two isomers, **5** and **9**. The head group modification of compound **9** was changed from amide to carbamate compared to compound **5**, and interestingly, the cytotoxicity profiles of both **5** and **9** were identical, as shown in Figures 1B and 2D. These results suggest that the hydrophobicity of the structure plays a major role in the structure–activity relationship. Finally, we focused on the two FALPs, **2** and **7**, to illustrate this effect. Although both **2** and **7** were very similar compounds, **2** had a hydrophilic polyethylene glycol (PEG) modification (Log P = −0.28), while compound **7** carried a C6 hydrocarbon chain (Log P = 4.02) of the same molecular weight. As shown in Figure 2E, compound **2** exhibited a much higher potency than **7** in a wide range of concentrations tested, and the $IC_{50}$ (**7**) = 3.75 ± 0.47 μM was 12 times higher than that of **2**. The live/dead cell imaging assays further validated this drastic difference in their in vitro anticancer activity, as shown in Figure 2F, where compound **2** effectively eliminated all drug-resistant A2780cis ovarian cancer cells, while compound **7** was deemed ineffective at the tested concentration ((Pt) = 1 μM). Overall, the combined evidence points out that the hydrophobicity of the head group modifications dictate the anticancer activities of the FALPs.

**Cell entry and mitochondrial accumulation of FALP derivatives with head groups of different hydrophobicities.** Our next objective was to gain a more detailed understanding of how modifications to the head group of FALPs affect their anticancer activities through hydrophobicity. Cellular uptake is widely recognized as a critical factor in the activity of metallodrugs. The influence of hydrophobicity on the cellular uptake of therapeutic molecules has been widely acknowledged, with an increase in hydrophobicity typically promoting cell entry. To this end, we conducted a graphite furnace atomic absorption spectroscopic (GFAAS) analysis of the cellular uptake for FALPs with various head group modifications, including **2** and **7**. The Log P values of compounds **2** and **7**, as determined via GFAAS, are 1.97 and 2.57, respectively. This indicates that compound **7** exhibits greater hydrophobicity than compound **2**. Surprisingly, our results in Figure 3A indicate that the hydrophilic modification of **2** (496.6 ± 16.09 pmol Pt/million cells) led to an uptake of over eight times greater than the hydrophobic modifications of **7** (60.03 ± 8.01 pmol Pt/million cells). We previously discovered that mitochondria play significant roles in the mechanism of action of FALPs, so we further investigated how the hydrophobicity of the head group modifications affects the mitochondrial accumulation of FALPs. As shown in Figure 3B, the mitochondrial Pt content of **2** (31.7 ± 5.1 pmol Pt/million cells) was three times higher than that of **7** (11.72 ± 2.76 pmol Pt/million cells). Nevertheless, all FALPs demonstrated a higher cellular uptake and mitochondrial accumulation than cisplatin, despite using a higher Pt concentration in the cisplatin sample. In summary, the introduction of a hydrophilic head group in FALPs promotes cell entry and mitochondrial accumulation.

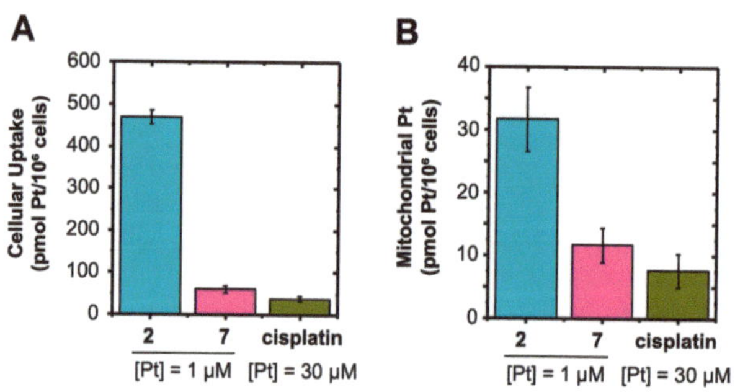


**Figure 3.** Cellular uptake (**A**) and mitochondrial accumulation (**B**) of FALPs (**2** and **7**) and cisplatin in A2780cis cells (24 h).

**Cellular Responses of FALP derivatives with head groups of different hydrophobicity.** Based on the observation that hydrophilic head group modifications lead to increased intracellular Pt levels, we formulated the hypothesis that such modifications would result in greater mitochondrial and DNA damage, as well as increased apoptosis. MitoSOX was utilized to assess the levels of mitochondrial ROS, while the γH2AX levels were analyzed to determine DNA damage. As mitochondrial and DNA damage are known to promote apoptosis, Annexin V/PI assays were also conducted to examine the apoptotic effects. A flow cytometric analysis was used to evaluate the mitochondrial ROS, γH2AX, and apoptosis in the cancer cells treated with FALPs (**2** and **7**) in the experiments. As illustrated in Figure 1A, compound **2** has a hydrophilic head group, while compound **7** has a hydrophobic head group. According to the flow cytometric results in Figure 4A, treatment with **2** (1 μM, 24 h) significantly increased the mitochondrial ROS levels compared to the control or **7**. Additionally, the treatment of cisplatin at a higher concentration (10 μM, 24 h) resulted in an insignificant change in the mitochondrial ROS levels, which is consistent with its mechanism of action. The treatment of **2** (0.25 μM, 24 h) also induced DNA damage in the treated A2780cis cells, as shown in Figure 4B. Furthermore, **7** at a higher concentration (1 μM, 24 h) triggered DNA damage, but to a lesser extent than **2**. Likewise, our flow cytometric analysis of Annexin V/PI showed that a larger population of cells were in the late stages of apoptosis (23.3% for **2**) compared to the effect of **7**, which induced only 3.58% of cells to undergo the late stages of apoptosis (Figure 4C). Based on the results, it can be inferred that the FALP derivative (**2**) with a hydrophilic head group modification induces mitochondrial and DNA damage as well as apoptosis more effectively than the one (**7**) with a hydrophobic modification.

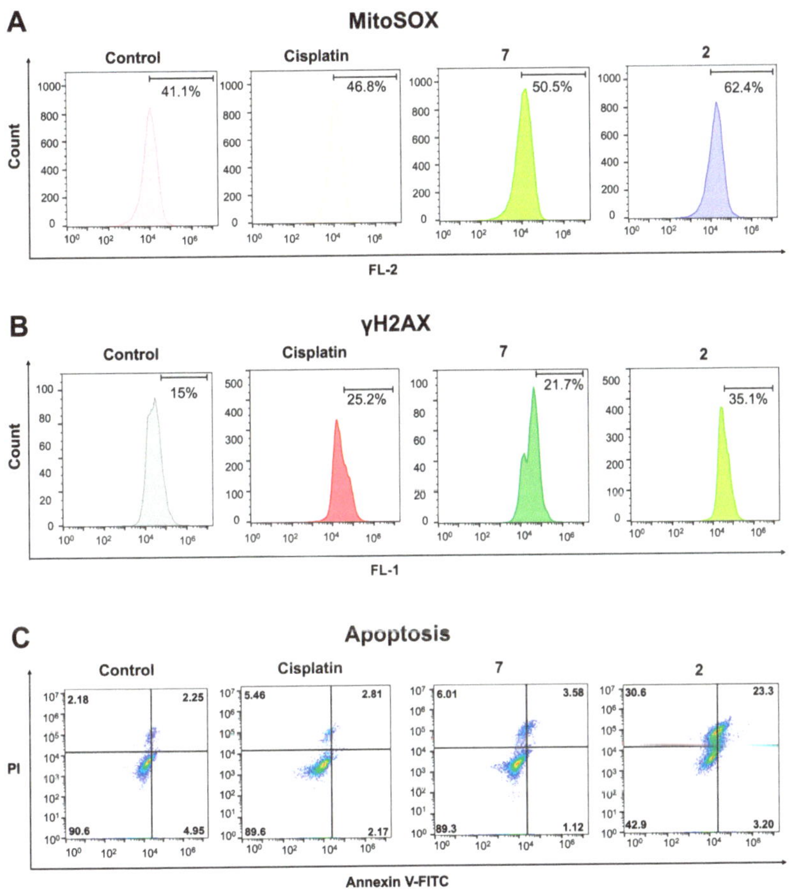


**Figure 4.** Cellular responses of A2780cis cells treated with FALPs and cisplatin. (**A**) Flow cytometric analysis of MitoSOX in the A2780cis cells treated with FALPs (**2** or **7**) or cisplatin for 24 h. (**B**) Flow cytometric analysis of γH2AX in the A2780cis cells treated with FALPs (**2** or **7**) or cisplatin for 24 h. (**C**) Flow cytometric analysis of apoptosis in the A2780cis cells treated with FALPs (**2** or **7**) or cisplatin for 48 h.

## 3. Materials and Methods

**General information**. All reagents were purchased from Strem, Aldrich, or Alfa and used without further purification. Compound **1** was synthesized according to the literature [36]. All reactions were carried out under normal atmospheric conditions. A Bruker 400 NMR was used for NMR data acquisition (frequency: 400 M Hz for $^{1}$H NMR; 100 MHz for $^{13}$C NMR). Chemical shifts in $^{1}$H and $^{13}$C{$^{1}$H} NMR spectra were internally referenced to solvent signals ($^{1}$H NMR: DMSO at $\delta$ = 2.50 ppm; $^{13}$C NMR: DMSO at $\delta$ = 40.45 ppm). The high-resolution mass spectra of created ions were recorded on an

Exactive Plus mass spectrometer (Thermo Scientific, Bremen, Germany). Analytical HPLC was conducted on an Agilent 1100 system using C18 reverse-phase columns (Hypersil GOLD; 100 mm × 3 mm; 5 μm). Graphite furnace atomic absorption spectroscopic (GFAAS) measurements were taken on a PinAAcle 900Z spectrometer (PerkinElmer, Shelton, CT, USA). Fluorescence spectra were taken on a FluoroMax-3 Fluorescence spectrophotometer (Horiba, Japan) using the software called FluorEssence. Fluorescence images were acquired using an IX70 (Olympus, Japan) inverted epifluorescence microscope equipped with a digital CCD camera (QImaging, Surrey, BC, Canada). Images were processed, and intensities were quantified with ImageJ software v1.53t. Live/dead cell assay was carried out using Invitrogen (Thermo Fisher Scientific) LIVE/DEAD$^{TM}$ Cell Viability Kit (Cat. No. L3224). Flow cytometry was carried out on a Accuri C6 flow cytometer (Becton, Dickinson and Company Biosciences, Lakes, NJ, USA).

**Synthesis of Compound 2**. An amount of 1 mL of anhydrous DMF was added to compound **1** (70 mg; 0.10 mmol) and HATU (46 mg; 0.12 mmol) in a vial under a stream of Ar and stirred at r.t. for 15 min. 2-(2-aminoethoxy) ethanol (28 μL; 0.28 mmol) was added to the mixture. After 20 min of stirring at r.t., DIPEA (70 μL; 0.41 mmol) was added. The reaction mixture was stirred in the dark at r.t. overnight, centrifuged, and the supernatant was added into 3 mL of brine. Then, the precipitation was collected via centrifugation, washed with water, and lyophilized overnight. Lyophilized product was dissolved in small amount of MeOH and purified with flash chromatography. Yield: 53 mg (67%). $^1$HNMR (400 MHz, DMSO-d$_6$): δ: 0.856 (NHCH2(CH2)14C*H3*, t, 3H), 1.23 (NHCH2(C*H2*)14CH3, m, 28H), 2.36 (CO(C*H2*)2CO, m, 4H), 2.88 (NHC*H2*(CH2)14CH3, q, 2H), 3.19 (NHC*H2*CH2O, q, 2H), 3.40 (CH2C*H2*OC*H2*CH2, m, 4H), 3.49 (OCH2C*H2*OH, q, 2H), 6.52 (N*H*CH2(CH2)14CH3, t, 1H), 6.62 (N*H3*, m, 6H), 7.90 (CON*H*(CH2)2O, t, 1H); $^{13}$C NMR(100 MHz, DMSO-d$_6$): δ: 180.4, 172.0, 164.4, 72.6, 69.5, 60.6, 41.4, 40.0, 31.8, 30.3, 29.5, 29.4, 29.2, 27.0, 22.6, 14.4; HR-MS (positive mode) for $[C_{25}H_{54}Cl_2N_4O_7Pt]^+$: *m/z* calc: 787.3088, obsd: 787.3064. Purity: 99% determined via HPLC.

**Synthesis of Compound 3**. An amount of 1 mL of anhydrous DMF was added to compound **1** (70 mg; 0.10 mmol) and HATU (46 mg; 0.12 mmol) in a vial under a stream of Ar and stirred at r.t. for 15 min. 2-methoxyethylamine (24 μL; 0.28 mmol) was added to the mixture. After 20 min of stirring at r.t., DIPEA (70 μL; 0.41 mmol) was added. The reaction mixture was stirred in the dark at r.t. overnight, centrifuged, and the supernatant was added into 3 mL of brine. Then, the precipitation was collected via centrifugation, washed with water, and lyophilized overnight. Lyophilized product was dissolved in small amount of MeOH and purified with flash chromatography. Yield: 59 mg (78%). $^1$H NMR (400 MHz, DMSO-d6): δ: 0.836 (NHCH2(CH2)14C*H3*, t, 3H); 1.23 (NHCH2(C*H2*)14CH3, m, 28H); 2.26 (CO(C*H2*)2CO, m, 4H), 2.85 (NHC*H2*(CH2)14CH3, q, 2H); 3.16 (NHC*H2*CH2O, 2H); 3.34 (C*H2*C*H2*OCH3, m, 4H); 3.23 (C*H2*CH2OC*H3*, s, 3H); 6.40 (N*H*CH2(CH2)14CH3 and N*H3*, m, 7H); 7.84 (CON*H*, 1H); 13C NMR (400MHz, DMSO-d$_6$): δ: 180.45,171.98, 164.49, 71.09, 58.33, 31.94, 31.73, 30.30, 29.48, 29.12, 26.92, 22.51, 14.36; HR-MS (positive mode) for $[C_{24}H_{52}Cl_2N_4O_6PtH]+$: *m/z* calc: 758.2986, obsd: 758.2985. Purity: 95% determined via HPLC.

**Synthesis of Compound 4**. An amount of 1 mL of anhydrous DMF was added to compound **1** (70 mg; 0.10 mmol) and HATU (46 mg; 0.12 mmol) in a vial under a stream of Ar and stirred at r.t. for 15 min. To the mixture, 0.5 mL anhydrous DMF solution of aminoacetonitrile bisulfate (43 mg; 0.28 mmol) was added. After 20 min of stirring at r.t., DIPEA (70 μL; 0.41 mmol) was added. The reaction mixture was stirred in the dark at r.t. overnight, centrifuged, and the supernatant was added into 3 mL of brine. Then, the precipitation was collected via centrifugation, washed with water, and lyophilized overnight to afford yellowish-white-colored solid. Yield: 47 mg (65%). $^1$H NMR (400 MHz, DMSO-d$_6$): δ: 0.856 (NHCH2(CH2)14C*H3*, t, 3H), 1.23 (NHCH2(C*H2*)14CH3, m, 28H), 2.42 (CO(C*H2*)2CO, m, 4H), 2.87 (NHC*H2*(CH2)14CH3, q, 2H), 4.10 (NHC*H2*CN, d, 2H), 6.51 (N*H*CH2(CH2)14CH3, t, 1H), 6.66 (N*H3*, m, 6H), 8.68 (CON*H*CH2CN, t, 1H); $^{13}$C NMR (100 MHz, DMSO-d$_6$): δ: 180.2, 172.5, 170.5, 118.2, 60.8, 41.1, 31.8, 30.3, 29.5, 29.4, 29.2,

26.9, 22.6, 14.4; HR-MS (positive mode) for $[C_{23}H_{47}Cl_2N_5O_5Pt]^+$: *m/z* calc: 739.2677, obsd: 739.2673. Purity: 95% determined via HPLC.

**Synthesis of Compound 5.** An amount of 1 mL of anhydrous DMF was added to compound **1** (70 mg; 0.1 mmol) and HATU (46 mg; 0.12 mmol) in a vial under a stream of Ar and stirred at r.t. for 15 min. To the mixture, 0.5 mL anhydrous DMF solution of glycine ethyl ester hydrochloride (39 mg; 0.28 mmol) was added. After 20 min of stirring at r.t., DIPEA (70 µL; 0.41 mmol) was added. The reaction mixture was stirred in the dark at r.t. overnight, centrifuged, and the supernatant was added into 3 mL of brine. Then, the precipitation was collected via centrifugation, washed with water, and lyophilized overnight to collect yellowish-white-colored solid. Yield: 47 mg (60%). $^1$HNMR (400 MHz, DMSO-d$_6$): δ: 0.853 (NHCH2(CH2)14C*H3*, t, 3H, J = 6.9 Hz), 1.15 (COOCH2C*H3*, t, 3H, J = 7.1 Hz), 1.23 (NHCH2(C*H2*)14CH3, m, 28H), 2.39 (CO(C*H2*)2CO, m, 4H), 2.87 (NHC*H2*(CH2)14CH3, q, 2H), 3.792 (NHC*H2*COOCH2, d, 2H), 4.08 (COOC*H2*CH3, q, 2H), 6.50 (N*H*CH2(CH2)14CH3, t, 1H), 6.69 (N*H3*, m, 6H), 8.37 (CON*H*CH2COOCH2, t, 1H); $^{13}$C NMR(100 MHz, DMSO-d$_6$): δ: 180.2, 172.4, 170.4, 60.8, 41.1, 31.6 30.3, 29.5, 29.4, 29.2, 26.9, 22.6, 14.6, 14.4; HR-MS (positive mode) for $[C_{25}H_{52}Cl_2N_4O_7Pt]^+$: *m/z* calc: 786.2936, obsd: 786.2933. Purity: 98% determined via HPLC.

**Synthesis of Compound 6.** An amount of 1 mL of anhydrous DMF was added to compound **1** (70 mg; 0.1 mmol) and HATU (46 mg; 0.12 mmol) in a vial under a stream of Ar and stirred at r.t. for 15 min. To the mixture, 0.5 mL anhydrous DMF solution of propylamine (23 µL; 0.28 mmol) was added. After 20 min of stirring at r.t., DIPEA (70 µL; 0.41 mmol) was added. The reaction mixture was stirred in the dark at r.t. overnight, centrifuged, and the supernatant was added into 3 mL of brine. Then, the precipitation was collected via centrifugation, washed with water, and lyophilized overnight to collect yellowish-white-colored solid. Lyophilized product was dissolved in small amount of MeOH and purified with flash chromatography. Yield: 53 mg (72%). $^1$H NMR (400 MHz, DMSO-d$_6$): δ: 0.814 (NHCH2(CH2)14C*H3* and NHCH2CH2C*H3* m, 6H); 1.23 (NHCH2(C*H2*)14CH3 and NHCH2C*H2*CH3 m, 28H); 2.25 (CO(C*H2*)2CO, m, 4H, J = 7.1, 42.4 Hz), 2.87 (NHC*H2*(CH2)14CH3, q, 2H); 6.34 (N*H* and N*H3*, 7H); 7.73 (N*H*, s, 1H); $^{13}$C NMR (100 MHz, DMSO-*d*$_6$): δ: 180.54,171.68, 164.46, 32.03, 31.73, 30.30, 29.49, 29.44, 29.35, 29.12, 26.93, 22.80, 14.37; HR-MS (positive mode) for $[C_{24}H_{52}Cl_2N_4O_5PtH]^+$: *m/z* calc: 742.3037, obsd: 742.3034. Purity: 95% determined via HPLC.

**Synthesis of Compound 7.** An amount of 1 mL of anhydrous DMF was added to compound **1** (70 mg; 0.1 mmol) and HATU (46 mg; 0.12 mmol) in a vial under a stream of Ar and stirred at r.t. for 15 min. To the mixture, 0.5 mL anhydrous DMF solution of hexylamine (37 µL; 0.28 mmol) was added. After 20 min of stirring at r.t., DIPEA (70 µL; 0.41 mmol) was added. The reaction mixture was stirred in the dark at r.t. overnight, centrifuged, and the supernatant was added into 3 mL of brine. Then, the precipitation was collected via centrifugation, washed with water, and lyophilized overnight to collect yellowish-white-colored solid. Lyophilized product was dissolved in small amount of MeOH and purified with flash chromatography. Yield: 54 mg (69%). $^1$H NMR (400 MHz, DMSO-d$_6$): δ: 0.83 (NHCH2(CH2)14C*H3* and NHCH2(CH2)4C*H3*, 6H); 1.23 (NHCH2(C*H2*)14CH3 and NHCH2(C*H2*)4CH3, 36H); 2.24 (CO(C*H2*)2CO, m, 4H), 2.87 (NHC*H2*(CH2)14CH3, q, 2H); 2.97 (NHC*H2*(CH2)4CH3, 4H); 6.48 (N*H* and N*H3*, 7H); 7.79 (N*H*, 1H); $^{13}$C NMR (100 MHz, DMSO-*d*$_6$): δ: 180.47,171.60, 164.41, 31.99, 31.75, 31.46, 29.51, 29.16, 26.94, 26.57, 22.55, 22.52, 14.42; HR-MS (positive mode) for $[C_{27}H_{58}Cl_2N_4O_5PtH]^+$: *m/z* calc: 784.3507, obsd: 784.3504. Purity: 95% determined via HPLC.

**Synthesis of Compound 8.** An amount of 1 mL of anhydrous DMF was added to compound **1** (70 mg; 0.1 mmol) and HATU (46 mg; 0.12 mmol) in a vial under a stream of Ar and stirred at r.t. for 15 min to obtain pale-yellow-colored solution. To the mixture, 0.5 mL anhydrous DMF solution of 1-adamantylamine (42 mg; 0.28 mmol) was added. After 20 min of stirring at R.T., DIPEA (70 µL; 0.41 mmol) was added. The reaction mixture was stirred in the dark at r.t. overnight. The solution turned into a golden yellow color. It was centrifuged, and the supernatant was added into 3 mL of brine. Then, the precipitation was collected via centrifugation, washed with water, and lyophilized overnight. Lyophilized product was dissolved in small amount of MeOH and purified with flash chromatography. Yield: 56 mg (67%). $^1$H NMR (400 MHz, DMSO-d$_6$): δ: 0.858 (NHCH2(CH2)14C*H3*, t, 3H 1.23 (NHCH2(C*H2*)14CH3, m, 28H), 1.60 (CHC*H2*CH, Adamantyl, t, 6H), 1.90 (CC*H2*CH, Adamantyl, d, 6H), 1.98 (CH2C*H*(CH2)2, Adamantyl, m, 3H), 2.31 (CO(C*H2*)2CO, m, 4H), 2.87 (NHC*H2*(CH2)14CH3, q, 2H), 6.52 (N*H*CH2(CH2)14CH3, t, 1H), 6.66 (N*H3*, m, 6H), 7.30 (CONH*C*(CH2)3, s, 1H); $^{13}$C NMR(100 MHz, DMSO-d$_6$): δ: 180.7, 171.2, 164.4, 51.0, 41.5, 41.4, 36.6, 31.8, 29.5, 29.4, 29.3, 26.9, 22.6, 14.4; HR-MS (positive mode) for $[C_{31}H_{60}Cl_2N_4O_5Pt]^+$: *m*/*z* calc: 834.3664, obsd: 834.3660. Purity: 96% determined via HPLC.

**Synthesis of Compound 9.** To PtC16 (80 mg; 0.122 mmol) and ethyl isocyanatoacetate (16 μL; 0.14 mmol) in a vial, 1.5 mL of anhydrous DMF was added under a stream of Ar and stirred at r.t. overnight. The product was extracted with $Et_2O$, washed with $H_2O$, and lyophilized overnight to obtain a yellowish-white-colored solid. Yield: 42 mg (44%). $^1$HNMR (400 MHz, DMSO-d$_6$): δ: 0.854 (NHCH2(CH2)14C*H3*, t, 3H), 1.18 (COOCH2C*H3*, t, 3H), 1.23 (NHCH2(C*H2*)14CH3, m, 28H), 2.34 (CO(C*H2*)2CO, m, 4H), 2.97 (NHC*H2*(CH2)14CH3, m, 2H), 3.72 (NHC*H2*COOCH2, d, 2H), 4.06 (COOC*H2*CH3, q, 2H), 6.647 (N*H3*, m, 6H), 6.83 (N*H*CH2(CH2)14CH3, t, 1H), 7.85 (CON*H*CH2COOCH2, t, 1H); $^{13}$C NMR(100 MHz, DMSO-d$_6$): δ: 180.4, 171.7, 171.6, 158.4, 60.6, 41.9, 40.0, 31.9, 31.8, 30.4, 29.5, 29.3, 29.2, 27.0, 22.6, 14.6, 14.4; HR-MS (positive mode) for $[C_{25}H_{52}Cl_2N_4O_7Pt]^+$: *m*/*z* calc: 786.2936, obsd: 786.2933. Purity: 95% determined via HPLC.

**GFAAS analysis of Log P values for 2 and 7.** The samples were first dissolved with DMSO to create 200 μM stocks. From these stocks, 50 μL was added to a $H_2O$:Octanol mixture with a 1:1 volume ratio. This mixture was vortexed for 5 min and subsequently centrifuged for 3 min at 3000 rpm. Following centrifugation, the $H_2O$ and octanol layers were isolated for analysis. The Pt content in each phase was quantified using GFAAS to calculate the Log P value.

**Cell culture.** A2780cis cell lines were purchased from Sigma-Aldrich and cultured in RPMI 1640 with L-glutamine (Corning, New York, NY, USA) supplemented with 10% FBS (Atlanta Biologicals, USA) and 1% penicillin-streptomycin (Corning). The MDA-MB-231 cell line was obtained via American Type Culture Collection, and cultured in DMEM 1 g/L glucose, with L-glutamine and sodium pyruvate (Corning) supplemented with 10% FBS and 1% penicillin-streptomycin (Corning). All cell lines were cultured at 37 °C under an atmosphere containing 5% $CO_2$. Cells were passaged upon reaching 80–90% confluence via trypsinization and split in a 1:5 ratio.

**MTT assays.** Cytotoxicity profiles of compounds **1–9** and cisplatin against different cell lines (A2780cis and MDA-MB-231) were evaluated using the MTT assays. A volume of 100 μL of a RPMI or DMEM containing $8 \times 10^4$ cells/mL was seeded in 96-well plates. The plates were incubated for 24 h at 37 °C with 5% $CO_2$ to allow for adherence of cells. A volume of 50 μL of RPMI or DMEM with various concentrations of cisplatin or compounds **1–9** were added to each well of the microplates. The Pt concentrations were determined via GFAAS. After 24 h, a volume of 30 μL of MTT (5.0 mg/mL in PBS, Alfa Aesar, Haverhill, MA, USA) was added to each well of the microplates. After 24 h, the medium was aspirated, and 200 μL of DMSO was added to each well. The plates were shaken gently on a shaker at r.t. for 10 min. Then, the absorbance of purple formazan was recorded at 562 nm with a BioTek ELx800 plate reader. $IC_{50}$ values were determined using Origin software v7.0. All experiments were performed in triplicate.

**LIVE/DEAD cell viability assays.** A2780cis cells were cultured in imaging disks (MatTek, Ashland, MA, USA) at a concentration of $5 \times 10^4$ cells with 2 mL of complete medium and incubated for 24 h at 37 °C with 5% $CO_2$. The cells were then treated with compound **2 or 7** ((Pt) = 1 μM) and incubated for 24 h at 37 °C with 5% $CO_2$. Before the assay, the cells were washed with 1 mL PBS and 1 mL dye-free RPMI to remove serum esterase activity that is generally present in serum-supplemented growth media. A 100 μL volume of LIVE/DEAD working solution (formed by mixing 2 μM of calcein AM and 2 μM ethidium homodimer-1 in PBS) was carefully added to the disk, which was then incubated at r.t. for 30 min. Images were acquired using an Olympus IX70 inverted epifluorescence microscope equipped with a digital CCD camera (QImaging, Surrey, BC, Canada). Images were processed, and intensities were quantified with ImageJ software (NIH).

**GFAAS analysis of cellular platinum contents in A2780cis cells**. A2780cis cells were seeded in a 6-well plate at a concentration of $5 \times 10^5$ cells per well and incubated at 37 °C with 5% $CO_2$ overnight. Next day, the cells were treated with compound **2** or **7** ((Pt) = 1 μM) or cisplatin ((Pt) = 30 μM) for 24 h at 37 °C with 5% $CO_2$. The remaining live cells were harvested via trypsinization and counted. The cells were then digested in 200 μL 65% $HNO_3$ at r.t. overnight. The Pt contents in the cells were analyzed via GFAAS. All experiments were performed in triplicate.

**Measurements of mitochondrial platinum contents in A2780cis cells**. A2780cis cells were seeded on a 6-well plate and incubated at 37 °C with 5% $CO_2$ overnight. The cells were treated with cisplatin ((Pt) = 30 μM) or compound **2** or **7** ((Pt) = 1 μM) for 24 h at 37 °C with 5% $CO_2$. Next, the wells were washed with PBS (1 mL) and harvested via trypsinization (1 mL) and counted. Mitochondrial fractions were isolated using the Thermo Scientific™ Mitochondria Isolation Kit for Mammalian Cells. The mitochondrial fraction was then dissolved in 200 μL 65% nitric acid and shaken at 400 rpm on an Eppendorf ThermoMixer™ F1.5 at r.t. overnight. Next, the fractions were diluted 4× in water and the platinum content was analyzed using GFAAS. All experiments were performed in triplicate.

**Flow cytometric analysis of MitoSOX**. A2780cis cells were seeded in 6-well plate at a concentration of $6 \times 10^4$ cells/mL and incubated overnight. Then, the cells were treated with cisplatin ((Pt) = 10 μM) or compound **2** or **7** ((Pt) = 1 μM) and incubated overnight. The medium was aspirated, and cells were washed with 1 mL PBS. Next, the cells were incubated with 5 μM MitoSOX reagent in fresh medium for 60 min at 37 °C with 5% $CO_2$ in the dark. Cells were trypsinized and collected. The cell pellet was washed 2 times with PBS. The cells were then re-suspended in PBS with 0.5% BSA to reach $10^6$ cells/mL and analyzed with BD Accuri C6 flow cytometer using FL-2 channel, and data were processed with FlowJo v10.

**Flow cytometric analysis of γH2AX**. A2780cis cells were seeded in a 6-well plate at a concentration of $4 \times 10^5$ cells/well. Cells were then incubated at 37 °C with 5% $CO_2$ for 24 h. Next, the cells were treated with compound **2** ((Pt) = 0.25 μM), **7** ((Pt) = 1 μM) or cisplatin ((Pt) = 30 μM) and incubated for 24 h. Live cells were collected and 250 μL BD Permeabilization solution was added to re-suspend the cells, which were then incubated for 20 min at 4 °C. Cell pellets were collected, washed twice with 1X BD Perm/Wash buffer, and resuspended in 50 μL of buffer. Alexa 488-anti γH2AX antibody solution was then added, and the samples were incubated in the dark for 60 min at r.t. The final cell pellets were suspended in 500 μL of PBS with 0.5% BSA and analyzed with BD Accuri C6 flow cytometer using FL-1 channel, and data were processed with FlowJo.

**Flow cytometric analysis of apoptosis**. A2780cis cells were seeded in a 6-well plate at a concentration of $3 \times 10^5$ cells/well. Cells were then incubated at 37 °C 5% $CO_2$ for 24 h. Next, compound **2** or **7** ((Pt) = 0.5 μM) or cisplatin ((Pt) = 7 μM) was added and incubated for 48 h. Both live and dead cells were collected, resuspended in 1mL PBS, and counted. A 1X binding buffer from the FITC Annexin V Apoptosis Detection Kit 1 (BD Biosciences, Franklin Lakes, NJ, USA) was then added to reach a concentration of $10^6$ cells/mL. An amount of 100 μL cell solution was transferred to a fresh 2 mL Eppendorf tube, and 5 μL of both Annexin V-FITC and PI solutions were added to cells. Cells were incubated for 15 min at r.t. in the dark and then brought to 400 μL volume by adding required volume of binding 1X buffer. Cells were then analyzed with FL-1 and FL-3 channels on a BD Accuri C6 flow cytometer and data were processed with FlowJo.

## 4. Conclusions

Our study represents the first comprehensive investigation of the structure–activity relationship of FALPs. We synthesized a small library of FALPs with diverse head group modifications and found that such modifications can greatly affect the cytotoxicity profiles of FALPs, ranging from low to highly potent. Interestingly, a further analysis revealed that only hydrophilic modifications led to a high potency, while hydrophobic moieties resulted in a much lower cytotoxicity. To explore the impact of hydrophobicity on the cytotoxicity of FALPs, we focused on two similar FALPs, one with a hydrophilic PEG head group and the other with a hydrophobic hydrocarbon modification of the same molecular weight. Using these model compounds, we evaluated cellular uptake and mitochondrial accumulation through GFAAS, as well as mitochondrial and DNA damage and apoptosis through flow cytometry. Our comprehensive findings reveal that FALPs incorporating hydrophilic modifications can readily penetrate cancer cells and mitochondria, initiating subsequent cellular responses that effectively eradicate cancer cells. Conversely, FALPs with hydrophobic modifications showed a notably lower uptake and weaker cellular responses. These combined results present an alternative perspective, differing from the conventional belief that increased hydrophobicity invariably enhances cellular uptake. These findings provide valuable new insights into the fundamental principles of developing metallodrugs. It underscores the significance of developing FALPs with hydrophilic modifications, which hold the potential to yield more potent and effective anticancer agents. This study lays the groundwork for future research endeavors aimed at optimizing the structural design of FALPs, with the objective of enhancing anticancer activity while minimizing side effects.

**Supplementary Materials:** The synthetic schemes and characterization of FALPs can be downloaded at https://www.mdpi.com/article/10.3390/ijms241713301/s1.

**Author Contributions:** Conceptualization, Y.-R.Z.; synthesis and characterization of the Pt compounds, M.K., M.C., P.D., A.S., T.B. and D.B.; cell-based studies, M.K., W.J., S.A. and P.D.; writing, M.K., W.J., S.A. and Y.-R.Z.; supervision, Y.-R.Z.; project administration, Y.-R.Z.; funding acquisition, Y.-R.Z. All authors have read and agreed to the published version of the manuscript.

**Funding:** This research was funded by the R15 grant (1R15CA249712-01A1) provided by the National Cancer Institute, NSF Award 2050873, and the Farris Family Innovation Fellowship and LaunchPad Award provided by Kent State University.

**Institutional Review Board Statement:** Not applicable.

**Informed Consent Statement:** Not applicable.

**Data Availability Statement:** Not applicable.

**Acknowledgments:** Y.-R.Z. is thankful for the support from the Research Council of Kent State University. W.J. is thankful for the financial support provided by the Healthy Communities Research Initiative at Kent State University.

**Conflicts of Interest:** The authors declare no conflict of interest.

## References

1. Wang, D.; Lippard, S.J. Cellular processing of platinum anticancer drugs. *Nat. Rev. Drug Discov.* **2005**, *4*, 307–320. [CrossRef]
2. Kelland, L. The resurgence of platinum-based cancer chemotherapy. *Nat. Rev. Cancer* **2007**, *7*, 573–584. [CrossRef] [PubMed]
3. Jamieson, E.R.; Lippard, S.J. Structure, recognition, and processing of cisplatin-DNA adducts. *Chem. Rev.* **1999**, *99*, 2467–2498. [CrossRef] [PubMed]
4. Todd, R.C.; Lippard, S.J. Inhibition of transcription by platinum antitumor compounds. *Metallomics* **2009**, *1*, 280–291. [CrossRef]
5. Jordan, C.T.; Guzman, M.L.; Noble, M. Cancer Stem Cells. *N. Engl. J. Med.* **2006**, *355*, 1253–1261. [CrossRef]
6. Gupta, P.B.; Chaffer, C.L.; Weinberg, R.A. Cancer stem cells: Mirage or reality? *Nat. Med.* **2009**, *15*, 1010–1012. [CrossRef] [PubMed]
7. Zhou, J.; Kang, Y.; Chen, L.; Wang, H.; Liu, J.; Zeng, S.; Yu, L. The Drug-Resistance Mechanisms of Five Platinum-Based Antitumor Agents. *Front. Pharmacol.* **2020**, *11*, 343. [CrossRef]
8. Jogadi, W.; Zheng, Y.R. Supramolecular platinum complexes for cancer therapy. *Curr. Opin. Chem. Biol.* **2023**, *73*, 102276. [CrossRef]

9. Xu, Z.; Wang, Z.; Deng, Z.; Zhu, G. Recent advances in the synthesis, stability, and activation of platinum(IV) anticancer prodrugs. *Coord. Chem. Rev.* **2021**, *442*, 213991. [CrossRef]
10. Johnstone, T.C.; Suntharalingam, K.; Lippard, S.J. The Next Generation of Platinum Drugs: Targeted Pt(II) Agents, Nanoparticle Delivery, and Pt(IV) Prodrugs. *Chem. Rev.* **2016**, *116*, 3436–3486. [CrossRef]
11. Konkankit, C.C.; Marker, S.C.; Knopf, K.M.; Wilson, J.J. Anticancer activity of complexes of the third row transition metals, rhenium, osmium, and iridium. *Dalton Trans.* **2018**, *47*, 9934–9974. [CrossRef] [PubMed]
12. Olelewe, C.; Awuah, S.G. Mitochondria as a target of third row transition metal-based anticancer complexes. *Curr. Opin. Chem. Biol.* **2023**, *72*, 102235. [CrossRef] [PubMed]
13. Vaidya, S.P.; Patra, M. Platinum glycoconjugates: "Sweet bullets" for targeted cancer therapy? *Curr. Opin. Chem. Biol.* **2023**, *72*, 102236. [CrossRef] [PubMed]
14. Northcote-Smith, J.; Suntharalingam, K. Targeting chemotherapy-resistant tumour sub-populations using inorganic chemistry: Anti-cancer stem cell metal complexes. *Curr. Opin. Chem. Biol.* **2023**, *72*, 102237. [CrossRef] [PubMed]
15. Elie, B.T.; Fernández-Gallardo, J.; Curado, N.; Cornejo, M.A.; Ramos, J.W.; Contel, M. Bimetallic titanocene-gold phosphane complexes inhibit invasion, metastasis, and angiogenesis-associated signaling molecules in renal cancer. *Eur. J. Med. Chem.* **2019**, *161*, 310–322. [CrossRef]
16. Boulet, M.H.C.; Bolland, H.R.; Hammond, E.M.; Sedgwick, A.C. Oxali(IV)Fluors: Fluorescence Responsive Oxaliplatin(IV) Complexes Identify a Hypoxia-Dependent Reduction in Cancer Cells. *J. Am. Chem. Soc.* **2023**, *145*, 12998–13002. [CrossRef]
17. Momeni, B.Z.; Abd-El-Aziz, A.S. Recent advances in the design and applications of platinum-based supramolecular architectures and macromolecules. *Coord. Chem. Rev.* **2023**, *486*, 215113. [CrossRef]
18. Li, H.; Cheng, S.; Zhai, J.; Lei, K.; Zhou, P.; Cai, K.; Li, J. Platinum based theranostics nanoplatforms for antitumor applications. *J. Mater. Chem. B* **2023**. [CrossRef]
19. Wang, T.; Wu, C.; Hu, Y.; Zhang, Y.; Ma, J. Stimuli-responsive nanocarrier delivery systems for Pt-based antitumor complexes: A review. *RSC Adv.* **2023**, *13*, 16488–16511. [CrossRef]
20. Zhong, T.; Yu, J.; Pan, Y.; Zhang, N.; Qi, Y.; Huang, Y. Recent Advances of Platinum-Based Anticancer Complexes in Combinational Multimodal Therapy. *Adv. Healthc. Mater.* **2023**, e2300253. [CrossRef]
21. Deng, Z.; Zhu, G. Beyond mere DNA damage: Recent progress in platinum(IV) anticancer complexes containing multi-functional axial ligands. *Curr. Opin. Chem. Biol.* **2023**, *74*, 102303. [CrossRef] [PubMed]
22. Li, Y.; Lin, W. Platinum-based combination nanomedicines for cancer therapy. *Curr. Opin. Chem. Biol.* **2023**, *74*, 102290. [CrossRef]
23. Giorgi, E.; Binacchi, F.; Marotta, C.; Cirri, D.; Gabbiani, C.; Pratesi, A. Highlights of New Strategies to Increase the Efficacy of Transition Metal Complexes for Cancer Treatments. *Molecules* **2022**, *28*, 273. [CrossRef] [PubMed]
24. Zhang, C.; Kang, T.; Wang, X.; Song, J.; Zhang, J.; Li, G. Stimuli-responsive platinum and ruthenium complexes for lung cancer therapy. *Front. Pharmacol.* **2022**, *13*, 1035217. [CrossRef] [PubMed]
25. Su, S.; Chen, Y.; Zhang, P.; Ma, R.; Zhang, W.; Liu, J.; Li, T.; Niu, H.; Cao, Y.; Hu, B.; et al. The role of Platinum(IV)-based antitumor drugs and the anticancer immune response in medicinal inorganic chemistry. A systematic review from 2017 to 2022. *Eur. J. Med. Chem.* **2022**, *243*, 114680. [CrossRef] [PubMed]
26. Lu, Y.; Ma, X.; Chang, X.; Liang, Z.; Lv, L.; Shan, M.; Lu, Q.; Wen, Z.; Gust, R.; Liu, W. Recent development of gold(I) and gold(III) complexes as therapeutic agents for cancer diseases. *Chem. Soc. Rev.* **2022**, *51*, 5518–5556. [CrossRef]
27. Alassadi, S.; Pisani, M.J.; Wheate, N.J. A chemical perspective on the clinical use of platinum-based anticancer drugs. *Dalton Trans.* **2022**, *51*, 10835–10846. [CrossRef]
28. Czarnomysy, R.; Radomska, D.; Szewczyk, O.K.; Roszczenko, P.; Bielawski, K. Platinum and Palladium Complexes as Promising Sources for Antitumor Treatments. *Int. J. Mol. Sci.* **2021**, *22*, 8271. [CrossRef]
29. Anthony, E.J.; Bolitho, E.M.; Bridgewater, H.E.; Carter, O.W.L.; Donnelly, J.M.; Imberti, C.; Lant, E.C.; Lermyte, F.; Needham, R.J.; Palau, M.; et al. Metallodrugs are unique: Opportunities and challenges of discovery and development. *Chem. Sci.* **2020**, *11*, 12888–12917. [CrossRef]
30. Gibson, D. Platinum(IV) anticancer agents; are we en route to the holy grail or to a dead end? *J. Inorg. Biochem.* **2021**, *217*, 111353. [CrossRef]
31. Zhang, C.; Xu, C.; Gao, X.; Yao, Q. Platinum-based drugs for cancer therapy and anti-tumor strategies. *Theranostics* **2022**, *12*, 2115–2132. [CrossRef] [PubMed]
32. Wu, T.; Liu, J.; Liu, M.; Liu, S.; Zhao, S.; Tian, R.; Wei, D.; Liu, Y.; Zhao, Y.; Xiao, H.; et al. A Nanobody-Conjugated DNA Nanoplatform for Targeted Platinum-Drug Delivery. *Angew. Chem. Int. Ed.* **2019**, *58*, 14224–14228. [CrossRef] [PubMed]
33. Kostrhunova, H.; Zajac, J.; Novohradsky, V.; Kasparkova, J.; Malina, J.; Aldrich-Wright, J.R.; Petruzzella, E.; Sirota, R.; Gibson, D.; Brabec, V. A Subset of New Platinum Antitumor Agents Kills Cells by a Multimodal Mechanism of Action Also Involving Changes in the Organization of the Microtubule Cytoskeleton. *J. Med. Chem.* **2019**, *62*, 5176–5190. [CrossRef] [PubMed]
34. Gorle, A.K.; Katner, S.J.; Johnson, W.E.; Lee, D.E.; Daniel, A.G.; Ginsburg, E.P.; von Itzstein, M.; Berners-Price, S.J.; Farrell, N.P. Substitution-Inert Polynuclear Platinum Complexes as Metalloshielding Agents for Heparan Sulfate. *Chemistry* **2018**, *24*, 6606–6616. [CrossRef] [PubMed]
35. Amarsy, I.; Papot, S.; Gasser, G. Stimuli-Responsive Metal Complexes for Biomedical Applications. *Angew. Chem. Int. Ed.* **2022**, *61*, e202205900. [CrossRef]

36. Zheng, Y.-R.; Suntharalingam, K.; Johnstone, T.C.; Yoo, H.; Lin, W.; Brooks, J.G.; Lippard, S.J. Pt(IV) Prodrugs Designed to Bind Non-Covalently to Human Serum Albumin for Drug Delivery. *J. Am. Chem. Soc.* **2014**, *136*, 8790–8798. [CrossRef]
37. Zhang, G.; Zhu, Y.; Wang, Y.; Wei, D.; Wu, Y.; Zheng, L.; Bai, H.; Xiao, H.; Zhang, Z. pH/redox sensitive nanoparticles with platinum(iv) prodrugs and doxorubicin enhance chemotherapy in ovarian cancer. *RSC Adv.* **2019**, *9*, 20513–20517. [CrossRef]
38. Jayawardhana, A.M.D.S.; Stilgenbauer, M.; Datta, P.; Qiu, Z.; Mckenzie, S.; Wang, H.; Bowers, D.; Kurokawa, M.; Zheng, Y.R. Fatty acid-like Pt(IV) prodrugs overcome cisplatin resistance in ovarian cancer by harnessing CD36. *Chem. Commun.* **2020**, *56*, 10706–10709. [CrossRef]
39. Jayawardhana, A.M.D.S.; Zheng, Y.R. Interactions between mitochondria-damaging platinum(IV) prodrugs and cytochrome c. *Dalton Trans.* **2022**, *51*, 2012–2018. [CrossRef]
40. Awuah, S.G.; Zheng, Y.R.; Bruno, P.M.; Hemann, M.T.; Lippard, S.J. A Pt(IV) Pro-drug Preferentially Targets Indoleamine-2,3-dioxygenase, Providing Enhanced Ovarian Cancer Immuno-Chemotherapy. *J. Am. Chem. Soc.* **2015**, *137*, 14854–14857. [CrossRef]
41. Wei, D.; Yu, Y.; Zhang, X.; Wang, Y.; Chen, H.; Zhao, Y.; Wang, F.; Rong, G.; Wang, W.; Kang, X.; et al. Breaking the Intracellular Redox Balance with Diselenium Nanoparticles for Maximizing Chemotherapy Efficacy on Patient-Derived Xenograft Models. *ACS Nano* **2020**, *14*, 16984–16996. [CrossRef] [PubMed]
42. Miller, M.; Zheng, Y.; Suresh, G.; Pfirschke, C.; Zope, H.; Engblom, C.; Kohler, R.; Iwamoto, Y.; Yang, K.; Askevold, B.; et al. Tumour-associated macrophages act as a slow-release reservoir of nano-therapeutic Pt(IV) pro-drug. *Nat. Commun.* **2015**, *6*, 8692. [CrossRef] [PubMed]
43. Zhou, F.; Feng, B.; Yu, H.; Wang, D.; Wang, T.; Ma, Y.; Wang, S.; Li, Y. Tumor Microenvironment-Activatable Prodrug Vesicles for Nanoenabled Cancer Chemoimmunotherapy Combining Immunogenic Cell Death Induction and CD47 Blockade. *Adv. Mater.* **2019**, *31*, e1805888. [CrossRef] [PubMed]
44. Kang, X.; Wang, Y.; Chen, Z.; Wu, Y.; Chen, H.; Yang, X.; Yu, C. Imidazole modified Pt(iv) prodrug-loaded multi-stage pH responsive nanoparticles to overcome cisplatin resistance. *Chem. Commun.* **2020**, *56*, 11271–11274. [CrossRef]
45. Ma, J.; Wang, Q.; Huang, Z.; Yang, X.; Nie, Q.; Hao, W.; Wang, P.G.; Wang, X. Glycosylated Platinum(IV) Complexes as Substrates for Glucose Transporters (GLUTs) and Organic Cation Transporters (OCTs) Exhibited Cancer Targeting and Human Serum Albumin Binding Properties for Drug Delivery. *J. Med. Chem.* **2017**, *60*, 5736–5748. [CrossRef]
46. Abu Ammar, A.; Raveendran, R.; Gibson, D.; Nassar, T.; Benita, S. A Lipophilic Pt(IV) Oxaliplatin Derivative Enhances Antitumor Activity. *J. Med. Chem.* **2016**, *59*, 9035–9046. [CrossRef]
47. Martinho, N.; Santos, T.C.B.; Florindo, H.F.; Silva, L.C. Cisplatin-Membrane Interactions and Their Influence on Platinum Complexes Activity and Toxicity. *Front. Physiol.* **2018**, *9*, 1898. [CrossRef]
48. Chin, C.F.; Tian, Q.; Setyawati, M.I.; Fang, W.; Tan, E.S.; Leong, D.T.; Ang, W.H. Tuning the activity of platinum(IV) anticancer complexes through asymmetric acylation. *J. Med. Chem.* **2012**, *55*, 7571–7582. [CrossRef]
49. Park, G.Y.; Wilson, J.J.; Song, Y.; Lippard, S.J. Phenanthriplatin, a monofunctional DNA-binding platinum anticancer drug candidate with unusual potency and cellular activity profile. *Proc. Natl. Acad. Sci. USA* **2012**, *109*, 11987–11992. [CrossRef]
50. Arzuman, L.; Beale, P.; Yu, J.Q.; Huq, F. Monofunctional Platinum-containing Pyridine-based Ligand Acts Synergistically in Combination with the Phytochemicals Curcumin and Quercetin in Human Ovarian Tumour Models. *Anticancer Res.* **2015**, *35*, 2783–2794.
51. Morstein, J.; Capecchi, A.; Hinnah, K.; Park, B.; Petit-Jacques, J.; Van Lehn, R.C.; Reymond, J.L.; Trauner, D. Medium-Chain Lipid Conjugation Facilitates Cell-Permeability and Bioactivity. *J. Am. Chem. Soc.* **2022**, *144*, 18532–18544. [CrossRef] [PubMed]
52. Annunziata, A.; Imbimbo, P.; Cucciolito, M.E.; Ferraro, G.; Langellotti, V.; Marano, A.; Melchiorre, M.; Tito, G.; Trifuoggi, M.; Monti, D.M.; et al. Impact of Hydrophobic Chains in Five-Coordinate Glucoconjugate Pt(II) Anticancer Agents. *Int. J. Mol. Sci.* **2023**, *24*, 2369. [CrossRef] [PubMed]
53. Desiatkina, O.; Anghel, N.; Boubaker, G.; Amdouni, Y.; Hemphill, A.; Furrer, J.; Păunescu, E. Trithiolato-bridged dinuclear ruthenium(II)-arene conjugates tethered with lipophilic units: Synthesis and Toxoplasma gondii antiparasitic activity. *J. Organomet. Chem.* **2023**, *986*, 122624. [CrossRef]

International Journal of *Molecular Sciences*

MDPI

*Article*

# Synthesis and Biological Activity of a New Indenoisoquinoline Copper Derivative as a Topoisomerase I Inhibitor

**Caroline Molinaro [1,†], Nathalie Wambang [2,†], Sylvain Pellegrini [2], Natacha Henry [2], Marc F. Lensink [1], Emmanuelle Germain [3], Till Bousquet [2], Jérôme de Ruyck [1], Katia Cailliau [1], Lydie Pélinski [2,*,‡] and Alain Martoriati [1,*,‡]**

1 Univ. Lille, CNRS, UMR 8576-UGSF-Unité de Glycobiologie Structurale et Fonctionnelle, F-59000 Lille, France; caroline.molinaro.fr@gmail.com (C.M.); marc.lensink@univ-lille.fr (M.F.L.); jerome.de-ruyck@univ-lille.fr (J.d.R.); katia.cailliau-maggio@univ-lille.fr (K.C.)

2 Univ. Lille, CNRS, Centrale Lille, Univ. Artois, UMR 8181-UCCS-Unité de Catalyse et Chimie du Solide, F-59000 Lille, France; wnath18@yahoo.fr (N.W.); sylvain.pellegrini@univ-lille.fr (S.P.); natacha.henry@univ-lille.fr (N.H.); till.bousquet@univ-lille.fr (T.B.)

3 Univ. Lille, Inserm U1003-PHYCEL-Physiologie Cellulaire, F-59000 Lille, France; emmanuelle.germain@univ-lille.fr

* Correspondence: lydie.pelinski@univ-lille.fr (L.P.); alain.martoriati@univ-lille.fr (A.M.)

† These authors contributed equally to this work.

‡ These authors contributed equally to this work.

**Abstract:** Topoisomerases are interesting targets in cancer chemotherapy. Here, we describe the design and synthesis of a novel copper(II) indenoisoquinoline complex, **WN198**. The new organometallic compound exhibits a cytotoxic effect on five adenocarcinoma cell lines (MCF-7, MDA-MB-231, HeLa, HT-29, and DU-145) with the lowest $IC_{50}$ (0.37 ± 0.04 μM) for the triple-negative MDA-MB-231 breast cancer cell line. Below 5 μM, **WN198** was ineffective on non-tumorigenic epithelial breast MCF-10A cells and *Xenopus* oocyte G2/M transition or embryonic development. Moreover, cancer cell lines showed autophagy markers including Beclin-1 accumulation and LC3-II formation. The DNA interaction of this new compound was evaluated and the dose-dependent topoisomerase I activity starting at 1 μM was confirmed using in vitro tests and has intercalation properties into DNA shown by melting curves and fluorescence measurements. Molecular modeling showed that the main interaction occurs with the aromatic ring but copper stabilizes the molecule before binding and so can putatively increase the potency as well. In this way, copper-derived indenoisoquinoline topoisomerase I inhibitor **WN198** is a promising antitumorigenic agent for the development of future DNA-damaging treatments.

**Keywords:** indenoisoquinoline; copper(II) complex; topoisomerase I; adenocarcinoma; molecular modeling; DNA intercalation

## 1. Introduction

Metal-based drugs have been designed and developed for their cytotoxic effects on cancer cells since the discovery of platinum's antitumor properties [1–4]. As they adopt a variety of coordination geometries, essential trace metals [5,6] are used for the implementation of metal-based complexed drugs in anticancer chemotherapies. In particular, numerous developed copper complexes have highly efficient antitumor activity [7–9] associated with various action mechanisms ranging from chelators [10], ionophores [11], and proteasome inhibitors [12] to inhibition of topoisomerase I and/or topoisomerase II that results in severe DNA damage, cell cycle arrest, and ultimately cancer cell death [13–15].

Novel topoisomerase I inhibitors, named indenoisoquinolines [16–19], were identified to have greater chemical stability compared to camptothecin and its derivatives currently used as a second-line treatment against endocrine-resistant breast cancer. The indenoisoquinoline derivatives indotecan (LMP400) and indimitecan (LMP776) have completed phase I/II clinical trials (Figure 1).

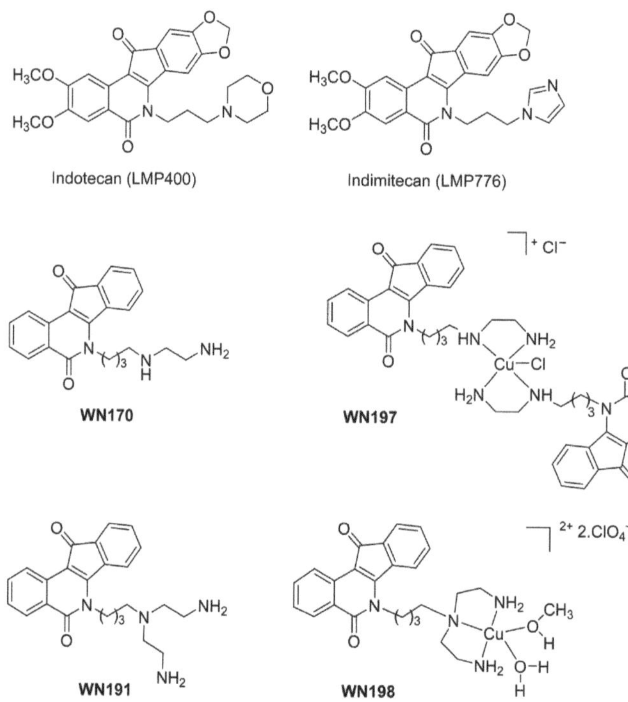


**Figure 1.** Biologically active indenoisoquinolines.

In previous studies, we have shown that copper complex **WN197**, derived from **WN170**, could be a new efficient drug to counteract cancer cells at low doses and exerts a specific cytotoxic effect at low concentration ($IC_{50}$ of 0.5 μM) on three adenocarcinoma cell lines from breast, cervix, and colon but not on non-cancerous breast MCF-10A cells and chemo-resistant pulmonary cancer H69AR cells (Figure 1) [15]. To our knowledge, no studies on the presence of branching on the indenoisoquinoline side chain have been reported. Our interest in the search for anticancer organometallic drugs has been focused on the synthesis of a novel triamine ligand **WN191** and its corresponding tridentate Cu(II) complex **WN198** (Figure 1). The DNA interaction of this new compound was evaluated and its cytotoxicity was tested on adenocarcinoma cell lines, revealing the lowest $IC_{50}$ for the triple-negative MDA-MB-231 breast cancer cell line.

## 2. Results

### 2.1. Synthesis and Characterization

The synthesis of the ligand **WN191** and its corresponding copper complex **WN198** are described in Figure 2. Indenoisoquinoline **WN191** was first obtained in a four-step reaction. Condensation of the commercially available benzo[*d*]indeno [1,2-*b*]pyran-5,11-dione **1** with a primary aminoalcohol was followed by tosylation of the alcohol function to lead to compound **2** in 85% yield. The substitution of the tosyl group by the diprotected triamine and the deprotection of the Boc group led to the indenoisoquinoline derivative **WN191** in 88% yield. Their complexation by copper(II) perchlorate in methanol afforded the copper indenosisoquinoline **WN198** in 70% yield.

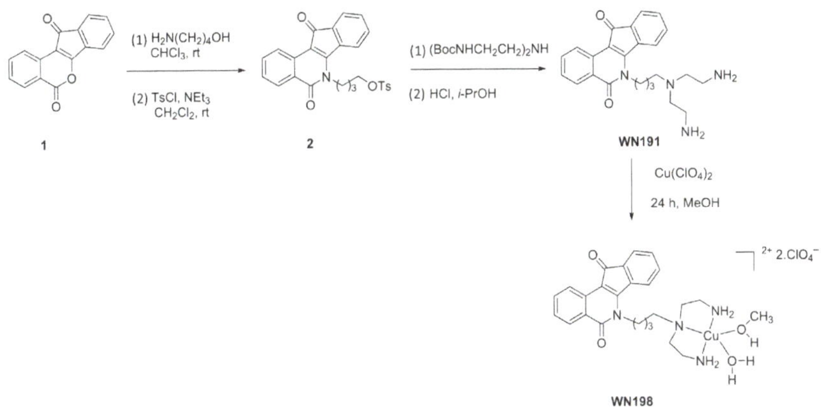


**Figure 2.** Synthesis of the ligand **WN191** and the Cu(II) complex **WN198**.

The structure of **WN198** was established by single-crystal X-ray analysis. The drawing is displayed in Figure 3. The Cu complex crystallizes in the triclinic space group $P\overline{1}$ with two formula units per unit cell. The XRD results show that the Cu(II) complex consists of a mononuclear $[Cu(II)L(H_2O)(CH_3OH)]^{2-}$ (where L is the indenoisoquinolinetriamine ligand) unit and two $(ClO_4)^-$. The five coordinated Cu(II) ions occupy the center of the distorted square base of the pyramid consisting of three nitrogen atoms from the indenoisoquinolinetriamine ligand and one oxygen atom of a water molecule (1.995 Å). One O atom of $(CH_3)OH$ completes the pyramidal environment with an axial bond distance of Cu–O = 2.333 (7) Å.

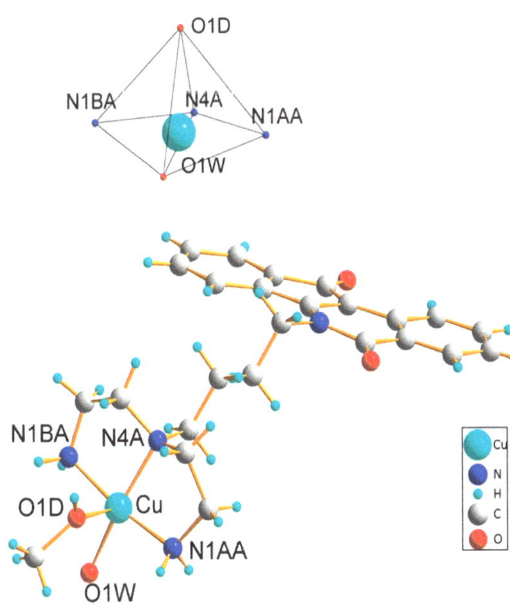


**Figure 3.** Molecular structure of Cu complex **WN198** showing the local geometry around the copper and ligand. Selected bond lengths (Å) and angles (°): Cu-N1BA 1.998(8); Cu-N4A 2.014(6); Cu-N1AA 1.990(7); Cu-O1W 1.980(7); Cu-O1D 2.333(7); N1BA Cu N4A 85.1(3); N1AA Cu N4A 86.2(3); O1W Cu N1AA 94.5(3); O1W Cu N1BA 94.6(3); N1BA Cu O1D 91.2(4); N4A Cu O1D 101.9(3); N1AA Cu O1D 88.1(3); O1W Cu O1D 94.0(3).

### 2.2. **WN191** *and* **WN198** *Display Cytotoxic Activity on Five Adenocarcinoma Cell Lines at Low Doses*

Cell viability was assayed on breast cancer cells (MCF-7), triple-negative breast cancer cells (MDA-MB-231), cervix cancer cells (HeLa), colorectal cancer cells (HT-29), and prostate cancer cells (DU-145), Table 1. The $IC_{50}$ values obtained for **WN191** were respectively 0.58 µM, 1.12 µM, 0.80 µM, 0.53 µM, and 1.09 µM. The copper metal significantly enhances the toxicity of the indenoisoquinoline on the triple-negative breast cancer line MDA-MB-231. For **WN198**, $IC_{50}$ values were lower by a factor of 3.02 for MDA-MB-231 (0.37 µM), or close to a factor of 1.11 for HeLa (0.72 µM) and 1.04 for DU-145 (1.04 µM), or slightly higher by a factor 1.53 for breast cancer hormone-dependent MCF-7 (0.89 µM) and by a factor of 2 for HT-29 (1.06 µM). All $IC_{50}$ values for compounds **WN191** and **WN198** were below the cisplatin values ranging from 2 to 40 µM. Compared to **WN170** [15], **WN198**'s $IC_{50}$ value was lower by a factor of 2.36 for MDA-MB-231 but not for all the other cell lines.

**Table 1.** Cell viability $IC_{50}$ (in µM).

| | MCF-7 | MDA-MB-231 | HeLa | HT-29 | DU-145 |
|---|---|---|---|---|---|
| **WN191** | 0.58 ± 0.02 | 1.12 ± 0.01 | 0.80 ± 0.09 | 0.53 ± 0.01 | 1.09 ± 0.05 |
| **WN198** | 0.89 ± 0.22 | 0.37 ± 0.04 | 0.72 ± 0.06 | 1.06 ± 0.02 | 1.04 ± 0.34 |
| **WN170** | 0.46 ± 0.17 | 0.875 ± 0.01 | 0.630 ± 0.09 | 0.479 ± 0.07 | 0.305 ± 0.04 |
| Cisplatin | 40.396 ± 11.9 | 33.802 ± 1.27 | 19.287 ± 5.323 | 21.313 ± 7.475 | 2.308 ± 0.04 |
| *Statistical difference* **(WN191/WN198)** | **** | **** | *** | **** | * |

Data are expressed as the mean ± SD of three independent experiments. Statistics were based on the Student's *t*-test of the difference between **WN191** and **WN198**. * $p < 0.05$, *** $p < 0.005$, and **** $p < 0.001$. Cisplatin and **WN170** were used as positive controls.

MTS viability assays were performed on the MCF-10A human non-tumorigenic epithelial cell line, most commonly used as a model for normal human breast cells (Figure 4A). The concentration of **WN198** required to inhibit 50% of the MCF-10A viability was significantly higher compared to the mean obtained from the other five adenocarcinomas' $IC_{50}$. **WN198**'s $IC_{50}$ was smaller compared to cisplatin but higher than that of **WN197**, a copper complex and a topoisomerase inhibitor [15]. Additionally, we tested the ability of **WN198** to alter *Xenopus* oocyte maturation and embryonic development, two useful single and highly organized cell systems to test metals and chemical effects [20–23]. *Xenopus* oocytes undergo a G2/M transition, after progesterone stimulation and concomitant microinjection and balneation with **WN198** (0.5, 1, and 5 µM). At those doses, oocytes treated with doxorubicin, a topoisomerase II inhibitor, displayed an inhibition of their meiosis progression (Figure 4B). *Xenopus* embryos could survive **WN198** but not doxorubicin treatment. For 1 and 5 µM of doxorubicin, the percentage of surviving embryos was lowered at segmentation (cell division), and only half of the embryos survived gastrulation (involving cell mobility to build a three-layered organized embryo). For all doxorubicin concentrations tested, the percentage of live embryos was drastically lowered at the time of neurulation and organogenesis (Figure 4C).

A

| MCF10A | $IC_{50}$ (μM) |
|---|---|
| **WN198** | 9.73 ± 0.75 |
| Cisplatin | 21.65 ± 1.2 |
| **WN197** | 1.08 ± 0.03 |

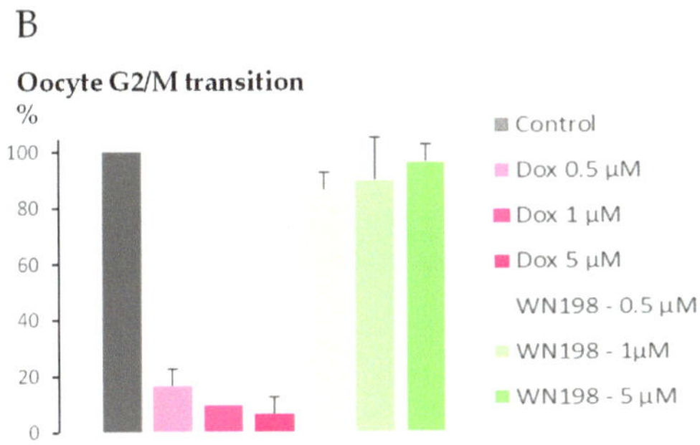


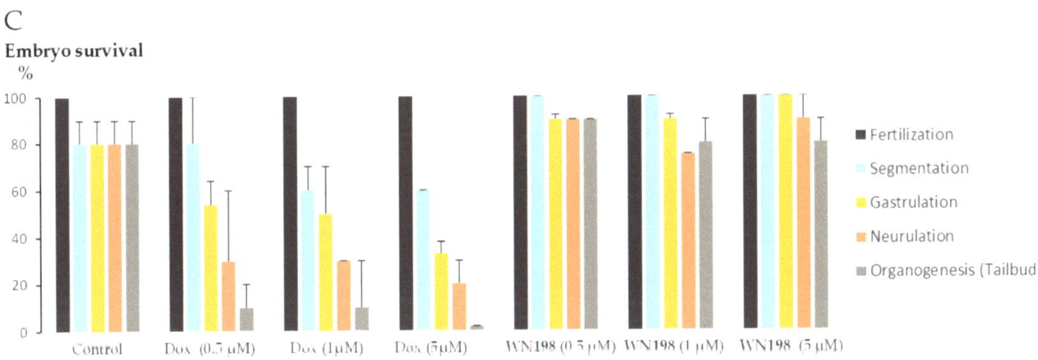


**Figure 4.** Non-cancerous cell viability under **WN198** treatment. (**A**) MCF-10A $IC_{50}$ values are expressed as the mean ± SD of three independent experiments. Cisplatin and **WN197** were used as control. (**B**) *Xenopus* oocyte G2/M transition was scored by the determination of a white spot at the animal pole showing the progression from prophase I to metaphase II (meiosis maturation) 12 h after microinjection and balneation with corresponding drugs (0.5, 1, 5 μM) and external stimulation by progesterone as a natural inducer (4 μM). Experiments were performed on 10 to 20 oocytes from three different females. (**C**) The viability of *Xenopus* embryos was followed after incubation with corresponding drugs (0.5, 1, 5 μM). Stages were identified using the Nieuwkoop and Faber table [24]: segmentation, (3 h 30 min after fertilization), gastrulation (9 h after fertilization), neurulation (19 h 45 min after fertilization), tailbud (24 h after fertilization). Experiments were performed on two independent fertilizations and 10 to 40 embryos in each condition.

### 2.3. **WN191** *and* **WN198** *Are Concentration-Dependent Topoisomerase Inhibitors*

To determine the effect of **WN191** and Cu(II) complex **WN198** on topoisomerase activity, in vitro tests were realized. Topoisomerase I assays rely on the relaxation of supercoiled DNA by an active topoisomerase I. With topoisomerase I, supercoiled DNA shows a relaxed profile (Figure 5). With a well-known topoisomerase I inhibitor, camptothecin (CPT), the DNA relaxation was disturbed and a part of the DNA remained supercoiled. With the addition of 1 and 2 μM of **WN198** in the reactional mixture, the quantity of relaxed DNA is decreased, indicating disruption of topoisomerase I activity. **WN191** disrupted topoisomerase I activity at 2 μM. DMSO (solvent control) and VP-16 (etoposide, a topoisomerase II inhibitor) have no effect on DNA relaxation induced by topoisomerase I, showing their lack of inhibitory effect on topoisomerase I activity.

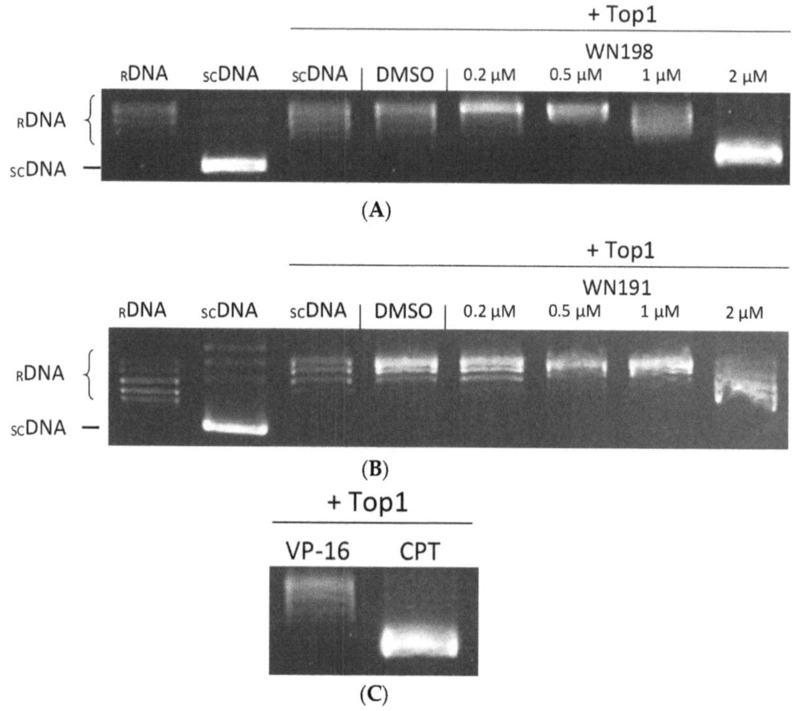


**Figure 5.** **WN191** and **WN198** inhibited human topoisomerase I activity in a dose-dependent manner. Topoisomerase I (Top1) activity is determined by in vitro assays after the addition of either **(A)** **WN198** or **(B)** **WN191** at different concentrations (0.2, 0.5, 1, and 2 µM, lanes 5–8). Relaxed DNA ($_{R}$DNA, lane 1) or supercoiled DNA ($_{SC}$DNA, lane 2) is used as migration control. The Top1 activity control allowing the relaxation of $_{SC}$DNA is in lane 3. DMSO is the solvent control (5%, lane 4). $_{SC}$DNA is used in all other reactions in the presence of Top1. **(C)** Etoposide (VP-16, 50 µM; topoisomerase II (Top2) poison, lane 1) is the negative control of Top1 activity inhibition, and camptothecin (CPT, 10 µM; Top1 poison, lane 2) the positive control of Top1 activity inhibition. After topoisomerase reactions, DNA is run in a 1% agarose gel, stained with ethidium bromide (0.5 µg/mL), and visualized under UV light.

### 2.4. **WN198** *Intercalates in DNA*

Melting curve and fluorescence measurements were performed to confirm the results shown in Table 2 and ascertain **WN191** and **WN198** intercalation in DNA.

**Table 2.** Melting curves and fluorescence measurements for **WN191** and **WN198**.

| Compound | Kapp ($10^7$ $M^{-1}$) | EtBr Displacement |
|---|---|---|
| **WN191** | 8.964 ± 0.964 | 91% |
| **WN198** | 10.791 ± 1.638 | 89% |

Apparent binding constants were measured by fluorescence using [EtBr]/[DNA] = 1.26. Data are the mean of at least three independent experiments.

The binding affinities, determined using a fluorescence-quenching assay based on DNA-binding competition between the intercalating drug ethidium bromide and the tested molecules, were used to gain insight into the DNA-binding affinity. The apparent DNA-binding constant Kapp value of the Cu(II) complex (10.791 ± 1.638 $10^7$ $M^{-1}$) is higher compared to the original ligand **WN191** value (8.964 ± 0.964 $10^7$ $M^{-1}$). The complexation of indenoisoquinoline ligand by copper allows a stronger interaction with DNA (Table 2). The EtBr displacement is in contrast only 2% different between the two compounds.

### *2.5. Modeling of the Interaction between* **WN170***,* **WN191***, and* **WN198** *and Top1*

As no experimental structures of complexes were available, we designed an approach to understand how our putative ligands could bind to Top1 and could intercalate DNA. The starting point was the crystal structure of human DNA Top1 in complex with topotecan poison (Figure 6A) (PDB ID: 1K4T, 2.1 Å). Based on the coordinates of the poison, we subsequently modeled the interaction of **WN170**, **WN191**, and **WN198** by molecular docking. The full procedure of molecular modeling is detailed in Section 4.7.

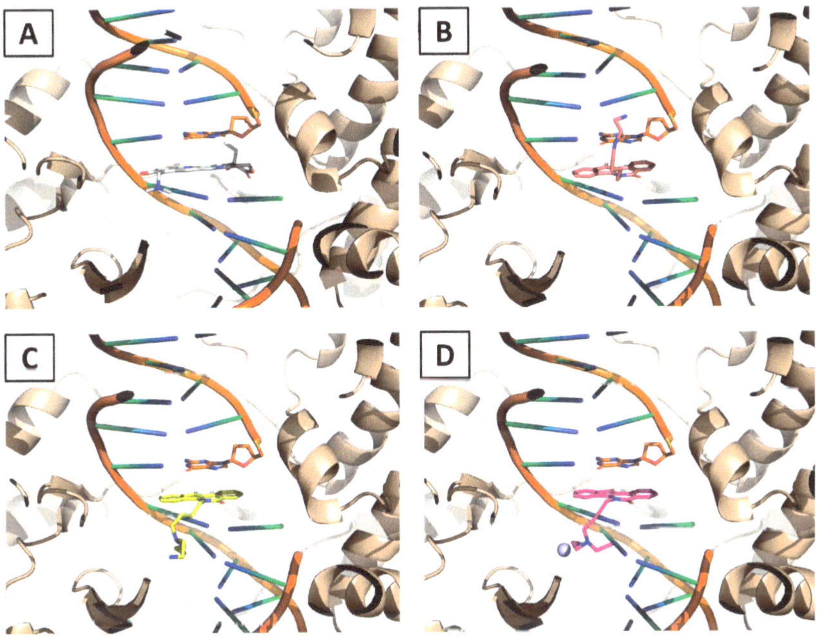


**Figure 6.** Interaction models of the studied ligands and Top1. (**A**) Crystal structure of the topotecan poison (grey sticks) fitting the DNA groove (orange sticks) of Top1 (brown helices) (PDB ID: 1K4T, 2.1 Å). (**B**) The binding mode of **WN170** (salmon sticks) is similar to topotecan's binding mode. The main interaction is taking place inside the DNA groove. (**C**) The interaction model of **WN191** (yellow sticks) is slightly different from **WN170** as the larger branched arm has to switch down. The main part of the infarction is occurring through the aromatic ring. (**D**) The binding mode of **WN198** (pink sticks with the copper molecule as a blue ball) is similar to the binding mode of **WN191**.

On one hand, based on these models, the previously described **WN170** adopts a similar conformation to the commercial poison topotecan which has an aromatic moiety intercalating the DNA (Figure 6B). The relatively small branched chain of **WN170** can easily fit the binding cavity. On the other hand, the complex between **WN191** and Top1 is similar to the previous one with the aromatic moiety totally fitting the DNA groove (Figure 6C). Nevertheless, one can observe that the larger branched arm of **WN191** had to switch to another position, putatively decreasing with a small impact on the affinity of this molecule. Moreover, **WN198** has a similar structure as **WN191** with a copper ion chelating a water molecule. This interaction was difficult to model by molecular mechanics as copper ions are not well parametrized in a classical forcefield. One model was proposed (Figure 6D) that can be easily superimposed on the ligand **WN191**. Although the binding modes of **WN191** and **WN198** are almost the same, their activities are slightly different, which may be explained by copper ion playing a role in the stabilization of the ligand before the binding to the DNA groove as discussed in Section 3.

### 2.6. **WN198** *Induces Autophagy but Not Apoptosis*

Apoptosis can be activated after DNA damage [25,26]. However, the early and late apoptosis markers, respectively, cleaved caspase 3 and cleaved PARP, were not detected after treatments with **WN198** at all concentrations tested, in three adenocarcinoma cell lines, MDA-MB-231, HeLa, HT-29, in contrast to doxorubicin treatment (Figure 7A). γH2AX, an indicator of DNA breaks, was detected after treatment with **WN198** and doxorubicin, indicating that **WN198** and doxorubicin could induce DNA damage. Untreated cells showed lower γH2AX signals. We then determined whether **WN198** could induce autophagy. Several autophagy markers [14] were present after 24h of treatment with 0.5, 1, and 5 μM of **WN198** and rapamycin, an inhibitor of the mTOR pathway known to trigger autophagy [27]. Beclin-1 was synthesized, and LC3-II (LC3-I in association with phosphatidyl-ethanolamine) was increased while control untreated cells did not show these markers (Figure 7B).

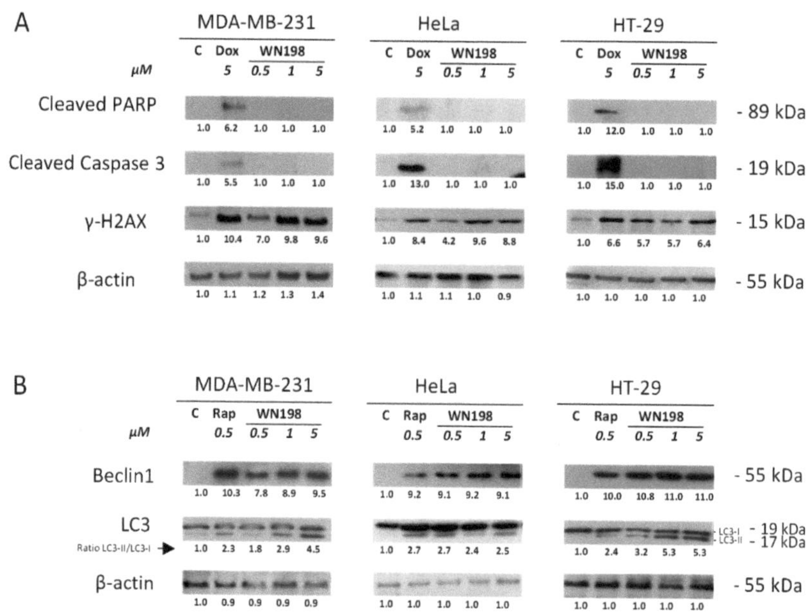


**Figure 7. WN198**-induced autophagy. MDA-MB-231, HeLa, and HT-29 cell lines were treated for 24 h with doxorubicin (Dox, 5 μM), rapamycin (Rap, 0.5 μM), or **WN198** (0.5, 1, 5 μM). (**A**) Western blots were performed with anticleaved caspase 3 and anticleaved PARP antibodies for detection of apoptosis. Anti-γH2AX antibody was used for detection of DNA breaks. (**B**) Western blot analysis of Beclin-1, and LC3 markers were used for detection of autophagy. LC3-II/LC3-I ratio was determined (arrow). β-actin level was used as a loading control. Relative protein levels were expressed by densitometry using software (Fiji Software, v1.52i).

## 3. Discussion

Topoisomerases are overexpressed in cancer cells which divide rapidly and have a high frequency of M phase in their cell cycle. This overexpression can be exploited by inhibition of topoisomerases by appropriate drugs that generate a high number of DNA breaks leading to cancer cell death [28–30]. Cells overexpressing topoisomerases have been shown to better respond to topoisomerase inhibitors [31,32].

Since the discovery of platinum's anticancer properties, platinum-derived drugs have become a mainstay of cancer therapy [2,33]. Other metal-based drugs have been designed and developed [3,4], like transition metals from the d-block of the periodic table (groups 3 to 12) [1,34,35], as they adopt a wide variety of coordination geometries. Indeed, previous studies have demonstrated organometallic compounds composed of metallic atoms such as copper [9], iron (e.g., ferrocifen/ferroquine [1,36]), ruthenium (e.g., indenoisoquinoline [37] and various complexes [38]), or platinum (e.g., cisplatin [39]), which can be used as efficient anticancer drugs. Among them, copper allows a modification in complexed ligand backbones and better DNA affinity [11,40,41]. Some copper derivatives interact with DNA, enhancing DNA damage and antitumor activity [42]. Other copper complexes inhibit topoisomerase activity and result in DNA damage, cell cycle arrest, autophagy, and death in cancer cells [13–15]. Recently, a new class of topoisomerase inhibitors derived from indenoisoquinoline were developed and selected for their capacity to escape efflux transporters implied in cell resistance and for their high stability [16,43]. These indenoisoquinoline derivatives are in phase I/II clinical trials [16,43].

In an attempt to develop new efficient organometallic compounds, a copper(II) indenoisoquinoline complex named **WN198** was synthesized with a branched triamine moiety attached to the indenoisoquinoline core. In this compound, the triamine ligand copper(II) was inserted at the end of the four-carbon chain as a linker because it was previously shown this moiety enhances the DNA-binding affinity due to many hydrogen bonds [36,44,45]. The Cu(II) complex structure displays a mononuclear unit of indenoisoquinolinetriamine and two $(ClO_4)^-$. At the center of a distorted square with three nitrogen atoms from the indenoisoquinolinetriamine and one oxygen atom from water are five coordinated Cu(II) ions.

In vitro tests reveal that **WN198** inhibits topoisomerase I in a dose-dependent manner starting at 1 μM and with optimum efficiency at 2 μM, while the initial compound (**WN191**) only starts inhibiting topoisomerase I at 2 μM. Furthermore, **WN198** could induce DNA breaks detected by a γH2AX signal. Because H2AX phosphorylation correlates with DNA lesions it is used as a marker. When DNA lesions are triggered, the activation of the DNA damage response (DDR) pathway leads to the phosphorylation of histone H2AX on serine 139 (γH2AX) by phosphoinositide 3-kinase-related kinase family members [46,47].

We could further determine that **WN198** interacts with DNA with an apparent DNA-binding constant value higher by a factor of 1.2 compared to the original indenoisoquinoline ligand value, showing that the complexation of the ligand with copper allows a slight improvement in the interaction with DNA. A catalytic mode of inhibition could also occur through the intercalation of **WN198** into the DNA as demonstrated by the melting curves and the fluorescence measurements. Indenoisoquinolines are composed of a planar skeleton, and the high affinity of the copper(II) complex for DNA could be the result of the π–cation interaction between the atom of Cu(II) coordinated with ligands and the base pairs [48,49]. DNA intercalation of **WN198** could impede topoisomerase's access to the DNA fixation sites as seen for other topoisomerase inhibitors, such as anthracyclines [50,51]. Due to a strong affinity for DNA duplexes, anthracycline compounds prevent topoisomerase binding to DNA [50,51]. Doxorubicin, one of the most effective chemotherapeutic drugs used against solid tumors in the treatment of several cancer types, displays a poison activity at low doses and an intercalating catalytic inhibitory action at high doses. Indenoisoquinoline copper derivative **WN197** also efficiently induced MDA-MB-231, HeLa, and HT-29 cell death below 0.5 μM. The planar indenoisoquinoline skeleton of **WN197** displays high intercalation into DNA [15]. Low doses of **WN197** inhibit topoisomerases while, at higher doses, the compound has DNA intercalation properties. However, **WN197** exhibits two complexed ligand backbones around a single Cu(II) atom, compared to a single one in **WN198**. While some of the topoisomerase I inhibitors developed do not use intercalation [52], and knowing DNA-binding and topoisomerase I poisoning activities can be viewed as separate mechanisms [53], our results nevertheless encourage the synthesis of series of indenoisoquinoline drugs composed of a backbone with a four-carbon side chain around a metallic center to keep topoisomerase inhibition and strong DNA interaction properties.

Based on existing structures, we demonstrate by molecular modeling that our drugs interfere with DNA and thus block the activity. Thanks to the structural approach, we can clearly see that the most important moiety is the aromatic intercalating one.

The viability assays showed that low doses between 0.37 and 1.6 μM could induce cell death in breast, cervix, colon, and prostate cancer cell lines, from five highly prevalent adenocarcinomas. The lowest $IC_{50}$ values were obtained for cervix and breast triple-negative cancer cell lines at doses below that of the original compound (respectively, by a factor of 1.1 and 3). These $IC_{50}$ values are under the values determined for many other topoisomerase I inhibitors [9]. Higher doses were necessary to affect the MCF-10A human non-tumorigenic epithelial breast line or *Xenopus* oocyte G2/M transition and embryonic development. The $IC_{50}$ of **WN198** on MCF-10A cells was higher compared to **WN197**, respectively, 9.73 and 1.080 μM, which makes **WN198** a better candidate to avoid side effects of chemotherapies on non-cancerous cells. The toxicity characteristics of **WN198** need further determination with respect to membrane permeability and the determination of cellular uptake.

Copper complexes or topoisomerase inhibitors arrest the cancer cell cycle in different phases and trigger cell death differently by apoptosis or senescence or have bimodal action through both apoptosis and autophagy [54–56]. Recently, another copper complex derived from the topoisomerase I inhibitor indenoisoquinoline was shown to arrest the cell cycle in the G2 phase through inhibition of Cdc25C phosphatase necessary to activate MPF (Cdk1/cyclin B) and to trigger cell death by autophagy [9,14,15]. **WN198** triggers autophagy, shown by the accumulation of Beclin-1 and the formation of LC3-II, but no apoptosis as cleaved markers such as PARP and caspase 3 were not detected in the cell lines MDA-MB-231, HeLa, and HT-29 for doses of **WN198** equal to and above the $IC_{50}$.

The influence of indenoisoquinoline amine complexations with other metals should be examined in future studies with the knowledge that low minimum and necessary doses of chemotherapeutic compounds could be useful in circumventing normal cell death and limiting cardiotoxicity, a strategy employed for anthracyclines [57,58].

## 4. Materials and Methods

### 4.1. Chemistry

All commercial reagents and solvents were used without further purification. Cisplatin was procured from Alfa Aesar (Heysham, UK); DMSO from Sigma-Aldrich (Saint-Quentin-Fallavier, France). Stock solutions were prepared in DMSO. Melting points were determined with a Barnstead Electrothermal (BI 9300) capillary melting point apparatus and are uncorrected. The $^{1}H$ and $^{13}C$ NMR spectra were recorded on a Bruker AC300 spectrometer at 300 and 75.5 MHz, respectively, using tetramethylsilane (TMS) as internal standard and DMSO-d6 as solvent. Elemental analyses were performed with a vario MICRO element analyzer. Thin layer chromatography (TLC) was carried out on aluminum-baked Macherey-Nagel silica gel 60. Column chromatography was performed on silica gel (230–400 mesh). The electronic absorption spectra were acquired on a SPECORD® PLUS UV–Vis double-beam spectrophotometer (Analytik Jena GmbH, Jena, Germany)). The molar conductance measurement was carried out using a CDRV 62 Tacussel electronic bridge, employing a calibrated $10^{-2}$ M KCl solution and $10^{-3}$ M solutions of compounds in DMSO. Purities of all tested compounds were ≥95%, as estimated by HPLC analysis. The high-resolution mass spectrum (HR-MS) was measured by REALACAT, University of Lille on a Synapt G2Si (Waters SAS, Saint-Quentin-en-Yvelines) equipped with an ion mobility cell, recorded in positive ion mode with an electrospray ionization (ESI) source.

#### 4.1.1. Synthesis of **WN191**

- 4-(5,11-Dioxo-5,11-dihydro-6*H*-indeno[1,2-*c*]isoquinolin-6-yl)butyl-4-methylbenzene sulfonate [59].

4-Aminobutan-1-ol (8 mmol, 0.74 mL) was added to a solution of indenopyranedione 1 (1.0 g, 4.0 mmol) in $CHCl_3$ (40 mL). After stirring at room temperature for 18 h, the reaction mixture was diluted in $CHCl_3$ (100 mL) and subsequently washed with distilled water (2 × 50 mL), HCl 0.1 M (1 × 25 mL), and brine (1 × 50 mL), dried over $MgSO_4$, filtered, and concentrated in vacuo to afford dark-orange needles (1.2 g, 96%). To the crude product in $CH_2Cl_2$ (8 mL) was added TsCl (7.8 g, 4.1 mmol). Triethylamine (1.5 mL, 11 mmol) was added and the mixture was stirred at room temperature for 16 h. The resulting mixture was diluted with $CH_2Cl_2$ (50 mL) and washed with distilled water (2 × 20 mL) and then brine (20 mL). The organic layer was dried over $Na_2SO_4$ and concentrated. Purification by flash column chromatography on silica gel ($CH_2Cl_2$/MeOH, 99:1, as eluent) afforded the desired product **2** as a dark-red solid (1.5 g, 85%). Mp: 154 °C. $^1$H NMR (300 MHz, $CDCl_3$) $\delta$ 8.60 (dd, *J* = 7.2, 0.3 Hz, 1H), 8.26 (dd, *J* = 6.8, 0.5 Hz, 1H), 7.75 (d, *J* = 8.3 Hz, 2H), 7.69 (td, *J* = 6.9, 1.3 Hz, 1H), 7.59 (dd, *J* = 6.1, 0.6 Hz, 1H), 7.45 (m, 4H), 7.30 (d, *J* = 7.9 Hz, 2H), 4.48 (t, *J* = 7.2 Hz, 2H), 4.13 (t, *J* = 5.6 Hz, 2H), 2.40 (s, 3H), 1.90 (m, 4H). ESI-MS (*m*/*z*): found: 474.1382 $[M + H]^+$, calculated: 474.1375 $[M + H]^+$.

- 6-(4-(Bis(2-aminoethyl)amino)butyl)-5*H*-indeno[1,2-*c*]isoquinoline-5,11(6*H*)-dione **WN191**

Compound 2 (800 mg, 1.7 mmol) was added to a solution of di-*tert*-butyl(azanediylbis (ethane-2,1-diyl))dicarbamate (2.6 mg, 8.4 mmol) in acetonitrile (25 mL). The resulting mixture was heated at 80 °C for 16 h. The solvent was evaporated in vacuo. The residue was taken up in $CH_2Cl_2$ (50 mL), washed with brine (2 × 25 mL), dried over $MgSO_4$, and concentrated in vacuo. Purification by flash column chromatography on silica gel ($CH_2Cl_2$/EtOAc, 7:3, as eluent) afforded the *N*-protected product as a red solid (663 mg, 65%). Mp: 94 °C. 1H NMR (300 MHz, $CDCl_3$) $\delta$ 8.79 (d, *J* = 7.8 Hz, 1H), 8.35 (d, *J* = 8.7 Hz, 1H), 7.72 (td, *J* = 6.9, 1.7 Hz, 1H), 7.64 (d, *J* = 6.5 Hz, 1H), 7.48–7.38 (m, 4H), 5.37 (m, 2H), 4.52 (t, *J* = 7.9 Hz, 2H), 3.18 (m, 4H), 2.55 (m, 6H), 1.98–1.18 (m, 2H), 1.70–1.65 (m, 2H), 1.36 (s, 18H). To a solution of protected amine (663 mg, 1.1 mmol) in $CHCl_3$ (5 mL) was slowly added a solution of HCl 5M in 2-propanol (40 mL) at 0 °C. After stirring at room temperature for 18 h, the mixture was neutralized by KOH, extracted with $CH_2Cl_2$ (3 × 25 mL), dried over $MgSO_4$, and concentrated in vacuo. Purification by recrystallization ($Et_2O$/EtOH) afforded compound **WN191** as an orange solid (394 mg, 88%). Mp: 172 °C. $^1$H-NMR (300 MHz DMSO-d6) $\delta$ 8.41 (d, *J* = 8.2 Hz, 1H), 8.08 (d, *J* = 8.2 Hz, 1H), 7.72–7.60 (m, 2H), 7.52–7.35 (m, 4H), 4.34 (t, *J* = 8.9 Hz, 2H), 2.85 (m, $CH_2$, 4H), 2.45 (m, 6H), 1.74 (q, *J* = 7.4 Hz, 2H), 1.53 (q, *J* = 8.9 Hz, 2H). $^{13}$C NMR (75 MHz DMSO-d6) $\delta$ 192.5, 162.3, 156.1, 136.5, 134.8, 134.0, 133.9, 131.9, 131.7, 131.2, 128.0, 126.9, 106.8, 53.5, 53.3, 43.9, 31.5, 29.3, 28.9, 27.3, 26.7, 23.3. ESI-MS (*m*/*z*): found: 405.2279 $[M + H]^+$, calculated: 405.2285 $[M + H]^+$.

4.1.2. Synthesis of Cu(II) Complex **WN198**

To a solution of **WN191** (298 mg, 0.74 mmol) in MeOH (8 mL) was slowly added a solution of $Cu(ClO_4)_2.6H_2O$ (273 mg, 0.74 mmol) in MeOH (7 mL). After stirring at room temperature for 24 h, the mixture was filtered and washed with methanol. The powder was recrystallized in a mixture of methanol/ether to give orange crystals (367 mg, 70%). Anal. calc. for $C_{25}H_{34}Cl_2CuN_4O_{12}$: C, 41.88; H, 4.78; N, 7.81; found: C, 41.61; H, 4.60; N, 7.90. UV-vis in DMSO-$H_2O$ (19/01) 5.10 at 5.0 M, [λmax, nm (ε, $M^{-1}$ $cm^{-1}$)]: 659 (200), 460 (2980), 372 (12320), 349 (11920), 327 (11120). 1.21. Λm ($\Omega^{-1}$.$cm^2$.$mol^{-1}$): 51. IR ($cm^{-1}$): 3326 (m) νas ($NH_2$), 3269 (m) νs ($NH_2$), 1656 (m) ν (C = O), 1548 (s) ν (C = C), 1077 (s), 622 (m) ν ($ClO_4^-$), 562 (s) ν (Cu-N). ESI-MS (*m*/*z*): found: 467.1507 $[M - (H_2O + MeOH)]^+$, calculated: 467.1508 $[M - (H_2O + MeOH)]^+$.

Crystal data, data collection, and structure refinement details are summarized in Table 3.

**Table 3.** Crystal data and structure refinement for Cu complex **WN198**.

| Empirical Formula | $C_{25}H_{34}Cl_2CuN_4O_{12}$ |
|---|---|
| Formula weight | 714.98 |
| Temperature/K | 296.15 |
| Crystal system | Triclinic |
| Space group | P-1 |
| a/Å | 9.0523(12) |
| b/Å | 11.6041(14) |
| c/Å | 14.3430(19) |
| $\alpha$/° | 89.459(8) |
| $\beta$/° | 87.775(8) |
| $\gamma$/° | 87.345(9) |
| Volume/Å$^3$ | 1503.8(3) |
| Z | 2 |
| $\rho_{calc}$g/cm$^3$ | 1.579 |
| $\mu$/mm$^{-1}$ | 0.972 |
| F(000) | 738.0 |
| Crystal size/mm$^3$ | 0.213 × 0.177 × 0.034 |
| Radiation | MoK$\alpha$ ($\lambda$ = 0.71073) |
| 2$\Theta$ range for data collection/° | 2.842 to 51.992 |
| Index ranges | $-11 \leq h \leq 12$, $-15 \leq k \leq 15$, $-19 \leq l \leq 19$ |
| Reflections collected | 24167 |
| Independent reflections | 5754 [$R_{int}$ = 0.0575, $R_{sigma}$ = 0.0819] |
| Data/restraints/parameters | 5754/3/350 |
| Goodness-of-fit on $F^2$ | 1.028 |
| Final R indexes [$I \geq 2\sigma$ (I)] | $R_1$ = 0.1042, $wR_2$ = 0.2914 |
| Final R indexes [all data] | $R_1$ = 0.1547, $wR_2$ = 0.3380 |
| Largest diff. peak/hole/e Å$^{-3}$ | 1.49/−1.39 |

### 4.2. Cell Culture

Human cell lines from cervix cancer (HeLa), breast cancer (MCF-7), triple-negative breast cancer (MDA-MB-231), colorectal cancer (HT-29), prostate cancer (DU-145), and a human breast epithelial cell line, arguably the most commonly used normal breast cell model (MCF-10A), were obtained from the American Type Culture Collection (ATCC, Manassas, VA, USA). Cells were cultured at 37 °C in a humid atmosphere containing 5% $CO_2$, in a DMEM culture medium supplemented with 10% fetal bovine serum (Dutscher, Dernolsheim, France), 1% Zell Shield (Dutscher), and 1% non-essential amino acids (Lonza, Basel, Switzerland). MCF-10A cells were maintained in MEBM (Lonza) supplemented with MEGM (Lonza) and 1% Zell Shield.

### 4.3. Cell Viability Assay

Cell viability was determined using the CellTiter 96® AQ$_{ueous}$ One Solution Cell Proliferation Assay (MTS test, Promega, Charbonnières-les-Bains, France). Cells were seeded in 96-well plates at a density of 2.10$^3$ for 24 h before treatment with 0 to 100 µM of **WN198, WN191, WN170, WN197,** or cisplatin for 72 h. Cells were incubated for 2 h with 20 µL of CellTiter solution at 37 °C in 5% $CO_2$, and the production of formazan from reduced 3-(4,5-dimethylthiazol-2-yl)-5-(3-carboxymethoxyphenyl)-2-(4-sulfophenyl)-2H-tetrazolium (MTS) was measured at 490 nm (SPECTROstar Nano, BMG LABTECH, Ortenberg, Germany). GraphPad Prism V6.0 software served to calculate $IC_{50}$. Statistical differences between **WN198** and **WN191** were ascertained by a Student's *t*-test (* $p < 0.05$, ** $p < 0.001$, *** $p < 0.0005$ and **** $p < 0.0001$).

### 4.4. Human Topoisomerase I In Vitro Activity

Topoisomerase I activity was determined with in vitro drug screening kits (TopoGEN, Inc., Buena Vista, CO, USA assays) based on the relaxation of supercoiled DNA into relaxed DNA as previously described [15]. Briefly, the assembled reaction mixture was composed of supercoiled pHOT1 DNA (250 ng), 10× TGS buffer (10 mM Tris-HCl pH 7.9, 1 mM EDTA), human topoisomerase I (5 units), the tested compound, and $H_2O$ to a final adjusted volume of 20 µL. **WN191** and **WN198** were tested at concentrations ranging from 0.2 to 2 µM. Camptothecin (10 µM) was used as a positive control (poison inhibitor of topoisomerase I activity), etoposide (100 µM) as negative control (inhibitor of topoisomerase II activity), 1% DMSO alone as vehicle control, and relaxed pHOT1 DNA (100 ng) alone as migration control. All reaction products were submitted to 1% agarose gel electrophoresis with EtBr (0.5 µg/mL) at 100 V for 1 h in TAE buffer (Tris-Acetate-EDTA; pH 8.3).

### 4.5. *Ethidium Bromide Competition Test*

Titrations of fluorescence were determined as described [36,45]. EtBr/**WN191** or EtBr/**WN198** with a molar ratio of 12.6/10 at concentrations ranging from 0.05 to 10 µM was used in a BPE buffer at pH 7.1. The excitation wavelength was 546 nm and the emission was monitored in the range of 560 to 700 nm (SPEX Fluorolog, Horiba-Jobin Yvon). EtBr displacement $IC_{50}$ values were calculated with a fitting function incorporated into GraphPad Prism V6.0. The apparent binding constants were calculated using the equation $K_{app} = (1.26 (K_{app}(EtBr)/IC_{50})$ with $K_{app}(EtBr) = 10^7\ M^{-1}$.

### 4.6. *Molecular Modeling*

Receptor structure was the crystal structure of human DNA Top1 in complex with topotecan poison (Figure 4A) (PDB ID: 1K4T, 2.1 Å). All ligands and water molecules were removed except DNA. Simulation of the binding modes of the receptor and the designed compounds was performed using the GOLD docking program [60]. GOLD is based on a genetic algorithm and considered the ligand as flexible, while side chains of most residues were kept rigid. For the search procedure, a sphere of 30 Å was centered on the X, Y, and Z coordinates of the topotecan poison (Origin = 22.7 −9.4 26.9). Twenty conformations were generated for each ligand using default parameters. Then, all the different binding poses were scored with the ChemPLP scoring function. Subsequent energy minimization was performed using the AMBER forcefield. Finally, all the molecular representations were sketched using PyMOL [61].

### 4.7. *Xenopus laevis Oocyte and Embryo Handling*

*Xenopus laevis* females were obtained from the CRB-University of Rennes I, France. They were housed at PHExMAR, University of Lille. *Xenopus* were maintained in accordance with the EU Directive 2010/63/EU and the French National Guidelines for Use of Animals for Scientific Purposes. The experimental protocols were approved by the "Comité d'Ethique CEEA-75 en Experimentation Animale Nord-Pas de Calais-France" (*Xenopus* project number: F59-00913).

Females were anesthetized by immersion in a solution of Tricaine (MS222, Sandoz) at 3 $g.L^{-1}$ for 1 h. The ovaries were surgically removed and placed in ND96 medium (96 mM NaCl, 2 mM KCl, 1.8 mM $CaCl_2$, 1 mM $MgCl_2$, 5 mM HEPES, pH 7.5) at 19 °C. Stage VI oocytes were harvested by using a 1 h collagenase A treatment (1 mg/mL, Boehringer Mannheim) for 45 min with a manual dissociation under a binocular microscope. Microinjections were performed under a binocular microscope with a Nishiryo positive displacement digital micropipette, in the equatorial region of the oocytes, with 60nl of control DMSO 0.1%, doxorubicin, or **WN198** at 0.5, 1, and 5 µM. Meiotic resumption was triggered by the addition of progesterone (4 µM) after microinjection and the addition of a corresponding concentration of doxorubicin or **WN198** to the external medium (0.5, 1, and 5 µM). The appearance of a white spot, resulting from the migration of the germinal vesicle at the apex of the pigmented hemisphere of the oocyte, indicated the meiosis process or G2/M transition. Oocytes arrested in prophase I progress to metaphase II and arrest [22,23,62].

For spawning, females were stimulated by a dorsal lymph sac injection of human chorionic gonadotropic hormone (900 U). Mature oocytes were obtained by a slight pressure on the animal's ovaries. Testicles were surgically removed from a male after anesthesia in 5 g.L$^{-1}$ MS222. In vitro fertilization was performed under gentle agitation in 0.22 μm filtered water with doxorubicin or **WN198** at 0.5, 1, and 5 μM or not. Eggs were deganguated with 2% L-cysteine with doxorubicin or **WN198** or not for 10 min and rinsed three times. Embryos were kept at 23 °C in daily renewed solution with doxorubicin or **WN198** or not. Embryo stages were identified using the Nieuwkoop and Faber table (Nieuwkoop and Faber, 1994) [24]: morula (segmentation, stage 6.5, 3 h 30 min after fertilization), gastrula (stage 10, 9 h after fertilization), neurula (stage 18, 19 h 45 min after fertilization), and tailbud (organogenesis, stage 22, 24 h after fertilization).

### 4.8. Electrophoresis and Western Blot

First, 7.5 × 10$^5$ cells were seeded for 24 h before treatment with 5 μM doxorubicin (a positive control for apoptosis), 0.5 μM rapamycin (a positive control for autophagy), **WN198** (0.5, 1, 5 μM), or 0.1% DMSO (solvent control). After 24 h, cells were lysed in RIPA buffer (150 mM EDTA; 150 mM NaCl; 50 mM NaF, 1% Triton X-100; NP40 2%; 0.4% Na-deoxycholate; 0.6% SDS; 50 mM TRIS-HCl pH 4) supplemented with 1% protease cocktail inhibitors (Sigma-Aldrich) and phosphatase inhibitors (Roche SAS by Merck).

After centrifugation for 10 min at 12,000× g, the protein concentration of sample supernatants was determined using a Bradford assay (BioRad, Marnes-la-Coquette, France) at 595 nm (SPECTROstar Nano, BMG LABTECH). After denaturation at 75 °C for 10 min in 2× Laemmli buffer (65.8 mM TRIS-HCl pH 6.8; 26.3% glycerol; 2.1% SDS; 0.01% bromophenol blue; 4% β-mercaptoethanol, BioRad), 15 μg of samples was separated on 4–20% SDS PAGE gels (mini protean TGX, BioRad, Marnes-la-Coquette, France) for 1 h at 200 V in denaturing buffer (0.3% TRIS base; 0.1% SDS; 1.44% glycine). Proteins were wet transferred in 0.32% TRIS; 1.8% glycine; 20% methanol (Sigma-Aldrich, Saint-Quentin-Fallavier, France) onto a nitrocellulose membrane (Amersham Hybond, Dutscher, Bernolsheim, France) for 1 h at 100 V. After saturation with 5% low-fat dry milk in TBS with 0.05% Tween (Sigma-Aldrich), membranes were incubated overnight at 4 °C with primary antibodies: rabbit polyclonal antibodies were against cleaved caspase 3 or Beclin-1 (Cell Signalling Technology (CST, by Ozyme, Saint-Cyr-L'École, France), 1/1000), phosphorylated H2AX (S139, CST, 1/750), LC3 (CST, 1/50); goat polyclonal antibodies against β-actin (SCB, 1/1200); and cocktail antibodies against cleaved PARP (Abcam, Cambridge, UK, cell cycle and apoptosis cocktail, 1/1500). After three TBS-Tween washes of 10 min, membranes were incubated for 1 h with appropriate horseradish peroxidase-labeled secondary antibodies: antirabbit or antimouse (Invitrogen, 1/30,000) or antigoat antibodies (SCB, 1/30,000). After three 10 min TBS-Tween washes, the signals were revealed using a chemiluminescent assay (ECL Select, GE Healthcare, Dutscher, Bernolsheim, France) on hyperfilms (Amersham hyperfilm MP, Dutscher). β-actin was used as a loading control. Signals were quantified with ImageJ (Fiji Software, v1.52i) and normalized to respective loading controls.

## 5. Conclusions

A new copper(II) complex containing an indenoisoquinoline scaffold was synthesized. The molecular structure of **WN198** was confirmed by single-crystal X-ray diffraction analysis. **WN198** displays a strong DNA interaction and kept a topoisomerase I inhibitory activity as detected by in vitro tests.

The compound exerts excellent cytotoxic activities against five adenocarcinoma cancer cell lines at a lower concentration compared to other classical topoisomerase drugs used in chemotherapies. It is particularly efficient against MDA-MB-231 (triple-negative breast cancer) cell line proliferation with an $IC_{50}$ of 0.37 μM. The $IC_{50}$ on non-cancerous cell line MCF-10A is significantly high compared to other copper complexes as topoisomerase inhibitors and no toxicity was detected below 5 μM for *Xenopus* oocyte maturation and embryo development.

**WN198** appears to be a new efficient drug to counteract cancer cells at low doses. **WN198** could benefit patients overexpressing topoisomerases, sensitize cancer cells to

DNA-damaging chemotherapies [63], bypass unwanted side effects, or be part of synthetic lethality or synergic strategies.

**Author Contributions:** Conceptualization, C.M., K.C., L.P. and A.M.; performing experiments, C.M., N.W., N.H., T.B., J.d.R., A.M. and L.P.; writing and editing, C.M., S.P., M.F.L., E.G., J.d.R., K.C., L.P. and A.M. All authors have read and agreed to the published version of the manuscript.

**Funding:** C.M. is a recipient of a doctoral fellowship from a French ministry. This work was supported by the CNRS, the University of Lille, Chevreul Institute (FR 2638), Ministère de l'Enseignement Supérieur et de la Recherche, Région Hauts de France and by grants from the "Ligue Contre le Cancer, Comités Nord Pas de Calais, Nord et Aisne" (A.M.).

**Institutional Review Board Statement:** Not applicable.

**Informed Consent Statement:** Not applicable.

**Data Availability Statement:** Data are contained within the article.

**Acknowledgments:** We are sincerely indebted to Arlette Lescuyer (UMR-CNRS 8576) for her help.

**Conflicts of Interest:** The authors declare that the research was conducted in the absence of any commercial or financial relationships that could be construed as a potential conflict of interest.

## References

1. Jaouen, G.; Vessières, A.; Top, S. Ferrocifen type anticancer drugs. *Chem. Soc. Rev.* **2015**, *44*, 8802–8817. [CrossRef] [PubMed]
2. Dilruba, S.; Kalayda, G.V. Platinum-based drugs: Past, present and future. *Cancer Chemother. Pharmacol.* **2016**, *77*, 1103–1124. [CrossRef] [PubMed]
3. Zhang, P.; Sadler, P.J. Advances in the design of organometallic anticancer complexes. *J. Organomet. Chem.* **2017**, *839*, 5–14. [CrossRef]
4. Anthony, E.J.; Bolitho, E.M.; Bridgewater, H.E.; Carter, O.W.L.; Donnelly, J.M.; Imberti, C.; Lant, E.C.; Lermyte, F.; Needham, R.J.; Palau, M.; et al. Metallodrugs are unique: Opportunities and challenges of discovery and development. *Chem. Sci.* **2020**, *11*, 12888–12917. [CrossRef]
5. Komeda, S.; Casini, A. Next-Generation Anticancer Metallodrugs. *Curr. Top. Med. Chem.* **2012**, *12*, 219–235. [CrossRef]
6. Mejía, C.; Ortega-Rosales, S.; Ruiz-Azuara, L. Mechanism of action of anticancer metallodrugs. In *Biomedical Applications of Metals*; Springer: Cham, Switzerland, 2018; pp. 213–234. [CrossRef]
7. Santini, C.; Pellei, M.; Gandin, V.; Porchia, M.; Tisato, F.; Marzano, C. Advances in copper complexes as anticancer agents. *Chem. Rev.* **2014**, *114*, 815–862. [CrossRef]
8. Kellett, A.; Molphy, Z.; McKee, V.; Slator, C. Recent advances in anticancer copper compounds. In *Metal-Based Anticancer Agents*; Vessieres, I.A., Meier-Menches, S.M., Casini, A., Eds.; Royal Society of Chemistry, RSC Metallobiology: London, UK, 2019; Volume 14, pp. 91–119.
9. Molinaro, C.; Martoriati, A.; Pelinski, L.; Cailliau, K. Copper complexes as anticancer agents targeting topoisomerases i and ii. *Cancers* **2020**, *12*, 2863. [CrossRef]
10. Baldari, S.; Di Rocco, G.; Toietta, G. Current biomedical use of copper chelation therapy. *Int. J. Mol. Sci.* **2020**, *21*, 1069. [CrossRef]
11. Denoyer, D.; Clatworthy, S.A.S.; Cater, M.A. *Copper Complexes in Cancer Therapy*; NIH: Bethesda, MD, USA, 2018; Volume 18, ISBN 9783110470734.
12. Zhang, Z.; Wang, H.; Yan, M.; Wang, H.; Zhang, C. Novel copper complexes as potential proteasome inhibitors for cancer treatment (Review). *Mol. Med. Rep.* **2017**, *15*, 3–11. [CrossRef]
13. Liang, X.; Wu, Q.; Luan, S.; Yin, Z.; He, C.; Yin, L.; Zou, Y.; Yuan, Z.; Li, L.; Song, X.; et al. A comprehensive review of topoisomerase inhibitors as anticancer agents in the past decade. *Eur. J. Med. Chem.* **2019**, *171*, 129–168. [CrossRef]
14. Molinaro, C.; Martoriati, A.; Cailliau, K. Proteins from the DNA Damage Response: Regulation, Dysfunction, and Anticancer Strategies. *Cancers* **2021**, *13*, 3819. [CrossRef] [PubMed]
15. Molinaro, C.; Wambang, N.; Bousquet, T.; Vercoutter-Edouart, A.S.; Pélinski, L.; Cailliau, K.M.A. A Novel Copper(II) Indenoisoquinoline Complex Inhibits Topoisomerase I, Induces G2 Phase Arrest, and Autophagy in Three Adenocarcinomas. *Front. Oncol.* **2022**, *12*, 837373. [CrossRef] [PubMed]
16. Pommier, Y.; Cushman, M.; Doroshow, J.H. Novel clinical indenoisoquinoline topoisomerase I inhibitors: A twist around the camptothecins. *Oncotarget* **2018**, *9*, 37286–37288. [CrossRef] [PubMed]
17. Thomas, A.; Pommier, Y. Targeting topoisomerase I in the era of precision medicine. *Clin. Cancer Res.* **2019**, *25*, 6581–6589. [CrossRef] [PubMed]
18. Pommier, Y.; Kiselev, E.; Marchand, C. Interfacial inhibitors. *Bioorganic Med. Chem. Lett.* **2015**, *25*, 3961–3965. [CrossRef]
19. Cushman, M. Design and Synthesis of Indenoisoquinolines Targeting Topoisomerase I and Other Biological Macromolecules for Cancer Chemotherapy. *J. Med. Chem.* **2021**, *64*, 17572–17600. [CrossRef]

20. Slaby, S.; Hanotel, J.; Marchand, G.; Lescuyer, A.; Bodart, J.; Leprêtre, A.; Lemière, S.; Marin, M. Maturation of Xenopus laevis oocytes under cadmium and lead exposures: Cell biology investigations. *Aquat. Toxicol.* **2017**, *193*, 110–115. [CrossRef]
21. Slaby, S.; Lemière, S.; Hanotel, J.; Lescuyer, A.; Demuynck, S.; Bodart, J.F.; Leprêtre, A.; Marin, M. Cadmium but not lead exposure affects Xenopus laevis fertilization and embryo cleavage. *Aquat. Toxicol.* **2017**, *177*, 1–7. [CrossRef]
22. Slaby, S.; Marin, M.; Marchand, G.; Lemiere, S. Exposures to chemical contaminants: What can we learn from reproduction and development endpoints in the amphibian toxicology literature? *Environ. Pollut.* **2019**, *248*, 478–495. [CrossRef]
23. Marchand, G.; Wambang, N.; Pellegrini, S.; Molinaro, C.; Martoriati, A.; Bousquet, T.; Markey, A.; Lescuyer-Rousseau, A.; Bodart, J.F.; Cailliau, K.; et al. Effects of Ferrocenyl 4-(Imino)-1,4-Dihydro-quinolines on *Xenopus laevis* Prophase I—Arrested Oocytes: Survival and Hormonal-Induced M-Phase Entry. *Int. J. Mol. Sci.* **2020**, *21*, 3049. [CrossRef]
24. Nieuwkoop, P.D.; Faber, J. *A Systematical and Chronological Survey of the Development from the Fertilized Egg till the End of Metamorphosis*; Faber, P.D.J., Ed.; Nieuwkoop, Garland Publishing Inc.: New York, NY, USA, 1994; ISBN 0-8153-1896-0.
25. Larsen, A.K.; Escargueil, A.E.; Skladanowski, A. From DNA Damage to G2 Arrest: The Many Roles of Topoisomerase II. *Prog. Cell Cycle Res.* **2003**, *5*, 295–300. [PubMed]
26. Pommier, Y. Drugging Topoisomerases: Lessons and Challenges. *ACS Chem. Biol.* **2013**, *8*, 82–95. [CrossRef] [PubMed]
27. Kim, K.W.; Mutter, R.W.; Cao, C.; Albert, J.M.; Freeman, M.; Hallahan, D.E.; Lu, B. Autophagy for cancer therapy through inhibition of pro-apoptotic proteins and mammalian target of rapamycin signaling. *J. Biol. Chem.* **2006**, *281*, 36883–36890. [CrossRef]
28. Heestand, G.M.; Schwaederle, M.; Gatalica, Z.; Arguello, D.; Kurzrock, R. Topoisomerase expression and amplification in solid tumours: Analysis of 24,262 patients. *Eur. J. Cancer* **2017**, *83*, 80–87. [CrossRef]
29. Villman, K.; Ståhl, E.; Liljegren, G.; Tidefelt, U.; Karlsson, M.G. Topoisomerase II-$\alpha$ expression in different cell cycle phases in fresh human breast carcinomas. *Mod. Pathol.* **2002**, *15*, 486–491. [CrossRef]
30. Lee, Y.C.; Lee, C.H.; Tsai, H.P.; An, H.W.; Lee, C.M.; Wu, J.C.; Chen, C.S.; Huang, S.H.; Hwang, J.; Cheng, K.T.; et al. Targeting of Topoisomerase I for Prognoses and Therapeutics of Camptothecin-Resistant Ovarian Cancer. *PLoS ONE* **2015**, *10*, e0132579. [CrossRef] [PubMed]
31. Ali, Y.; Abd Hamid, S. Human topoisomerase II alpha as a prognostic biomarker in cancer chemotherapy. *Tumor Biol.* **2016**, *37*, 47–55. [CrossRef] [PubMed]
32. Zhong, W.; Yang, Y.; Zhang, A.; Lin, W.; Liang, G.; Ling, Y.; Zhong, J.; Yong, J.; Liu, Z.; Tian, Z.; et al. Prognostic and predictive value of the combination of TOP2A and HER2 in node-negative tumors 2 cm or smaller (T1N0) breast cancer. *Breast Cancer* **2020**, *27*, 1147–1157. [CrossRef]
33. Rosenberg, B.; VanCamp, L.; Trosko, J.E.; Mansour, V.H. Platinum Compounds: A New Class of Potent Antitumour Agents. *Nature* **1969**, *5191*, 385–386. [CrossRef]
34. Szczepaniak, A.; Fichna, J. Organometallic Compounds and Metal Complexes in Current and Future Treatments of Inflammatory Bowel Disease and Colorectal Cancer—A Critical Review. *Biomolecules* **2019**, *9*, 398. [CrossRef]
35. Ndagi, U.; Mhlongo, N.; Soliman, M.E. Metal Complexes in Cancer Therapy—An Update from Drug Design Perspective. *Drug Des. Devel. Ther.* **2017**, *11*, 599–616. [CrossRef] [PubMed]
36. Kondratskyi, A.; Kondratska, K.; Vanden Abeele, F.; Gordienko, D.; Dubois, C.; Toillon, R.A.; Slomianny, C.; Lemière, S.; Delcourt, P.; Dewailly, E.; et al. Ferroquine, the next generation antimalarial drug, has antitumor activity. *Sci. Rep.* **2017**, *7*, 15896. [CrossRef] [PubMed]
37. Wambang, N.; Schifano-Faux, N.; Martoriati, A.; Henry, N.; Baldeyrou, B.; Bal-Mahieu, C.; Bousquet, T.; Pellegrini, S.; Meignan, S.; Cailliau, K.; et al. Synthesis, Structure, and Antiproliferative Activity of Ruthenium(II) Arene Complexes of Indenoisoquinoline Derivatives. *Organometallics* **2016**, *35*, 2868–2872. [CrossRef]
38. Pragti; Kundu, B.K.; Mukhopadhyay, S. Target based chemotherapeutic advancement of ruthenium complexes. *Coord. Chem. Rev.* **2021**, *448*, 214169. [CrossRef]
39. Brown, A.; Kumar, S.; Tchounwou, P.B. Cisplatin-Based Chemotherapy of Human Cancers. *J. Cancer Sci. Ther.* **2019**, *11*, 97.
40. Ceramella, J.; Mariconda, A.; Iacopetta, D.; Saturnino, C.; Barbarossa, A.; Caruso, A.; Rosano, C.; Sinicropi, M.S.; Longo, P. From coins to cancer therapy: Gold, silver and copper complexes targeting human topoisomerases. *Bioorg. Med. Chem. Lett.* **2020**, *30*, 126905. [CrossRef]
41. Marzano, C.; Pellei, M.; Tisato, F.; Santini, C. Copper Complexes as Anticancer Agents. *Anticancer Agents Med. Chem.* **2012**, *9*, 185–211. [CrossRef]
42. Shobha Devi, C.; Thulasiram, B.; Aerva, R.R.; Nagababu, P. Recent Advances in Copper Intercalators as Anticancer Agents. *J. Fluoresc.* **2018**, *28*, 1195–1205. [CrossRef]
43. Marzi, L.; Sun, Y.; Huang, S.Y.N.; James, A.; Difilippantonio, S.; Pommier, Y. The indenoisoquinoline LMP517: A novel antitumor agent targeting both TOP1 and TOP2. *Mol. Cancer Ther.* **2020**, *19*, 1589–1597. [CrossRef]
44. Beck, D.E.; Agama, K.; Marchand, C.; Chergui, A.; Pommier, Y.; Cushman, M. Synthesis and biological evaluation of new carbohydrate-substituted indenoisoquinoline topoisomerase I inhibitors and improved syntheses of the experimental anticancer agents indotecan (LMP400) and indimitecan (LMP776). *J. Med. Chem.* **2014**, *57*, 1495–1512. [CrossRef]
45. Wambang, N.; Schifano-Faux, N.; Aillerie, A.; Baldeyrou, B.; Jacquet, C.; Bal-Mahieu, C.; Bousquet, T.; Pellegrini, S.; Ndifon, P.T.; Meignan, S.; et al. Synthesis and biological activity of ferrocenyl indeno[1,2-c]isoquinolines as topoisomerase II inhibitors. *Bioorganic Med. Chem.* **2016**, *24*, 651–660. [CrossRef]

46. Sharma, A.; Singh, K.; Almasan, A. Histone H2AX Phosphorylation: A Marker for DNA Damage. *Methods Mol. Biol.* **2012**, *920*, 613–626. [CrossRef] [PubMed]
47. Smith, H.L.; Southgate, H.; Tweddle, D.A.; Curtin, N.J. DNA Damage Checkpoint Kinases in Cancer. *Expert. Rev. Mol. Med.* **2020**, *8*, e2. [CrossRef] [PubMed]
48. Deng, J.H.; Luo, J.; Mao, Y.L.; Lai, S.; Gong, Y.N.; Zhong, D.C.; Lu, T.B. Π-π stacking interactions: Non-negligible forces for stabilizing porous supramolecular frameworks. *Sci. Adv.* **2020**, *6*, eaax9976. [CrossRef] [PubMed]
49. Thakuria, R.; Nath, N.K.; Saha, B.K. The Nature and Applications of π-π Interactions: A Perspective. *Cryst. Growth Des.* **2019**, *19*, 523–528. [CrossRef]
50. Atwal, M.; Swan, R.L.; Rowe, C.; Lee, K.C.; Armstrong, L.; Cowell, I.G.; Austin, C.A.; Lee, D.C. Intercalating TOP2 Poisons Attenuate Topoisomerase Action at Higher Concentrations. *Mol. Pharmacol.* **2019**, *96*, 475–484. [CrossRef] [PubMed]
51. Marinello, J.; Delcuratolo, M.; Capranico, G. Anthracyclines as Topoisomerase II Poisons: From Early Studies to New Perspectives. *Int. J. Mol. Sci.* **2018**, *19*, 3480. [CrossRef]
52. Bailly, C.; Dassonneville, L.; Colson, P.; Houssier, C.; Fukasawa, K.; Nishimura, S.; Yoshinari, T. Intercalation into DNA is not required for inhibition of topoisomerase I by indolocarbazole antitumor agents. *Cancer Res.* **1999**, *59*, 2853–2860.
53. Grossman, S.; Fishwick, C.W.G.; McPhillie, M.J. Developments in Non-Intercalating Bacterial Topoisomerase Inhibitors: Allosteric and ATPase Inhibitors of DNA Gyrase and Topoisomerase IV. *Pharmaceuticals* **2023**, *16*, 261. [CrossRef]
54. Hu, J.; Liao, C.; Guo, Y.; Yang, F.; Sang, W.; Zhao, J. Copper (II) complexes inducing apoptosis in cancer cells, and demonstrating DNA and HSA interactions. *Polyhedron* **2017**, *132*, 28–38. [CrossRef]
55. Rezaei, A.; Falahati-Pour, S.K.; Mohammadizadeh, F.; Hajizadeh, M.R.; Mirzaei, M.R.; Khoshdel, A.; Fahmidehkar, M.A.; Mahmoodi, M. Effect of a Copper (II) complex on the induction of apoptosis in human hepatocellular carcinoma cells. *Asian Pac. J. Cancer Prev.* **2018**, *19*, 2877–2884. [CrossRef] [PubMed]
56. Gul, N.S.; Khan, T.M.; Chen, M.; Huang, K.B.; Hou, C.; Choudhary, M.I.; Liang, H.; Chen, Z.F. New copper complexes inducing bimodal death through apoptosis and autophagy in A549 cancer cells. *J. Inorg. Biochem.* **2020**, *213*, 111260. [CrossRef] [PubMed]
57. McGowan, J.V.; Chung, R.; Maulik, A.; Piotrowska, I.; Walker, J.M.; Yellon, D.M. Anthracycline Chemotherapy and Cardiotoxicity. *Cardiovasc. Drugs Ther.* **2017**, *31*, 63–75. [CrossRef] [PubMed]
58. Barrett-Lee, P.J.; Dixon, J.M.; Farrell, C.; Jones, A.; Leonard, R.; Murray, N.; Palmieri, C.; Plummer, C.J.; Stanley, A.; Verrill, M.W. Expert opinion on the use of anthracyclines in patients with advanced breast cancer at cardiac risk. *Ann. Oncol.* **2009**, *20*, 816–827. [CrossRef] [PubMed]
59. Park, E.-J.; Kondratyuk, T.P.; Morrell, A.; Kiselev, E.; Conda-Sheridan, M.; Cushman, M.; Ahn, S.; Choi, Y.; White, J.J.; van Breemen, R.B.; et al. Induction of retinoid X receptor activity and consequent upregulation of p21WAF1/CIP1 by indenoisoquinolines in MCF7 cells. *Cancer Prev. Res.* **2011**, *4*, 592–607. [CrossRef]
60. Jones, G.; Willett, P.; Glen, R.C.; Leach, A.R.; Taylor, R. Development and validation of a genetic algorithm for flexible docking. *J. Mol. Biol.* **1997**, *267*, 727–748. [CrossRef]
61. *The PyMOL Molecular Graphics System*; Version 2.0; Schrödinger, LLC: New York, NY, USA, 2015.
62. Meneau, F.; Dupré, A.; Jessus, C.; Daldello, E.M. Translational Control of Xenopus Oocyte Meiosis: Toward the Genomic Era. *Cells* **2020**, *9*, 1502. [CrossRef]
63. Topatana, W.; Juengpanich, S.; Li, S.; Cao, J.; Hu, J.; Lee, J.; Suliyanto, K.; Ma, D.; Zhang, B.; Chen, M.; et al. Advances in synthetic lethality for cancer therapy: Cellular mechanism and clinical translation. *J. Hematol. Oncol.* **2020**, *13*, 118. [CrossRef]

International Journal of *Molecular Sciences*

MDPI

*Article*

# Copper(II) Complexes with 1-(Isoquinolin-3-yl)heteroalkyl-2-ones: Synthesis, Structure and Evaluation of Anticancer, Antimicrobial and Antioxidant Potential

**Łukasz Balewski [1], Tomasz Plech [2], Izabela Korona-Głowniak [3], Anna Hering [4], Małgorzata Szczesio [5], Andrzej Olczak [5], Patrick J. Bednarski [6], Jakub Kokoszka [1] and Anita Kornicka [1,*]**

1 Department of Chemical Technology of Drugs, Faculty of Pharmacy, Medical University of Gdansk, Gen. J. Hallera 107, 80-416 Gdańsk, Poland; lukasz.balewski@gumed.edu.pl (Ł.B.); jakub.kokoszka@gumed.edu.pl (J.K.)
2 Department of Pharmacology, Medical University of Lublin, Radziwiłłowska 11, 20-080 Lublin, Poland; tomasz.plech@umlub.pl
3 Department of Pharmaceutical Microbiology, Medical University of Lublin, Chodźki 1, 20-093 Lublin, Poland; iza.glowniak@umlub.pl
4 Department of Biology and Pharmaceutical Botany, Faculty of Pharmacy, Medical University of Gdansk, Gen. J. Hallera 107, 80-416 Gdańsk, Poland; anna.hering@gumed.edu.pl
5 Institute of General and Ecological Chemistry, Faculty of Chemistry, Lodz University of Technology, Żeromskiego 116, 90-924 Łódź, Poland; malgorzata.szczesio@p.lodz.pl (M.S.); andrzej.olczak@p.lodz.pl (A.O.)
6 Department of Pharmaceutical and Medicinal Chemistry, Institute of Pharmacy, University of Greifswald, F.-L. Jahn Strasse 17, D-17489 Greifswald, Germany; bednarsk@uni-greifswald.de
* Correspondence: anita.kornicka@gumed.edu.pl

**Abstract:** Four copper(II) complexes, **C1–4**, derived from 1-(isoquinolin-3-yl)heteroalkyl-2-one ligands **L1–4** were synthesized and characterized using an elemental analysis, IR spectroscopic data as well as single crystal X-ray diffraction data for complex **C1**. The stability of complexes **C1–4** under conditions mimicking the physiological environment was estimated using UV-Vis spectrophotometry. The antiproliferative activity of both ligands **L1–4** and copper(II) compounds **C1–4** were evaluated using an MTT assay on four human cancer cell lines, A375 (melanoma), HepG2 (hepatoma), LS-180 (colon cancer) and T98G (glioblastoma), and a non-cancerous cell line, CCD-1059Sk (human normal skin fibroblasts). Complexes **C1–4** showed greater potency against HepG2, LS180 and T98G cancer cell lines than *etoposide* ($IC_{50}$ = 5.04–14.89 μg/mL vs. $IC_{50}$ = 43.21–>100 μg/mL), while free ligands **L1–4** remained inactive in all cell lines. The prominent copper(II) compound **C2** appeared to be more selective towards cancer cells compared with normal cells than compounds **C1**, **C3** and **C4**. The treatment of HepG2 and T98G cells with complex **C2** resulted in sub-G1 and G2/M cell cycle arrest, respectively, which was accompanied by DNA degradation. Moreover, the non-cytotoxic doses of **C2** synergistically enhanced the cytotoxic effects of chemotherapeutic drugs, including *etoposide*, *5-fluorouracil* and *temozolomide*, in HepG2 and T98G cells. The antimicrobial activities of ligands **L2–4** and their copper(II) complexes **C2–4** were evaluated using different types of Gram-positive bacteria, Gram-negative bacteria and yeast species. No correlation was found between the results of the antiproliferative and antimicrobial experiments. The antioxidant activities of all compounds were determined using the DPPH and ABTS radical scavenging methods. Antiradical tests revealed that among the investigated compounds, copper(II) complex **C4** possessed the strongest antioxidant properties. Finally, the ADME technique was used to determine the physicochemical and drug-likeness properties of the obtained complexes.

**Keywords:** isoquinoline derivatives; copper(II) complexes; synthesis; structural; stability studies; antitumor; antimicrobial; antioxidant properties; ADME analysis

## 1. Introduction

Since metal-based compounds play remarkable roles as therapeutic and diagnostic agents, the search for novel metallopharmaceuticals represents an area of significant interest in the field of medicinal chemistry [1,2]. The therapeutic potential of metal complexes has a long history; however, the discovery of the bacteriostatic and anticancer activity ruthenium(II) phenanthroline complexes by Francis Dwyer, followed by the discovery of the anticancer properties of *cis*-diaminodichloridoplatinum(II)—*cisplatin* by Barnet Rosenberg, was a milestone in the development of metal-containing drugs [1]. Among clinically approved anticancer chemotherapeutics, platinum agents such as *cisplatin* and its derivatives, e.g., *carboplatin oxaliplatin* and *picoplatin*, are still the most prominent drugs used in the treatment of solid cancers [3]. Nonetheless, despite the evident success of *cisplatin* and its analogues in medicine, their progress in clinical application is limited due to their well-known drawbacks, such as low solubility, severe side effects, including nephrotoxicity and neurotoxicity, and the intrinsic or acquired resistance of cancer cells to platinum-containing drugs [4–7]. Consequently, with the emergence of many biomedical applications of other transition metal complexes, including anticancer and antimicrobial agents, attention is shifting beyond platinum-based compounds [1,8–11]. The preclinical studies provide evidence that non-platinum agents have the potential to circumvent the problem of chemoresistance and the toxicity of platinum-based agents by exhibiting different specific modes of action, reduced undesirable effects and the ability to overcome drug-resistance mechanisms [10,12,13].

Antitumor agents based on endogenous metals such as cobalt, zinc, iron or copper were found to be less toxic compared to platinum analogues [14]. Among them, copper(II)-containing coordination compounds have attracted considerable interest due to the significant role of copper in cancer [15,16]. The copper ion is involved in essential processes of cancer progression such as cell proliferation, angiogenesis or metastasis, and it provides various mechanisms of antitumor action [17–19]. For example, the antiproliferative effect of copper(II) complexes results from the inhibition of the activities of enzymes, which play a pivotal role in cancer therapy, e.g., protein disulfide isomerase (PDI) [20], topoisomerases I and II [21–23], telomerase [24] or proteasome [25–27], as well as DNA intercalation [28–30] and DNA degradation [31,32]. Moreover, the antitumor activity of copper compounds may also be the consequence of their ability to induce apoptosis [33,34] and non-apoptotic cell death—paraptosis [25,35]—reactive oxygen species (ROS) formation that triggers tumor cell death [36,37], and antiangiogenic properties [38]. In turn, the copper(II) complex with *disulfiram*, a drug used in humans to treat alcoholism with great potential for the treatment of human cancers [39], was found to be capable of reversing the drug resistance of *doxorubicin* (ADM)-resistant acute leukemia cell lines by the induction of apoptosis [40]. Additionally, it is well known that the copper coordination in organic compounds may lead to higher antitumor activity [23,41], selectivity [23] and reduced toxicity [42] compared to the free ligands. In this line, it is worth noting that the uptake of copper(II) complexes by cancer cells is higher than that by normal cells [38,43]. In addition, copper(II) complexes may exhibit a different response to cancer cells than to non-cancerous cells [44]. Overall, copper(II)-containing coordination compounds have emerged as a promising class of therapeutic anticancer agents with various mechanisms of action and the potential to overcome drug resistance [17,20,45].

It should be noted that the therapeutic potency of copper(II) complexes is not limited to anticancer activity. These compounds have also gained much interest due to their anti-inflammatory and antioxidant properties [46–48], antiviral properties [49,50] or antibiofilm and antimicrobial [51,52] activities. The latter is a multi-faceted process, although the main mechanism of the bactericidal effect is the formation of ROS, causing irreversible damage to membranes [51].

Isoquinoline-containing compounds are of scientific interest due to their fluorescent properties and broad spectrum of biological activities including antihypertensive [53], anti-inflammatory and analgesic [54] or antioxidant [55] activities, as well as their ability to act as antidepressant and antipsychotic [56] agents. Moreover, the importance of isoquinoline scaffold in drug design [57,58] has also led to the development of bioactive compounds with antimalarial [59], antifungal and antibacterial effects [60,61]. In addition, isoquinoline derivatives constitute an important source of novel anticancer agents that may exert their biological activities through various mechanisms such as apoptosis, DNA fragmentation, the inhibition of tubulin polymerization, induced cell cycle arrest and the interruption of cell migration [62,63]. On the other hand, the research on the biological activities of their metal complexes, especially copper(II)-coordination compounds, is not very extensive.

Our previous work indicated that the newly synthesized 1-(isoquinolin-3-yl)heteroalkyl-2-ones of type **A** (Figure 1) possess promising fluorescent properties [64]. In the present work, as a continuation of a research program on the chemistry and biological activity of copper(II) complexes undertaken in our laboratory [64–68], we wish to report the results of studies on the reactions of the above-mentioned isoquinoline derivatives **A** with copper(II) chloride, an X-ray structure determination of the complexes obtained of type **B** (Figure 1), as well as the results of the evaluation of their anticancer, antimicrobial and antioxidant potential. Furthermore, to confirm the importance of copper coordination in organic ligands, the biological properties of the free ligands **A** were also evaluated.

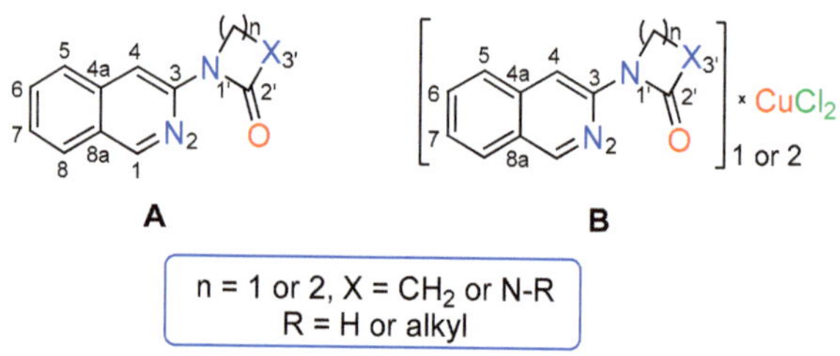


**Figure 1.** 1-(isoquinolin-3-yl)heteroalkyl-2-one ligands (**A**) and copper(II) complexes of 1-(isoquinolin-3-yl)-heteroalkyl-2-ones (**B**).

## 2. Results and Discussion

### *2.1. Chemistry*

#### 2.1.1. Synthesis of 1-(Isoquinolin-3-yl)heteroalkyl-2-one Ligands **L1–4**

Ligand 1-(isoquinolin-3-yl)azetidin-2-one (**L1**) was synthesized according to a previously described procedure involving copper-catalyzed Goldberg–Ullmann-type coupling of 3-bromoisoquinoline with azetidin-2-one [64]. These reactions were carried out in anhydrous dioxane or *n*-butyl alcohol in the presence of a base, copper(I) iodide and *N,N*-dimethylethylenediamine (Scheme 1).

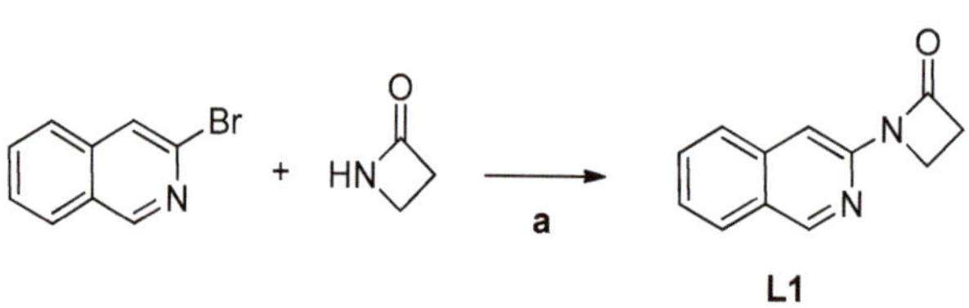


**a**: $K_2CO_3$, CuI, DMEDA, anh. dioxane, 90°C

**Scheme 1.** Synthesis of 1-(isoquinolin-3-yl)azetidin-2-one ligand (**L1**).

In turn, the use of isoquinoline *N*-oxide and 2-chloroimidazoline as substrates allowed us to obtain the ligand-containing cyclic urea moiety—1-(isoquinolin-3-yl)imidazolidin-2-one (**L2**)—as a major product [69,70].

The reaction leading to compound **L2** proceeds exothermically and spontaneously in a polar aprotic solvent (dichloromethane or chloroform) at an ambient temperature, and it involves a few steps (Scheme 2). Firstly, isoquinoline *N*-oxide attacks 2-chloroimidazoline to form 2-((4,5-dihydro-1*H*-imidazol-2-yl)oxy)isoquinolin-2-ium chloride (**Ia**), which, in the mesomeric form **Ib**, possesses a nucleophilic carbon atom in position 3 that is susceptible to the intramolecular attack of the nitrogen atom of the imidazoline ring. In the next step, the simultaneous re-aromatization of intermediate—3,4*a*-dihydro-2*H*-imidazo[1′,2′:4,5][1,2,4]oxadiazolo[2,3-*b*]isoquinoline (**II**) followed by the elimination of hydrogen chloride leads to the formation of the desired 1-(isoquinolin-3-yl)imidazolidin-2-one (Scheme 2). Upon the treatment of compound **L2** with methyl or ethyl iodide in the presence of sodium hydroxide, the corresponding *N*-alkylated ligands **L3** and **L4** were synthesized in acceptable yields (Scheme 2).

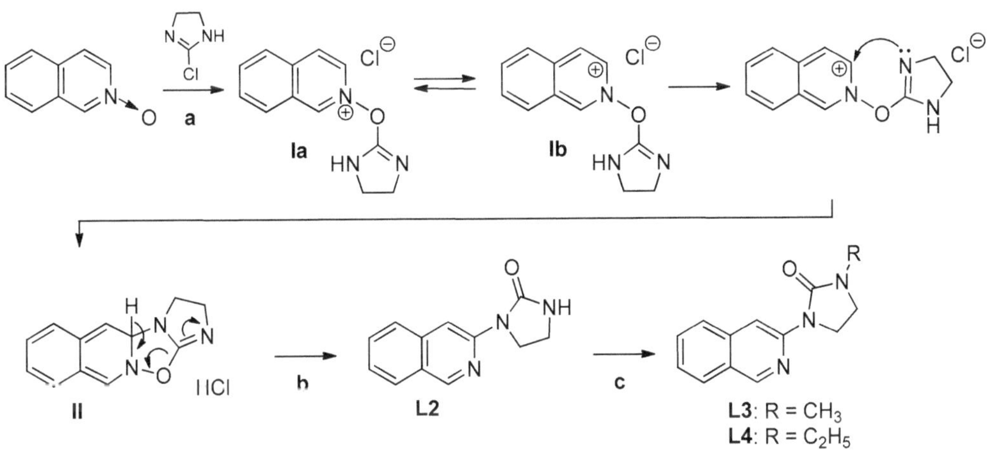


**Scheme 2.** Synthesis of ligand **L2** in the reaction of isoquinoline *N*-oxide with 2-chloroimidazoline and *N*-alkylated analogous **L3** and **L4**.

2.1.2. Synthesis of Copper(II) Complexes of 1-(Isoquinolin-3-yl)heteroalkyl-2-ones **C1–4**

Copper(II) complexes **C1–4** were prepared through the reaction of copper(II) chloride dihydrate with previously described 1-(isoquinolin-3-yl)heteroalkyl-2-ones **L1–4** [64]. For the preparation of metal complexes that are stable under physiological conditions, reactions were carried out in dimethylformamide (DMF) containing 0.5% water and the dihydrate of copper(II) salt. The use of dimethylformamide as a solvent had a positive effect on the purity and further isolation of the desired copper(II) complexes. The formation of precipitate or crystals of green or brown copper(II) complexes was observed at room temperature over a period of 3 to 12 days. After the required time, the products—copper(II) complexes—were separated by filtration.

Firstly, we began by studying the equimolar ratio of ligands **L1–4** and copper(II) salt in 99.5% dimethylformamide. It was found that for the copper(II) complexes **C3** and **C4**, a stoichiometric ratio of the ligands **L3** and **L4** in terms of yields was the most important factor. In the case of copper(II) complexes **C1** and **C2**, two molecules of a neutral bidentate ligand **L1** or **L2** can coordinate with the copper(II) ion. Thus, a 2-fold excess of ligands **L1** or **L2** resulted in the creation of tetra-coordinate mononuclear copper(II) complexes **C1** and **C2**, respectively.

Mononuclear complex **C1** ($L_2CuCl_2$) was obtained in the reaction of ligand **L1** with copper(II) chloride dihydrate in a 1:1 molar ratio through the slow evaporation of the solvent at room temperature (Scheme 3).

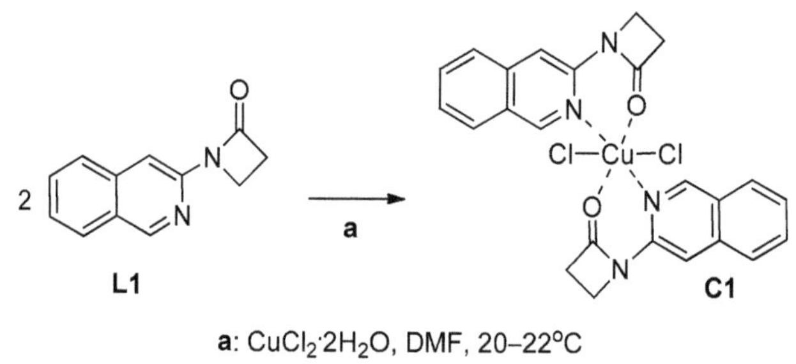


**Scheme 3.** Synthesis of copper(II) complex **C1** in the reaction of ligand **L1** with copper(II) chloride dihydrate in dimethylformamide.

The coordination compounds **C2, C3** and **C4** were obtained in acceptable yields with analogous reaction conditions and copper(II) salt stoichiometries (Scheme 4). It should be emphasized that the ligand unsubstituted at the nitrogen atom in position 3 of the imidazolidin-2-one; derivative **L2** formed a neutral mononuclear chelate **C2** ($L_2CuCl_2$), while the $N^3$-substituted ligands **L3** and **L4** allowed the preparation of brown-green bidentate *N,O*-chelates **C3** and **C4** with the structure $LCuCl_2$.

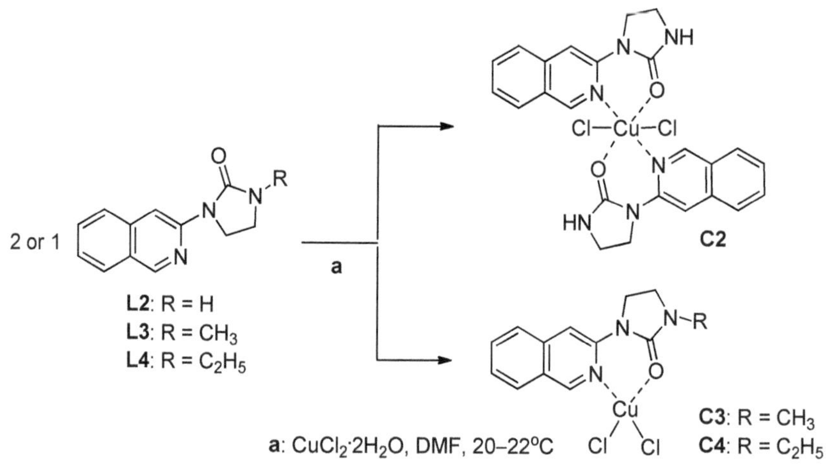


**Scheme 4.** Synthesis of copper(II) complexes **C2, C3** and **C4** in the reaction of **L2, L3** and **L4** with copper(II) chloride dihydrate in dimethylformamide.

Summing up, the efficiency of the complexation reactions, calculated as the ratio of the achieved yield to the theoretical yield, was approximately twice higher (61–68%) in the case of the bidentate *N,O*-chelates **C3** and **C4** ($LCuCl_2$) than the mononuclear complexes **C1** and **C2** with the $L_2CuCl_2$ structure (27–33%).

### 2.2. *Structural Analysis of Copper(II) Complexes* **C1–4**

The structures of copper(II) complexes **C1–4** were confirmed using an elemental analysis and infrared spectroscopic data. Moreover, the crystal structure of copper(II) complex **C1** was determined using X-ray crystallography. It should be mentioned that the presence of an unpaired electron attributed to the copper(II) ion in complexes **C1–4** results in their paramagnetic properties; thus, the nuclear magnetic resonance (NMR) spectra of compounds **C1–4** cannot be recorded.

### 2.2.1. Infrared Spectra

In the infrared spectra of copper(II) complexes **C1**, **C2**, **C3** and **C4** stretching vibrations of the carbonyl group (C=O) were observed in the range of 1639 to 1751 $cm^{-1}$. It should be noted that the IR spectra of the copper(II) complexes showed the characteristic shifts of functional group absorptions, confirming their involvement in the chelation of a metal ion. Hence, in the case of synthesized complexes, shifted stretching bands were observed for the C=O group of the lactam (**C1**) or cyclic urea ring (**C2**, **C3** and **C4**), and the C=N moiety of the isoquinoline ring (Figures S1–S8, Supplementary Materials).

For example, the IR spectrum registered for dichloro[1-ethyl-3-(isoquinolin-3-yl)imidazolidin-2-one]copper(II) (**C4**)—superimposed on the IR spectrum of the parent ligand **L4**—indicated a shift in the carbonyl stretching band by a value of 50 $cm^{-1}$ towards lower values of wavelengths (1689 $cm^{-1}$ → 1639 $cm^{-1}$) (Figure 2).

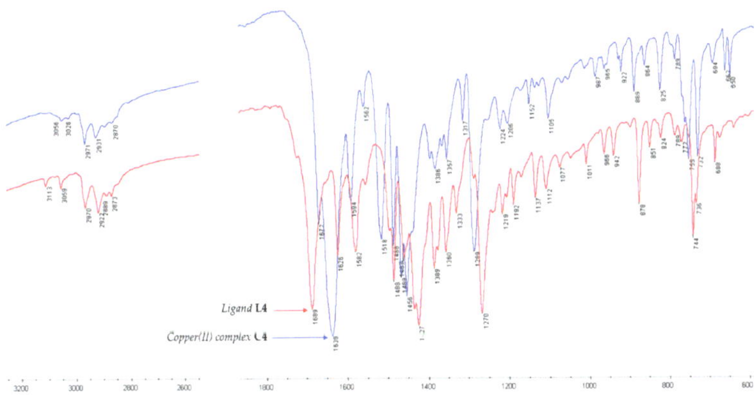


**Figure 2.** FTIR spectra of ligand **L4** (red line) and corresponding copper(II) complex ($LCuCl_2$) **C4** (blue line).

### 2.2.2. X-ray Crystallographic Studies

The crystal data, data collection and structure refinement details are summarized in Table 1.

**Table 1.** X-ray diffraction data and structure refinement details for copper(II) complex **C1**.

| Compound | (C1_1_auto) |
|---|---|
| Chemical formula | $C_{24}H_{20}Cl_2CuN_4O_2$ |
| $M_r$ | 530.89 |
| Crystal system, space group | Triclinic, $P\bar{1}$ |
| Temperature (K) | 100 |
| $a, b, c$ (Å) | 7.8288 (2), 8.2173 (2), 9.5081 (2) |
| $\alpha, \beta, \gamma$ (°) | 106.092 (2), 111.334 (3), 97.897 (2) |
| $V$ (Å$^3$) | 527.68 (2) |
| $Z$ | 1 |
| Radiation type | Mo $K\alpha$ |
| $\mu$ ($mm^{-1}$) | 1.32 |
| Crystal size (mm) | 0.22 × 0.2 × 0.09 |
| Diffractometer | XtaLAB Synergy, Dualflex, Pilatus 300 K |
| $T_{min}$, $T_{max}$ | 0.586, 1.000 |
| No. of measured, independent and observed [$I > 2\sigma(I)$] reflections | 15,833, 3057, 2752 |
| $R_{int}$ | 0.040 |
| $(\sin\theta/\lambda)_{max}$ (Å$^{-1}$) | 0.748 |
| $R[F^2 > 2\sigma(F^2)]$, $wR(F^2)$, $S$ | 0.029, 0.072, 1.09 |
| No. of reflections | 3057 |
| No. of parameters | 151 |
| H-atom treatment | H-atom parameters constrained |
| $\Delta_{max}$, $\Delta_{min}$ (e Å$^{-3}$) | 0.91, −0.35 |

In the crystal structure of complex **C1**, the organic ligand **L1** is coordinated in the bidentate manner via the oxygen atom of the carbonyl group and the nitrogen atom of the isoquinoline ring, forming a six-membered chelate cycle (Figure 3).

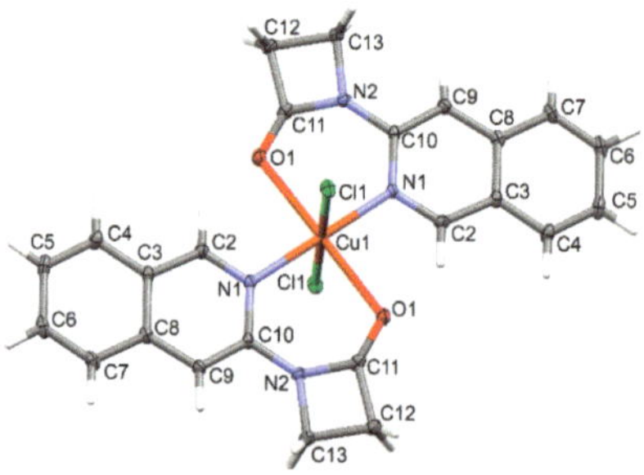


**Figure 3.** Structural representation and atom-numbering scheme. Thermal ellipsoids are drawn at the 50% probability level.

Coordinated ligands are in *trans* position to each other. The copper ion is in a distorted octahedral environment with two N atoms and two O atoms from two ligands in the equational plane and two Cl donors in the opposite axial sites (Figure 4). The copper ions reside in the center of the octahedron, in which the bond lengths are Cu1 − N1 = 2.0666 (12)Å, Cu1 − O1 = 2.4210 (11) Å and Cu1 − Cl1 = 2.3051 (4) Å.

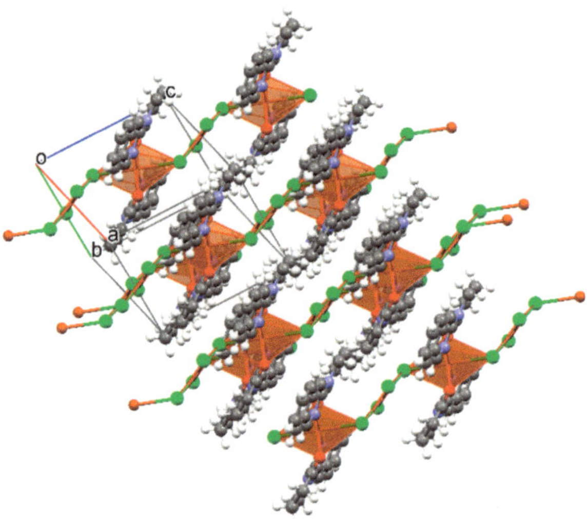


**Figure 4.** Crystal packing of copper(II) complex **C1** in the unit cell.

There are no strong hydrogen bonds in the **C1** structure. There are hydrogen bonds of the C-H...Cl(O) type. The C-H...O bond stabilizes the coordination fragment (Table 2). However, the C-H...Cl hydrogen bond stabilizes the packing.

**Table 2.** Hydrogen bond geometry for copper(II) complex **C1**.

| *D*—H···*A* | *D*—H (Å) | H···*A* (Å) | *D*···*A* (Å) | *D*—H···*A* (º) |
|---|---|---|---|---|
| C9—H9···Cl1 [i] | 0.95 | 2.86 | 3.7205 (15) | 152 |
| C2—H2···O1 [ii] | 0.95 | 2.21 | 2.9512 (19) | 135 |

Symmetry codes: [i] $x, y+1, z$; [ii] $-x+1, -y, -z+1$.

Some additional geometrical details can be found in the Supplementary Materials (Tables S1–S3).

### 2.2.3. Molecular Modeling Studies of Ligands **L1** and **L2**

Our previous X-ray studies indicated that in the crystal state, the compound **L2** adopts the *E* conformation, which is probably stabilized by intramolecular C-H···O (2.26 Å). The molecules of ligand **L2** are connected by pairs of N-H···O hydrogen bonds (1.99 Å) between two imidazolidin-2-one ring fragments with the formation of centrosymmetric dimers [64]. The many possible rotamers of the representative ligand—1-(isoquinolin-3-yl)imidazolidin-2-one (**L2**)—can be generated from the rotation around the bond axis $C3_{(isoquinoline)}$-$N1'_{(imidazolin-2-one)}$ (Figure 5).

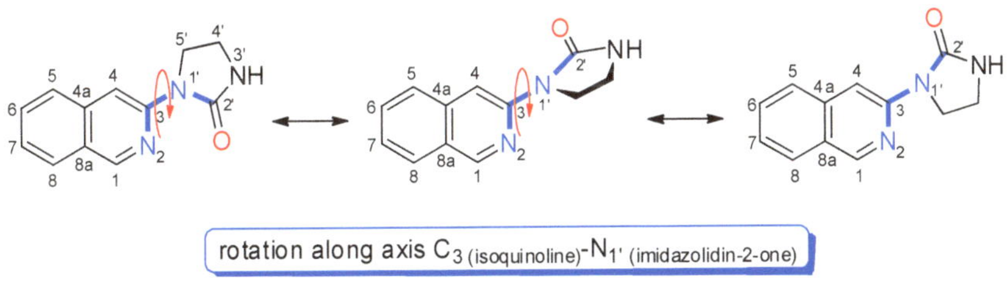


**Figure 5.** Structures of possible rotamers of ligand **L2** obtained by rotation along the $C_3$-$N_{1'}$ axis.

In this study, the X-ray diffraction analysis results obtained for copper(II) complex **C1** showed that two molecules of ligand **L1** exist in the conformation *E*. For this reason, we decided to perform quantum chemical calculations to gain a better understanding of the structure of the ligands **L1–4**. We assumed that the formation of the copper(II) complexes of ligands **L1–4** requires a rotation of their conformation from *E* to *Z*.

The structures of the selected two ligands **L1** and **L2** were optimized in a polar solvent (DMF) by using the Spartan program suite (Spartan version 14 V 1.1.4.). The possible conformers of compounds **L1** and **L2** were calculated at the B3LYP/6.31G** level of theory [71–73]. It should be mentioned that B3LYP—the so-called 'hybrid functional'—is one of the most popular DFT functionals used for the prediction of the physicochemical properties of molecules in in silico drug design [74].

In the case of ligand **L1**, the structure with a torsion angle ($N_2$-$C_3$-$N_{1'}$-$C_{2'}$, $\Phi = 180°$) was proven to be the lowest energy rotameric form in the DMF solution (*E* conformation of molecule) (Figure 6). The energy difference between the *E* conformation and its rotamer in the *Z* conformation ($N_2$-$C_3$-$N_{1'}$-$C_{2'}$, $\Phi = 0$) was calculated to be $\Delta E = 8.146$ kcal/mol. Based on this, it may be concluded that the barrier to the $C_3$-$N_{1'}$ bond rotation is low, and it is easy to overcome the energy barrier under normal conditions. This suggests that rotamers having different torsion angles may exist together in the solution at room temperature. In the *Z* conformation of 1-(isoquinolin-3-yl)azetidin-2-one (**L1**), the position of the nitrogen and oxygen atoms of the two heterocyclic rings is favored for the chelation of copper(II) ions.

In the case of ligand **L1** in its *E* conformation with torsion angle $\Phi = 0$, the highest occupied molecular orbital ($E_{HOMO} = -5.67$ eV) is confined to carbon atoms $C_3$–$C_8$ of the isoquinoline ring, and nitrogen ($N_{1'}$) and carbon ($C_{2'}$) atoms of the azetydin-2-one system, while the frontier orbital LUMO ($E_{LUMO} = -1.52$ eV) is located mostly on the entire isoquinoline scaffold (Figure 7). The calculated HOMO-LUMO energy gap for ligand **L1** in conformation with the torsion angle $\Phi = 0$ ($E_g = 4.15$ eV) is lower than the energy gap obtained for its $\Phi = 180$ rotamer ($E_g = 4.27$ eV). This may suggest the higher reactivity of conformer **L1** with the torsion angle $\Phi = 0$ towards bonding with transient metals such as copper.

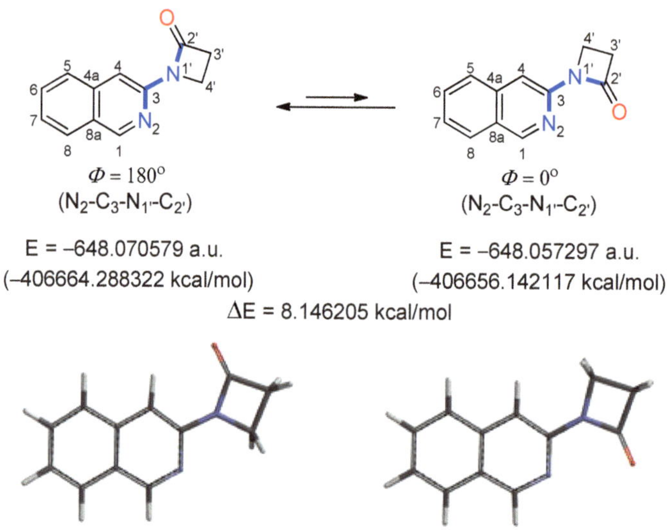


**Figure 6.** Structures of two possible rotamers of ligand **L1** and corresponding electronic energies (E, a.u.) and relative energy (ΔE, kcal/mol) calculated in DMF at B3LYP/6.31G** level of theory.

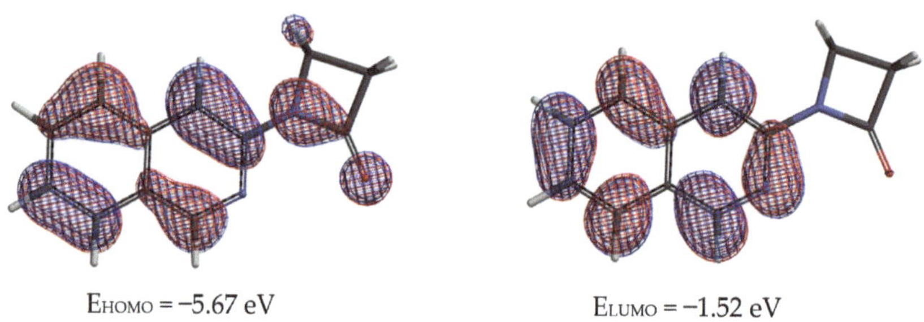


**Figure 7.** Orbital diagrams of HOMO (**left**) and LUMO (**right**) for optimized structure of conformer ligand L1 with torsion angle $\Phi = 0°$.

Similarly, the in silico data for ligand **L2** revealed that conformer with torsion angle $N_2$-$C_3$-$N_{1'}$-$C_{2'}$ at about $\Phi = 172$ was calculated to be slightly lower in energy ($E = -441{,}256.812142$ kcal/mol) than the conformer with torsion angle $\Phi = 0$ ($E = -441{,}250.96698$ kcal/mol) (Figure 8). The energy difference between these two rotamers was estimated to be $\Delta E = 5.845$ kcal/mol. Moreover, based on calculated dipole moments, the conformer with torsion angle $\Phi = 0$ ($\mu = 6.60$ debye)—favored for the binding of copper(II) ions—would be predicted to prevail over the second one ($\Phi = 172$, $\mu = 3.23$ debye) in a polar solvent such as dimethylformamide.

Based on this, it may be concluded that for ligands **L1** and **L2**, the barrier to the $C_3$-$N_{1'}$ bond rotation is low, and it is easy to overcome the energy barrier under normal conditions (at room temperature) and in polar solvents (dimethylformamide). This suggests that rotamers of ligands **L1–4** having different torsion angles may exist in the solution at room temperature. In their *Z* conformations, the position of the atoms having donating properties is favored for the binding of copper(II) ions. Therefore, forming a six-membered chelate ring requires energy, which can be compensated for by creating novel bonds involving copper and nitrogen or oxygen atoms: $Cu^{2+}$---N=C and $Cu^{2+}$---O=C.

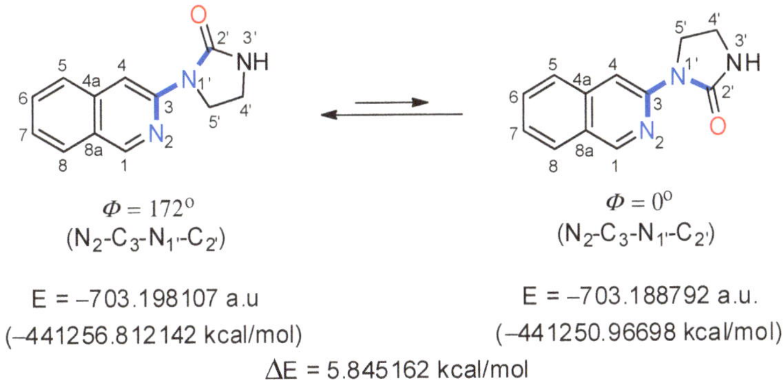


**Figure 8.** Structures of two possible rotamers of ligand **L2**, and corresponding electronic energies (E, a.u.) and relative energy ($\Delta$E, kcal/mol) calculated in DMF at B3LYP/6.31G** level of theory.

### 2.3. *Stability Studies of Copper(II) Complexes* **C1–4** *in Aqueous Buffer*

The stability testing of the synthesized compounds must validate the biological results by ensuring that the compound remains biologically active over time. It should be emphasized that the limited stability of drug candidates in physiological pH ranges prevents their use in vivo. In the case of metal complexes, this is a crucial aspect due to the fact that they may dissociate under physiological conditions, releasing the metal ions and the free ligands.

To address this question and confirm the validity of the biological results, we performed UV-vis stability measurements of copper(II) complexes **C1–4** under conditions that mimic the physiological environment (phosphate buffered aqueous solution, pH = 7.4, 37 °C). An increase in absorbance during the measurements may indicate the release of free ligands, whereas a decrease in absorbance shows the precipitation of the complex from the buffer solution.

Firstly, copper(II) complexes were dissolved in 99.5% DMF at a concentration of 4 mM. These solutions were diluted into a 100 mM potassium phosphate buffer with pH 7.4 to a final concentration of 40 μM of copper(II) complex in a quartz cuvette at a temperature of 37 °C. To identify very small changes in the UV-vis spectra, difference spectra between $\lambda$ = 250–600 nm were recorded every 10 min with a diode array photometer over the course of 3 h at 37 °C. The difference spectra were obtained by subtracting the spectrum at time = 0 from each of the following recorded spectra between $\lambda$ = 250 and 600 nm.

The tested copper(II) complexes **C1–4** did not show noticeable changes in their time-dependent difference spectra over 3 h of incubation in the phosphate-buffered aqueous solution (pH 7.4, 37 °C). It was observed that the intensity did not change during the experiments. It is also worth noting the lack of isosbestic points in the range of 250 and 600 nm. The presence of isosbestic points indicates a possibility of reaction in the buffered solution, for example, ligand exchange. Thus, the complexes **C1–4** appear stable under biologically similar conditions with no apparent loss of the Cu(II) from the ligand.

Figure 9 presents the time-dependent UV-Vis spectra of the representative copper(II) complex **C1**. The UV-vis difference spectra of copper(II) complexes **C2**, **C3** and **C4** are shown in the supporting information (Figures S9–S12, Supplementary Materials).

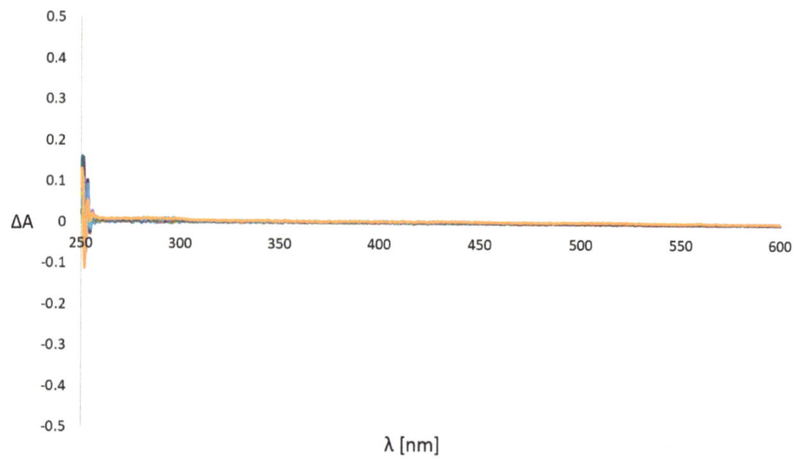


**Figure 9.** Time-dependent UV-vis difference spectra of dichloro{*bis*[1-(isoquinolin-3-yl)azetidin-2-one]}copper(II) (**C1**) over 3 h at pH = 7.4 and 37 °C.

### *2.4. Biological Evaluation*

#### 2.4.1. In Vitro Cytotoxic Activity

The in vitro cytotoxic activities of free ligands **L1–4** and the corresponding copper(II) complexes **C1–4** were evaluated against four human cancer cell lines, namely melanoma A375, hepatoma HepG2, colon cancer LS-180 and glioblastoma T98G. To establish the selectivity towards cancer cell lines, the investigated compounds were also tested on a non-cancerous human skin fibroblast cell line—CCD-1059-Sk.

From the results presented in Table 3, it is apparent that free ligands **L1**, **L2**, **L3** and **L4** were inactive in all cell lines up to 200 μg/mL, while their copper(II) complexes **C1**, **C2**, **C3** and **C4** exhibited remarkable growth inhibitory potency towards cancer cell lines ($IC_{50}$ values ranging from 5.04 μg/mL to 37.97 μg/mL) compared with the widely used anticancer agent—*etoposide* ($IC_{50}$ values between 10.20 and >100 μg/mL).

**Table 3.** Cytotoxic activity of the investigated free ligands **L1–4**, Cu(II) complexes **C1–4** and reference compound against human cancer cell lines and non-cancerous cells determined by MTT assay after 24 h incubation.

| Ligands/Complexes | $IC_{50} \pm SD$ (μg/mL) | | | | |
|---|---|---|---|---|---|
| | Cell Line | | | | |
| | A375 | HepG2 | LS180 | T98G | CCD-1059-Sk |
| L1 | >200 | >200 | >200 | >200 | >200 |
| C1 | 37.97 ± 3.01 | 14.89 ± 0.56 | 11.46 ± 0.69 | 8.25 ± 0.36 | 18.83 ± 0.91 |
| L2 | >200 | >200 | >200 | >200 | >200 |
| C2 | 28.91 ± 1.79 | 5.04 ± 0.42 | 5.88 ± 0.40 | 6.97 ± 0.52 | 21.86 ± 1.03 |
| L3 | >200 | >200 | >200 | >200 | >200 |
| C3 | 22.78 ± 1.14 | 12.00 ± 0.36 | 7.09 ± 0.53 | 10.55 ± 0.74 | 25.17 ± 0.85 |
| L4 | >200 | >200 | >200 | >200 | >200 |
| C4 | 37.80 ± 3.11 | 6.72. ± 0.12 | 5.92 ± 0.54 | 9.27 ± 0.11 | 18.25 ± 1.37 |
| Etoposide * | 10.20 ± 0.83 | 43.21 ± 2.75 | >100 | >100 | 83.53 ± 3.19 |

* *Etoposide* was used as a positive control; $IC_{50}$—the concentration that inhibits 50% of cell viability. The values shown are mean ± SD from three repetitions in two independent experiments.

It should be noted that copper(II) compounds **C1–4** showed several times greater effectiveness against cancer cell lines than the reference drug ($IC_{50}$ = 5.04–14.89 μg/mL vs. $IC_{50}$ = 43.21–>100 μg/mL); the exception to this was the melanoma A375 cell line, which was the least susceptible to the effect of the tested complexes ($IC_{50}$ = 22.78–37.97 μg/mL vs. $IC_{50}$ = 10.20 μg/mL). Among the tested complexes, compound **C2** bearing imidazolidin-2-one moiety was found to be the most potent on the HepG2 and T98G cancer cell lines (Table 3). Its analogues bearing the methyl or ethyl substituent at the R position of the imidazolidin-2-one scaffold (**C3**: R = $CH_3$; **C4**: R = $C_2H_5$) displayed slightly weaker antiproliferative activities ($IC_{50}$ = 5.04–6.97 μg/mL vs. $IC_{50}$ = 6.72–12.00 μg/mL). Furthermore, complex **C1** with an azetidin-2-one functionality showed a comparable level of ability to inhibit the growth of the HepG2 and T98G cancer cell lines with the complexes **C3** and **C4** featuring the imidazolidin-2-one moiety ($IC_{50}$ = 8.25–14.89 μg/mL vs. $IC_{50}$ = 6.72–12.00 μg/mL), although compound **C1** turned out to be the least active in all cancer cell lines ($IC_{50}$ = 14.89–37.97 μg/mL) (Table 3).

Summing up, the data presented here confirmed the hypothesis that the introduction of a metal ion into organic ligands may have a beneficial effect on the anticancer potential [11,23,38,43,75].

As evidenced in Table 3, the cytotoxic potency of the investigated copper(II) complexes **C1, C2, C3** and **C4** was also observed in a non-cancerous cell line, CCD-1059-Sk ($IC_{50}$ values ranging from 18.25 μg/mL to 25.17 μg/mL). Nevertheless, when individual cell lines such as HepG2, LS180 and T98G were compared with CCD-1059-Sk, a moderate degree of selectivity for compounds **C2, C3** and **C4** became apparent ($IC_{50}$ = 5.04–12.00 μg/mL vs. $IC_{50}$ = 15.03–25.17 μg/mL). In this regard, the most active Cu(II) complex **C2** was characterized by the greatest selective effect on the HepG2, LS180 and T98G cancer cell lines, with the selectivity index (SI) ranging from 3.14 (T98G) to 4.33 (HepG2).

#### 2.4.2. Cell Cycle Analysis

Since copper(II) complex **C2** most effectively inhibited the growth of HepG2 and T98G cancer cells, its effect on cell cycle progression was examined by using the cytometry method (Figure 10). Interestingly, this compound revealed a distinct effect on the cell cycle progression of HepG2 and T98G cells. The growth inhibition of HepG2 cells was associated with cell cycle arrest in the sub-G1 phase, showing low-molecular-weight fragments of DNA as the evidence of apoptosis (Figure 10A). When analyzing the DNA content in the sub-G1 phase, a significant increase ($p < 0.0001$) from 10% in the control cells to 29% in the **C2(CX)**-treated HepG2 cells was observed. The identification of the occurrence of apoptosis on the basis of the elevated number of cells in the sub-G1 phase relies on the principle that degraded DNA fragments (i.e., early signs of apoptosis) are released from cells, which results in the increased number of cells possessing a reduced DNA content. Consecutively, the cytotoxic effect of **C2 (CX)** on the T98G cells resulted from cell cycle arrest in the G2/M phase, which indicates considerable DNA damages that are unable to be repaired before mitosis (Figure 10B).

#### 2.4.3. Interaction of Copper(II) Complex **C2** with Clinically Used Anticancer Agents

Possible interactions between anticancer drugs should be an important consideration among patients undergoing antineoplastic therapy since they are exposed to several types of treatments, each including a number of drugs. As most of them have a narrow therapeutic index, it is important to examine the possible new strategies that can increase the clinical outcomes by using lower doses of the currently available anticancer drugs. Such co-treatments can also be an effective strategy for overcoming resistance in cancer therapy. In our studies, copper(II) complex **C2**, which exhibited the most potent effect against HepG2 and T98G cancer cells, was selected to examine the possible synergism with clinically used anticancer agents. The combinations of **C2 (CX)** and *etoposide* (**ETO**), *cisplatin* (**CIS**) and *5-fluorouracil* (**5-FU**) were evaluated against both of the cell lines (Figures 11 and 12). Additionally, the combination of **C2 (CX)** and *temozolomide* (**TMZ**)—a chemotherapeutic agent being used as a first-line treatment for glioblastoma—was tested on the T98G glioblastoma cell line (Figure 11).

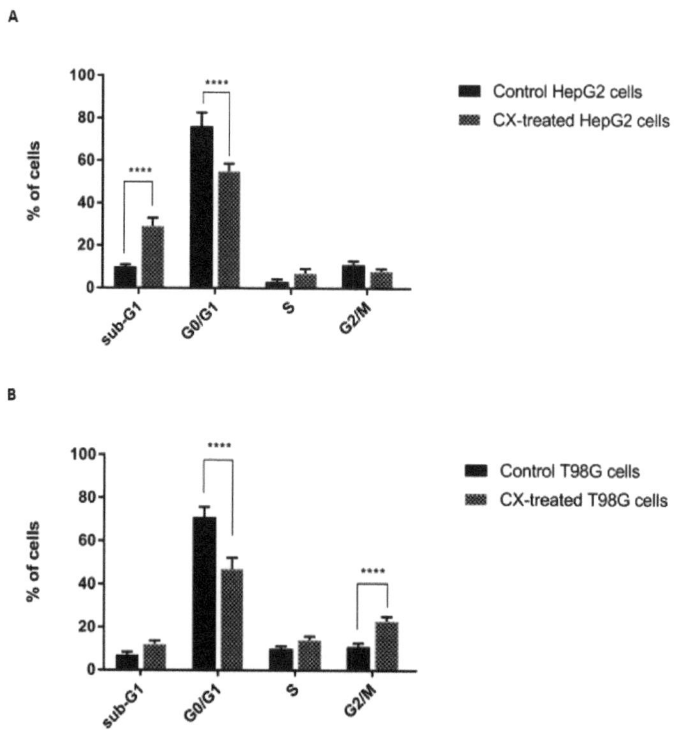


**Figure 10.** Cell cycle analysis of HepG2 (**A**) and T98G cells (**B**) incubated for 24 h with copper(II) complex **C2** (**CX**) at its $IC_{50}$ concentration. Results are expressed as means ± SEM. Statistical significance was designated as **** when $p < 0.0001$ (vs. control cells) using ANOVA analysis followed by Tukey's post hoc test.

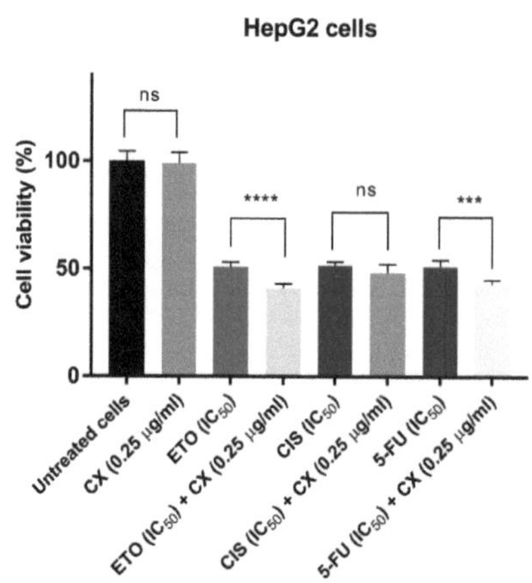


**Figure 11.** Interactions between copper(II) complex **C2** (**CX**) and anticancer drugs, *etoposide* (**ETO**), *cisplatin* (**CIS**) and *5-fluorouracil* (**5-FU**), examined on HepG2 cells using MTT assay. **CX** was tested at the highest concentration that did not affect the viability of HepG2 cells (i.e., 0.25 μg/mL). Chemotherapeutics (**ETO, CIS, 5-FU**) were tested at their $IC_{50}$ concentrations. Statistical analysis: one-way ANOVA with Tukey's post hoc test; ns—not significant; *** $p < 0.001$; **** $p < 0.0001$.

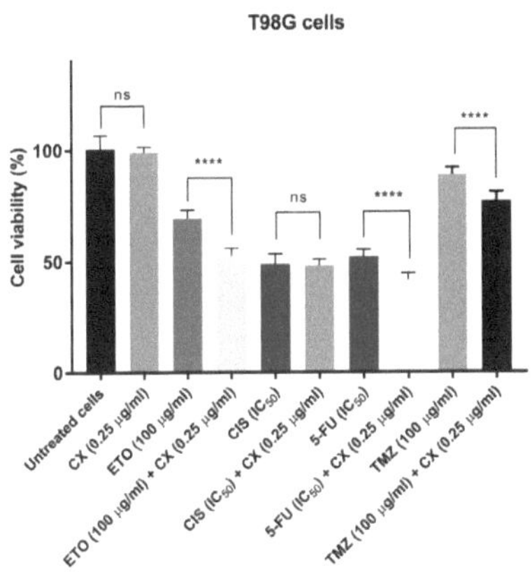


**Figure 12.** Interactions between copper(II) complex **C2** (**CX**) and anticancer drugs, *etoposide* (**ETO**), *cisplatin* (**CIS**) and *5-fluorouracil* (**5-FU**), examined on T98G cells using MTT assay. **CX** was tested at the highest concentration that did not affect the viability of T98G cells (i.e., 0.25 μg/mL). Chemotherapeutics (**ETO**, **CIS**, **5-FU**) were tested at their $IC_{50}$ concentrations. Statistical analysis: one-way ANOVA with Tukey's post hoc test; ns—not significant; **** $p < 0.0001$.

The investigated drugs represent different mechanisms of anticancer activity, including human DNA topoisomerase IIα inhibitors (**ETO**) [76], alkylating agents (**CIS**, **TMZ**) [77,78], thymidylate synthase inhibitors and antimetabolites (**5-FU**) [79]. Compound **C2** (**CX**), at the concentration that did not inhibit the growth of cancer cells (i.e., 0.25 μg/mL), statistically significantly improved the cytotoxic effects of **ETO**, **5-FU** and **TMZ** against HepG2 and T98G cells (Figures 11 and 12).

To better understand the mechanism of synergism between **C2** and the clinically used drugs, it is important to gain insight into the mechanism of action of **C2** alone. Usually, beneficial drug–drug interactions (synergism) can be expected when the components of the drug mixture act via different mechanisms.

In this context, it should be mentioned that metal complexes with isoquinoline derivatives may exert anticancer properties through S-phase cell cycle arrest by the up-regulation of p53, p27 and p21 proteins and the down-regulation of cyclin A and cyclin E [80], mitochondrial (intrinsic) pathway-dependent apoptosis [81,82], caspase-3 activation triggering apoptosis [83], the inhibition of telomerase activity [75] as well as DNA intercalation [81,83].

In our studies, the cytotoxic activity of copper(II) complexes **C2** against the tested cancer cell lines appears to be the result of cell cycle arrest in the sub-G1 phase (HepG2 cells) or G1/M phase (T98G cells) associated with DNA damage. Nevertheless, more work is needed to clarify the molecular mechanism of action of compound **C2** leading to apoptosis.

### 2.4.4. Antimicrobial Activity

The in vitro antimicrobial activities of three ligands, **L2–4**, and their complexes, **C2–4**, were investigated on reference strains of Gram-positive bacteria, namely *Staphylococcus aureus* ATCC 25923, *Staphylococcus aureus* ATCC BAA-1707, *Staphylococcus epidermidis* ATCC 12228, *Micrococcus luteus* ATCC 10240 and *Bacillus cereus* ATCC 10876 and Gram-negative bacterial strains of *Salmonella Typhimurium* ATCC 14028, *Escherichia coli* ATCC 25922, *Proteus mirabilis* ATCC 12453, *Klebsiella pneumoniae* ATCC 13883 and *Pseudomonas aeruginosa* ATCC 9027, as well as three strains of yeasts such as *Candida albicans* ATCC 102231, *Candida parapsilosis* ATCC 22019 and *Candida glabrata* ATCC 90030.

The tested compounds did not show antibacterial activity against Gram-positive nor Gram-negative bacteria (MIC > 1000 mg/L,). As revealed by the data in Table 4, they displayed mild or no bioactivity towards the yeasts that were tested, except for ligand **L3**, which demonstrated moderate anti-*Candida* activity (MIC = 125–250 mg/L).

**Table 4.** Antimicrobial activity of free ligands, **L2**, **L3** and **L4**, and their complexes, **C2**, **C3** and **C4**, against reference strains of yeasts.

| Microorganism/Compounds | L2 | C2 | L3 | C3 | L4 | C4 |
|---|---|---|---|---|---|---|
| | MIC (mg/L) * | | | | | |
| Yeasts | | | | | | |
| *C. albicans* ATCC 102231 | 500 | 1000 | 250 | 1000 | 1000 | 500 |
| *C. parapsilosis* ATCC 22019 | 1000 | 1000 | 125 | 1000 | >1000 | >1000 |
| *C. glabrata* ATCC 90030 | >1000 | >1000 | 250 | 1000 | 1000 | >1000 |

* MIC—minimum inhibitory concentration in milligrams per liter.

Considering the promising anticancer potential of the tested complexes, it is notable that no antimicrobial activity is a beneficial property of these complexes as they do not cause harm to human microbiota during treatment. The gut microbiota plays a significant role in maintaining normal gut physiology and body health. It includes protection from pathogens by colonizing mucosal surfaces, producing different antimicrobial substances and enhancing the immune system, playing a significant role in digestion and metabolism, as well as influencing brain–gut communication and thus impacting the mental and neurological functions of the host [84].

### 2.5. Determination of Free Radical Scavenging Capacity

The disturbed redox balance between reactive oxygen species (ROS) and the antioxidant system is a critical factor in cancer development. One of the strategies for reducing tumors is targeting the redox metabolism by increasing the antioxidant capacity of cancer cells. In this way, antioxidants are being studied to develop more effective anticancer therapy [85]. Furthermore, it has been reported that some oxidants may act as the inducers of DNA damage response, which may result in cell death [86,87].

The free radical scavenging abilities of the free ligands, **L1**, **L2**, **L3** and **L4**, and the corresponding copper(II) complexes, **C1**, **C2**, **C3** and **C4**, were analyzed with two colorimetric methods, DPPH (2,2′-diphenyl-1-picrylhydrazyl) and ABTS (2,2′-azinobis(3-ethylbenzothiazoline-6-sulphonic acid) assays, which were previously used to study the antioxidant activities of the Cu(II) complexes [88]. The results are expressed as $IC_{50}$—the sufficient concentration to obtain 50% of the maximum scavenging activity—and they are shown in Table 5. Ascorbic acid, a well-known antioxidant, was used as a positive control.

**Table 5.** Antiradical activity (DPPH and ABTS) of ligands **L1**, **L2**, **L3** and **L4** and their copper(II) complexes **C1**, **C2**, **C3** and **C4** expressed as $IC_{50}$ (µg/mL) with standard deviation (±SD). Ascorbic acid was used as a positive control.

| Ligands/ Complexes | DDPH | ABTS |
|---|---|---|
| **L1** | NR * | 183.21 ± 2.45 |
| **C1** | 37.45 ± 0.66 | 112.67 ± 1.8 |
| **L2** | NR * | 82.08 ± 2.77 |
| **C2** | 401.52 ± 2.48 | 107.14 ± 1.42 |
| **L3** | NR * | 96.67 ± 2.84 |
| **C3** | 380.65 ± 2.74 | 106.19 ± 2.55 |
| **L4** | NR * | 108.59 ± 1.51 |
| **C4** | 26.46 ± 1.04 | 72.5 ± 0.97 |
| **Ascorbic acid** | 11.65 ± 0.54 | 20.15 ± 0.33 |

* NR—the $IC_{50}$ value was not reached.

From our results, copper(II) complexes **C1–4** exhibited moderate to good DPPH scavenging ability compared with the ascorbic acid ($IC_{50}$ = 26.46–401.52 µg/mL vs. $IC_{50}$ = 11.65 µg/mL), while ligands **L1–4** showed no activity in the DPPH assay; they did not reach the $IC_{50}$ value despite the increasing concentration until 2 mg/mL (higher concentrations resulted in precipitation). The highest DPPH antiradical activity was found for Cu(II) complex **C4** containing 3-ethylimidazolidin-2-one moiety (R = $C_2H_5$, $IC_{50}$ = 26.46 µg/mL). The Cu(II) complex **C1** bearing an azetidin-2-one ring system was observed to display slightly weaker activity ($IC_{50}$ = 37.45 µg/mL). In turn, the analogue of **C4** with 3-methylimidazolidin-2-one functionality (complex **C3**, R = $CH_3$) showed a significant decrease in potency ($IC_{50}$ = 380.65 µg/mL). A further decrease in the radical scavenging capability was observed for complex **C2** ($IC_{50}$ = 401.52 µg/mL) with imidazolidin-2-one scaffold (R = H).

On the other hand, in the ABTS assay, all tested compounds displayed antioxidant capacity with $IC_{50}$ values in the range from 72.5 µg/mL to 183.21 µg/mL (Table 5). The free ligands **L2** and **L3** demonstrated slightly higher ABTS quenching ability when compared to their complexes **C2** and **C3** ($IC_{50}$ = 82.08 and 96.76 µg/mL vs. $IC_{50}$ = 107.14 and 106.19 µg/mL, respectively), while ligands **L1** and **L4** proved to be less potent than the corresponding complexes **C1** and **C4** ($IC_{50}$ = 183.21 and 108.59 µg/mL vs. $IC_{50}$ = 112.67 and 72.7 µg/mL, respectively). As in the DPPH assay, the most promising antiradical properties for ABST radical scavenging ability were presented by complex **C4** ($IC_{50}$ = 72.7 µg/mL).

It could be concluded that in the DPPH assay, the coordination of ligands to the copper(II) metal center appears to be beneficial for antiradical potency as was previously reported [88–90]. In general, no similar correlation was found for complexes **C1–4** compared with their ligands **L1–4** in the ABTS analysis. Nevertheless, it should be pointed out that the highest scavenging activity on both the DPPH and ABTS radicals was exhibited by Cu(II) complex **C4**. This observation suggests that the presence of the electron-donating ethyl group at the N-3 position of the imidazolidin-2-one moiety (R = $C_2H_5$) facilitates antioxidant activity in the complex **C4** by increasing the electron density at the central ion, leading to improved radical scavenging ability in the molecule [91]. However, it was not possible to derive a correlation between antioxidant and antiproliferative activities with the only exception of Cu(II) complex **C4**, which demonstrated remarkable activity on the cancer cell lines, especially HepG2, LS180 and T98G ($IC_{50}$ = 5.92–9.27 µg/mL), and the strongest antioxidant properties within the tested group ($IC_{50}$ = 26.46 µg/mL in DDPH and 72.7 µg/mL in ABTS). On the contrary, the most potent complex against cancer cells, copper(II) complex **C2** ($IC_{50}$ = 5.04–6.97 µg/mL), exhibited less antiradical potency ($IC_{50}$ = 401.52 µg/mL in DDPH, and 107.13 µg/mL in ABTS).

It is worth noting that due to the redox activity of the copper(II) complexes, some of the previously reported copper(II) compounds combine both antioxidant and pro-oxidant modes of action, inducing apoptosis in tumor cells [87]. However, regarding the results of our studies, further work will be needed to clarify this.

### 2.6. In Silico Physicochemical, Pharmacokinetic and Drug-Likeness Predictions

The basic features of a drug molecule that determine whether it can be absorbed and transported inside the body include its solubility, lipophilicity, charge and size. Lipinski's rules dictate that undissociated substances with molecular weights below 500 Da, and a lipophilicity level in the range of 1–3 will have the best absorption rate.

The estimation of drug likeliness and the prediction of the physicochemical and pharmacokinetic properties—ADME (absorption, distribution, metabolism and excretion)—of copper(II) complexes **C1**, **C2**, **C3** and **C4** were carried out using the free available web tool SwissADME [92]. The prediction of the principal properties of the molecules was carried out by using Lipinski's filter, which confirmed the drug likeness of the synthesized copper(II) complexes. The results of the calculated basic parameters of Cu(II) complexes **C1**, **C2**, **C3** and **C4** are presented in Table 6 and Figures 13 and 14 (for more details, see Table S4 in the Supplementary Materials).

**Table 6.** Predicted physicochemical, pharmacokinetic and drug-likeness properties of copper(II) complexes **C1**, **C2**, **C3** and **C4**.

| | Physicochemical Properties | | | | | Lipophilicity | Water Solubility | Pharmacokinetics | | Drug Likeness | |
|---|---|---|---|---|---|---|---|---|---|---|---|
| | mol. wt. (g/mol) | ROTB (n) | HBA (n) | HBD (n) | TPSA | CLogP o/w | Solubility Class | GI Absorption | BBB Permeant | Lipinski Filter | BS |
| Rule | <500 | <10 | <10 | <5 | - | <5 | - | - | - | - | - |
| C1 | 530.89 | 2 | 4 | 0 | 66.40 | 3.04 | Soluble(p) | High | Yes | Yes(1) | 0.55 |
| C2 | 560.92 | 2 | 4 | 2 | 90.46 | 2.48 | Soluble(p) | High | No | Yes(1) | 0.55 |
| C3 | 361.71 | 1 | 2 | 0 | 36.44 | 1.87 | Soluble(m) | High | Yes | Yes(0) | 0.55 |
| C4 | 375.74 | 2 | 2 | 0 | 36.44 | 2.14 | Soluble(m) | High | Yes | Yes(0) | 0.55 |

mol. wt.—molecular weight; n—number; ROTB—stable bonds; HBA—hydrogen bond acceptors; HBD—hydrogen bond donors; TPSA—topological polar surface area calculated in Å$^2$; CLogP o/w—consensus logarithm of partition coefficient between *n*-octanol and water; m—moderate; p—poor; Lipinski filter with number of violations in bracket; GI—gastrointestinal absorption; BBB—blood–brain barrier; BA—bioavailability score.

The topological polar surface area (TPSA) of a molecule can be defined as the sum of the polar atoms, namely oxygen and nitrogen, as well as their hydrogen atoms attached. A heightened TPSA rate (>140 Å$^2$) may be attributed to poor membrane permeability and blood–brain barrier accessibility. Thus, it can be said that a TPSA is a metric for describing the ability of compounds to permeate living cells [93].

As can be seen from the data in Table 6, copper(II) complexes **C1–4** are characterized by reasonable polarity. Their TPSA values are in the range of 33.20–66.40 Å$^2$, except for compound **C4**, which has an estimated value equaling 90.46 Å$^2$. All complexes possess a suitable lipophilicity estimated as a partition coefficient between *n*-octanol and water, with consensus logP (ClogP) values ranging from 1.87 to 3.04. Moreover, three copper(II) complexes, **C1**, **C3** and **C4**, are expected to be moderately soluble in water.

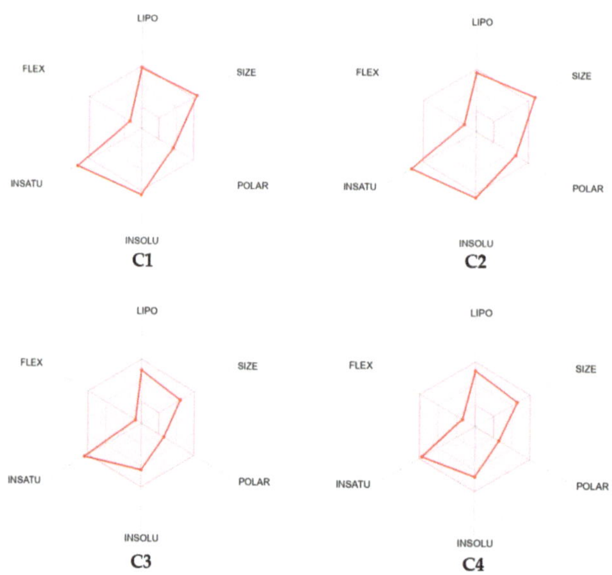


**Figure 13.** Oral bioavailability radar charts for the studied compounds **C1**, **C2**, **C3** and **C4**. In bioavailability radar, the pink area represents the optimal range for each physicochemical property of oral bioavailability, while the red lines represent compounds **C1**, **C2**, **C3** and **C4** (LIPO—lipophilicity SIZE—size; POLAR—polarity; INSOLU—solubility; INSATU—saturation; FLEX—flexibility).

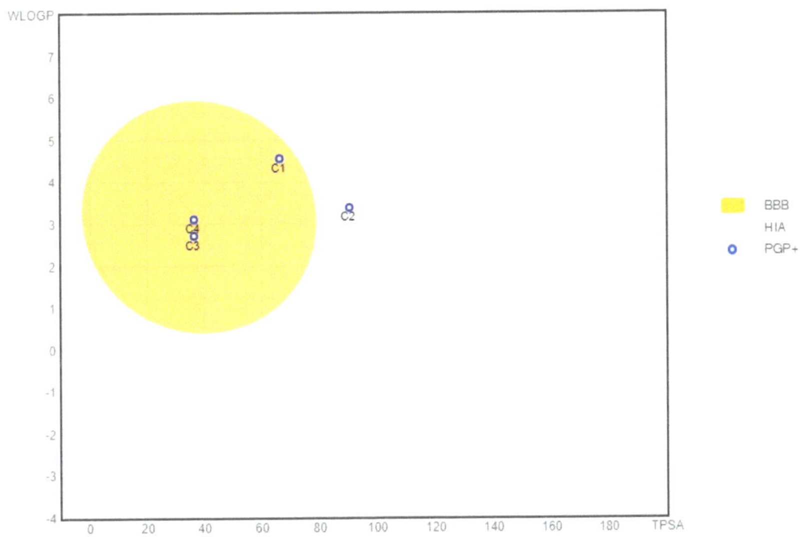


**Figure 14.** BOILED−egg plot for the studied copper(II) complexes **C1**, **C2**, **C3** and **C4**.

Lipinski's rule of five indicates that the lead compound should not contain more than 5 hydrogen bond donors (HBD), while hydrogen bond acceptors (HBAs) should not exceed 10. The calculated copper(II) complexes **C1–4** exhibited two or four HBAs, and the aforementioned standard was congregated. According to the Veber's rule (rotatable bonds must be equivalent to 10 and PSA must be lower than 140 Å), the complexes are also expected to possess high oral bioavailability.

Moreover, according to Table 6, the bioavailable radar charts in Figure 13 and the "BOILED-egg" plot in Figure 14, the investigated copper(II) complexes **C1**, **C2**, **C3** and **C4** are predicted to possess high gastrointestinal tract (GI) absorption and blood–brain barrier BBB permeant. In this regard, all of the tested molecules show the same bioavailability score of 0.55, which suggests desirable pharmacokinetic properties.

In light of Lipinski's "rule of five", copper(II) complexes **C1** and **C2** slightly exceed the molecular weight and violate this criterion, while copper(II) complexes **C3** and **C4** meet all of the criteria as one of the key drug-likeness characteristics. Furthermore, according to Table 6, the ADME properties of copper(II) complexes are favorable and indicate that designed compounds may be considered drug-likeness molecules.

## 3. Materials and Methods

### 3.1. Chemistry

#### 3.1.1. General Information

Melting points were determined with a Boëtius apparatus and are uncorrected. The infrared spectra were obtained on KBr pellets using a Nicolet 380 FT-IR 1600 spectrophotometer (Thermo Fisher Scientific Inc., Waltham, MA, USA). The elemental analyses for C, H and N were within 0.4% of the theoretical values. Thin layer chromatography (TLC) of ligands **L1**, **L2**, **L3** and **L4** was performed on silica gel precoated 60 F254 Merck plates (Merck KGaA, Darmstadt, Germany) using the following eluents: chloroform and ethyl acetate (8:2, *v*/*v*), chloroform and methanol (99:1, *v*/*v*) or chloroform, ethyl acetate and methanol (8:1.5:0.5, *v*/*v*/*v*). The developed chromatograms were viewed under UV light at 254 nm.

The mass spectrum of ligand **L4** was recorded on a Shimadzu LCMS-2010 EV (Tokyo, Japan) spectrometer equipped with an electrospray source, and the ESI-MS spectrum was registered in positive ion mode.

The difference spectra of copper complexes **C1**, **C2**, **C3** and **C4** from λ = 250 nm to 600 nm were recorded at 37 °C with a UV-VIS spectral photometer, Specord S600 (Analytik Jena, Jena, Germany), over the course of 3 h, with spectra being recorded every 10 min. Copper(II) complexes were dissolved in dimethylformamide (DMF) (Sigma-Aldrich Chemie GmbH, Steinheim, Germany). Freshly prepared solution of copper(II) complex at concentration 4 mM (25 µL) was added to 2.475 mL of 100 mM phosphate-buffered solution (pH 7.4), giving the complex a final concentration of 40 µM. Baseline correction was carried out by subtracting the mean absorption between λ = 250 and 600 nm from each spectrum.

The ligands **L1** and **L3**, and the new ligand **L4**, were prepared as previously reported [64]. Ligand **L2** was synthesized using a modified method [69].

#### 3.1.2. Synthesis of 1-(Isoquinolin-3-yl)imidazolidin-2-one (**L2**)

Isoquinoline 2-oxide (0.726 g, 0.005 mol) was added in one portion to freshly prepared solution of 2-chloro-4,5-dihydro-1*H*-imidazole (2.5 g, 0.025 mol) in chloroform (30 mL) at room temperature (20–22 °C) [94]. The stirring was continued until the exothermic reaction subsided (30–60 min.), and then the mixture was cooled. The precipitated crude product was isolated by suction, washed with chloroform (2 × 5 mL) and made alkaline with aqueous 20% potassium carbonate. Then, the oily residue was extracted with chloroform (3 × 20 mL). The collected organic layers were dried with anhydrous magnesium sulfate, filtered and concentrated to dryness. Upon addition of acetone to the oily residue, the product **L2** was precipitated, collected by filtration and washed with acetone. Ligand **L2** was purified on silica gel using preparative thin-layer chromatography (chromatotron); eluent: chloroform/ethyl acetate (4:1, *v*/*v*), yield: 43%, m.p. 234–236 °C (m.p. 233–236 °C [64]); IR (KBr) $v$ ($cm^{-1}$): 3201, 3111, 3053, 2996, 2965, 2902, 1717, 1625, 1583, 1488, 1454, 1415, 1365, 1268, 877, 750. Calculated for $C_{12}H_{11}N_3O$ (213.24): C, 67.59; H, 5.20; N, 19.71. Found: C, 67.78; H, 5.08; N, 19.96.

#### 3.1.3. Synthesis of 1-Ethyl-3-(isoquinolin-3-yl)imidazolidin-2-one (**L4**)

To a stirred suspension of ligand **L2** (0.213 g, 1 mmol) in 1–2 mL of anhydrous DMF, solid NaOH (0.1 g, 2.5 mmol) and ethyl iodide (0.9358 g, 0.4823 mL, 6 mmol) were added. After 48 h, a mixture was dissolved with chloroform (15 mL) and evaporated to dryness. Compound **L4** was separated using preparative thin-layer chromatography (chromatotron); eluent: dichloromethane/ethyl acetate/methanol, (8:1.5:0.5, *v*/*v*/*v*); yield: 62%; m.p. 151–153 °C; IR (KBr) $v$ ($cm^{-1}$): 3114, 3059, 2970, 2922, 2890, 1688, 1626, 1582, 1488, 1426, 1389, 1360, 1270, 878; $^1$H-NMR (DMSO-$d_6$, 400 MHz) $\delta$ (ppm): 1.12 (t, 3H, $CH_3$), 3.30 (q, 2H, $CH_2$), 3.47–3.51 (m, 2H, $CH_2$), 4.04–4.08 (m, 2H, $CH_2$), 7.47 (t, 1H, $CH^7$-isoquin.), 7.66 (t, 1H, $CH^6$-isoquin.), 7.83 (d, $J$ = 8.3 Hz, 1H, $CH^5$-isoquin.), 8.00 (dd, $J_1$ = 1.2 Hz, $J_2$ = 8.3 Hz, 1H, $CH^8$-isoquin.), 8.48 (s, 1H, $CH^4$-isoquin.), 9.12 (s, 1H, $CH^1$-isoquin.); $^{13}$C-NMR (DMSO-$d_6$, 100 MHz) $\delta$ (ppm): 12.81, 38.40, 40.97, 41.77, 105.36, 125.19, 125.39, 126.50, 127.96, 131.11, 137.53, 148.63, 151.11, 157.00; MS (ESI, $CH_3OH$:$CH_3CN$+0,1% $CH_3COOH$, 1:1, *v*/*v*) $m/z$ = 242 $[M+H]^+$, $m/z$ = 264 $[M + Na]^+$, and $m/z$ = 305 $[M + Na + CH_3CN]^+$. Calculated for $C_{14}H_{15}N_3O$ (241.29): C, 69.69; H, 6.27; N, 17.41. Found: C, 69.51; H, 6.02; N, 17.72.

#### 3.1.4. Synthesis of Copper(II) Complexes of 1-(Isoquinolin-3-yl)heteroalkyl-2-ones **C1–4** (General Method)

To a solution of appropriate ligand (**L1–4**) in 99.5% DMF (2–3 mL) at a temperature of 60–70 °C, copper(II) chloride dihydrate ($CuCl_2 \cdot 2H_2O$) dissolved in 1 mL of DMF was added dropwise in a molar ratio of 1:1 (**C3**, and **C4**) or 2:1 (**C1**, **C2**). The mixture was then left at ambient temperature (20–22 °C) for slow evaporation of the solvent (3–12 days). The green precipitate of the copper(II) complex was filtered off, washed with a small amount of 99.5% dimethylformamide and dried. Using the above procedure, the following copper(II) complexes were obtained:

Dichloro{bis[1-(Isoquinolin-3-yl)azetidin-2-one]}copper(II) (**C1**)

For the reaction, 0.132 g (0.664 mmol) of 1-(isoquinolin-3-yl)azetidin-2-one (**L1**) and 0.057 g (0.332 mol) copper(II) chloride dihydrate were used. After 12 days, 0.048 g of copper(II) complex **C1** was obtained, yield 27%, m.p. 188–190 °C; **IR** (KBr) $v$ (cm$^{-1}$): 3110, 3060, 3031, 2959, 2894, 1751, 1630, 1594, 1472, 1387, 1287, 1247, 1154, 1087, 1027, 968, 883, 775, 464. Calculated for $C_{24}H_{20}Cl_2CuN_4O_2$ (530.89): C, 54.30; H, 3.80; N, 10.55. Found: C, 54.05; H, 3.97; N, 10.40.

Dichloro{bis[1-(Isoquinolin-3-yl)imidazolidin-2-one]}copper(II) (**C2**)

For the reaction, 0.15 g (0.702 mmol) of 1-(isoquinolin-3-yl)imidazolidin-2-one (**L2**) and 0.06 g (0.351 mol) copper(II) chloride dihydrate were used. After 10 days, 0.052 g of copper(II) complex **C2** was obtained, yield 33%, m.p. > 350 °C (decomp.); **IR** (KBr) $v$ (cm$^{-1}$): 3106, 3022, 2827, 1670, 1633, 1596, 1464, 1434, 1368, 1285, 1149, 757, 657, 564, 476. Calculated for $C_{24}H_{22}Cl_2CuN_6O_2$ (560.92): C, 51.39; H, 3.95; N, 14.98. Found: C, 51.12; H, 3.84; N, 14.71. Found: C, 52.25; H, 4.05; N, 14.65.

Dichloro[1-(Isoquinolin-3-yl)-3-methylimidazolidin-2-one]copper(II) (**C3**)

For the reaction, 0.09 g (0.396 mmol) of 1-(isoquinolin-3-yl)-3-methylimidazolidin-2-one (**L3**) and 0.068 g (0.396 mol) copper(II) chloride dihydrate were used. After 3 days, 0.045 g of copper(II) complex **C4** was obtained, yield 61%, m.p. 315–316 °C; **IR** (KBr) $v$ (cm$^{-1}$): 3464, 3389, 3072, 2947, 2891, 1655, 1635, 1596, 1519, 1465, 1410, 1396, 1294, 1286, 757, 466. Calculated for $C_{13}H_{13}Cl_2CuN_3O$ (361.71): C, 43.17; H, 3.62; N, 11.62. Found: C, 43.11; H, 3.57; N, 11.23.

Dichloro[1-ethyl-3-(isoquinolin-3-yl)imidazolidin-2-one]copper(II) (**C4**)

For the reaction, 0.09 g (0.373 mmol) of 1-ethyl-3-(isoquinolin-3-yl)imidazolidin-2-one (**L4**) and 0.064 g (0.373 mol) copper(II) chloride dihydrate were used. After 5 days, 0.1 g of copper(II) complex **C5** was obtained, yield 68%, m.p. 270–274 °C; **IR** (KBr) $v$ (cm$^{-1}$): 3043, 2967, 2923, 2852, 1639, 1518, 1458, 1289, 732, 460. Calculated for $C_{14}H_{15}Cl_2CuN_3O$ (375.74): C, 44.75; H, 4.02; N, 11.18. Found: C, 44.64; H, 4.22; N, 11.34. Found: C, 43.11; H, 3.57; N, 11.23.

3.1.5. Single Crystal X-ray Diffraction Studies

Single crystals of copper(II) complex **C1** suitable for X-ray diffraction were obtained by slow solvent evaporation at room temperature from DMF. Diffraction measurements were performed using an XtaLAB Synergy, Dualflex diffractometer (CrysAlisPro (Rigaku Oxford Diffraction, Tokyo, Japan, 2023) with a Pilatus 300 K detector at low temperature (100.0(2) K) using MoK$\alpha$ radiation for complex **C1**. Diffraction data were processed using CrysAlisPRO (version 1.171.42.86a) software (Rigaku Oxford Diffraction, 2023). Solving and refinement of the crystal structure were performed with SHELX (version 2018/2) [95] and SHELXL (version 2019/3) [96] using full-matrix least-squares minimization on $F^2$. All H atoms were optimized using constraints with riding model and distances suitable for 100 K temperature and with $U_{iso}(H) = 1.2\ U_{eq}(C)$. ShelXle (version 1143) software [97] was used to visualize the molecular structure. Graphical representation of the crystal structures was performed using the Mercury program (version 2022.3.0) [98]. OLEX2 program [99] was used in data preparation. The tables were prepared using PublCIF software (version 1.9.21_c) [100].

## 3.2. *Aqueous Stability Studies*

The stability of copper(II) complexes **C1**, **C2**, **C3** and **C4** was determined in 100 mM phosphate-buffered solution (pH 7.4) at 37 °C by using a Specord S600 (Analytic Jena, Jena, Germany) UV-vis diode-array photometer. Difference spectra were recorded every 10 min to help identify very small changes in the UV-vis spectra over 3 h incubations.

### 3.3. Biological Studies

#### 3.3.1. Examination of the Cytotoxic Activity with Assessment by MTT Assay

Cytotoxic activity of the investigated compounds was evaluated against T98G (glioblastoma), HepG2 (hepatoma), LS-180 (colon cancer) and A375 (melanoma) cell lines. Human normal skin fibroblasts, CCD-1059Sk (CRL-2072), were used as reference (non-cancerous) cells. All of the cell lines were obtained from American Type Culture Collection (ATCC; Manassas, VA, USA). A375, T98G and CCD-1059Sk cells were cultured in Dulbecco's Modified Eagle's Medium (DMEM) (Sigma Aldrich, St. Louis, MO, USA) supplemented with 10% FBS, penicillin (100 U/mL) and streptomycin (100 μg/mL). LS-180 and HepG2 cells were cultured in Eagle's Minimum Essential Medium supplemented with 10% FBS, penicillin (100 U/mL) and streptomycin (100 μg/mL). All of the cells were maintained at 37 °C in a humidified atmosphere of 5% $CO_2$. The compounds were dissolved in DMSO in order to obtain stock solutions. At the day of the experiment, the suspension of cells ($1 \times 10^5$ cells/mL) was distributed onto 96-well plates at a volume of 100 μL/well. After attachment, the cells were treated with the increased concentrations of the tested compounds in medium containing 2% FBS and incubated for 24 h. Then, the medium was removed from wells, and cells were rinsed with PBS. Afterwards, 15 μL of MTT working solution (5 mg/mL in PBS) was added to each well, and the plates were incubated for 3 h. Subsequently, 100 μL of 10% SDS solution was added to each well in order to dissolve the precipitated formazan crystals. After overnight incubation at 37 °C, the absorbance of the obtained solution was measured at $\lambda$ = 570 nm using a microplate reader (Epoch, BioTek Instruments, Inc., Winooski, VT, USA). At least two independent experiments were performed in triplicate. DMSO in the concentrations present in the dilutions of stock solutions did not influence the viability of the tested cells. $IC_{50}$ values of the investigated derivatives and positive control (*etoposide*) were calculated using the $IC_{50}$ online calculator [101].

#### 3.3.2. Cell Cycle Analysis

Cell cycle analysis of T98G and HepG2 cells pretreated with copper(II) complex **C2 (CX)** (at $IC_{50}$ concentration) was performed on NucleoCounter NC-3000 Image Cytometer (ChemoMetec, Lillerod, Denmark). The investigated cells were seeded on 6-well culture plates (Corning Inc., New York, NY, USA) at the density of $1 \times 10^5$ cells/mL and cultured in the respective medium at 37 °C in a humidified atmosphere of 5% $CO_2$. When the cells reached approximately 80% confluence, they were treated with **C2 (CX)**, at a concentration of $IC_{50}$, for 24 h. Subsequently, the cells were detached with trypsine, suspended in 250 μL of lysis buffer (Solution 10) supplemented with DAPI (10 μg/mL) and incubated for 5 min at 37 °C. Following this, stabilization buffer (Solution 11) was added, and the obtained cell suspension was applied on NC-slide and analyzed using NucleoCounter NC-3000 Image Cytometer equipped with NucleoView NC-3000™ software (ChemoMetec A/S, Lillerod, Denmark). Experiments were repeated three times, and the measurements in each experiment were run in duplicate.

#### 3.3.3. Analysis of Interaction of Copper(II) Complex **C2 (CX)** with Clinically Used Anticancer Agents

Possible interactions between copper(II) complex **C2 (CX)** and anticancer drugs including *etoposide* (**ETO**), *cisplatin* (**CIS**), *5-fluorouracil* (**5-FU**) and *temozolomide* (**TMZ**) were examined on the most sensitive cancer cell lines, i.e., T98G and HepG2. Firstly, the highest concentrations of **C2 (CX)** that did not affect the viability of T98G and HepG2 cells, as well as the $IC_{50}$ values for **ETO, CIS, 5-FU** and **TMZ,** were established using MTT assay (as described above). Chemotherapeutics whose $IC_{50}$ values were higher than 100 μg/mL were tested at the maximal concentration of 100 μg/mL. Next, T98G or HepG2 cells were incubated for 24 h with medium containing a mixture of the respective drugs mixed with the highest non-toxic concentration of **C2 (CX)**. The viability of cells was evaluated using MTT assay, as described above. ANOVA analysis (with Tukey's post hoc test) was performed in order to examine the possible interactions between compound **C2** (**CX**) and **ETO**, **CIS**, **5-FU** and **TMZ**.

#### 3.3.4. In Vitro Antimicrobial Activity

Antibacterial and antifungal activities of the free ligands **L2**, **L3** and **L4** and the corresponding copper(II) complexes **C2**, **C3** and **C4** were screened using the two-fold microdilution broth method. Minimal inhibitory concentrations (MICs) of tested compounds for the panel of reference Gram-positive bacteria, including *Staphylococcus aureus* ATCC 25923, *Staphylococcus aureus* ATCC BAA-1707, *Staphylococcus epidermidis* ATCC 12228, *Micrococcus luteus* ATCC 10240 and *Bacillus cereus* ATCC 10876; Gram-negative bacteria, including *Salmonella Typhimurium* ATCC 14028, *Escherichia coli* ATCC 25922, *Proteus mirabilis* ATCC 12453, *Klebsiella pneumoniae* ATCC 13883 and *Pseudomonas aeruginosa* ATCC 9027; and yeasts, including *Candida albicans* ATCC 102231, *Candida parapsilosis* ATCC 22019 and *Candida glabrata* ATCC 90030 were determined. The procedure has been described in detail before [102]. Briefly, the solutions of tested compounds dissolved in dimethylosulfoxide (DMSO) were suspended in Mueller–Hinton broth for bacteria or Mueller–Hinton broth with 2% glucose for fungi. Then, the series of two-fold dilutions were carried out in the sterile Nunc™ MicroWell™ 96-Well Microplates (ThermoFisher Scientific Inc.), obtaining concentrations from 1000 to 7.8 mg/L in the appropriate medium. Simultaneously, the inocula of 24 h cultures of microorganisms in sterile physiological saline (0.5 McFarland standard density) were prepared and added to each well, obtaining final density of $5 \times 10^5$ CFU/mL for bacteria and $5 \times 10^4$ CFU/mL for yeasts. Proper positive (inoculum without tested compound) and negative (compound without inoculum) controls were added in each microplate. After incubation (35 °C, 24 h), the growth of microorganisms was measured spectrophotometrically at 600 nm (BioTEK ELx808, Bio-Tek Instruments, Inc., Winooski, VT, USA). MICs were marked at the lowest concentration of the compound without the growth of bacteria or fungi.

### 3.4. *Antioxidant*

#### 3.4.1. Materials

Ascorbic acid, oleanolic acid, DPPH (2,2-diphenyl-1-picrylhydrazyl), ABTS (2,2′-azino-bis(3-ethylbenzothiazoline-6-sulfonic acid) diammonium salt, potassium persulfate and DMSO (dimethyl sulfoxide) were sourced from Sigma Chemical Co. (St. Louis, MO, USA). TRIS HCl (0.2 M, pH 8) and HPLC-grade methanol were sourced from P.O.Ch. (Gliwice, Poland).

#### 3.4.2. DPPH Assay

The DPPH radical scavenging assay of samples was performed with ascorbic acid as a positive control [88]. Briefly, 100 μL of different concentrations of the complexes, dissolved in DMSO, was mixed with 100 μL of 0.06 mM DPPH methanolic solution and incubated at room temperature in the dark for 30 min. The change in absorbance at $\lambda = 517$ nm was analyzed with the use of a 96-well microplate reader (Epoch, BioTek System, Winooski, VT, USA). The control was composed of DPPH and DMSO.

DPPH inhibition was calculated according to the following equation:

$$\text{DPPH Inhibition (\%)} = [(\text{Acontrol} - \text{Asample})/\text{Acontrol}] \times 100\%$$

The radical scavenging activity of the samples was shown as the $IC_{50}$ value (the concentration of the analyzed samples that caused a decrease in the non-reduced form of the DPPH radical by 50%).

#### 3.4.3. ABTS Assay

The ABTS radical scavenging assay of samples (ligands or complexes) was performed with ascorbic acid as a positive control [88]. Briefly, 30 μL of different concentrations of the samples, dissolved in DMSO, was mixed with 170 μL of ABTS solution (2 mM ABTS diammonium salt, 3.5 mM potassium persulfate) and completed with water to a final volume of 300 μL. ABTS solution with DMSO was used as a control. After 10 min of incubation at 30 °C in the dark, a change in absorbance was observed at $\lambda = 750$ nm by a 96-well microplate reader (Epoch, BioTek System, Santa Clara, CA, USA).

ABTS inhibition was calculated according to the following equation:

$$\text{ABTS Inhibition (\%)} = [(\text{Acontrol} - \text{Asample})/\text{Acontrol}] \times 100\%$$

The radical scavenging activity of the samples was shown as the $IC_{50}$ value (the concentration of the analyzed samples that caused a decrease in the non-reduced form of the ABTS radical by 50%).

### *3.5. ADME/Drug-Likeness Calculation*

The physicochemical, pharmacokinetic and drug-likeness properties of copper(II) complexes **C1–4** were predicted using the SwissADME web tool, which is available online [103].

## 4. Conclusions

Four new copper(II) complexes, **C1–4**, derived from 1-(isoquinolin-3-yl)heteroalkyl-2-one ligands, **L1–4**, were prepared, and their coordination modes were established using an elemental analysis, infrared spectra as well as X-ray crystallographic study for **C1**. The structures of copper(II) complexes **C1** and **C2** were determined as mononuclear species incorporating two molecules of the neutral 1-(isoquinolin-3-yl)heteroalkyl-2-one ligand bound to the central copper ion via a bidentate manner. Interestingly, ligands **L3** and **L4** containing an alkyl group at the N-3 position of the imidazolidin-2-one moiety form tetracoordinate mononuclear copper(II) complexes **C3** and **C4** with the central atom chelated by one molecule of the neutral bidentate ligand.

The results of UV-vis spectrophotometry confirmed the stability of complexes **C1–4** under conditions that mimic the physiological environment.

The in vitro cytotoxicity studies showed that the coordination of 1-(isoquinolin-3-yl)heteroalkyl-2-one ligands **L1–4** with a copper(II) ion results in metal complexes **C1–4** with remarkable growth inhibitory properties against tested human cancer cell lines, especially hepatoma HepG2, colon cancer LS-180 and glioblastoma T98G cells. The cytotoxic effect of the synthesized copper(II) complexes towards HepG2, LS-180 and T98G cancer cells was higher than the known antitumor agent *etoposide*. Among these compounds, dichloro{bis[1-(isoquinolin-3-yl)imidazolidin-2-one]}copper(II) (**C2**) was found to be the most promising agent with the greatest selective effect on HepG2, LS180 and T98G cancer cell lines compared with the non-cancerous CCD-1059-Sk cell line. The complex **C2** induced sub-G1 cell cycle arrest in the HepG2 cells and induced G1/M cell cycle arrest in the T98G cells, which was accompanied by DNA degradation. Furthermore, the treatment of HepG2 and T98G cells with the tested copper(II) compound **C2** at the concentration that did not inhibit the growth of cancer cells resulted in a significant increase in the cytotoxic effects of chemotherapeutics such as *etoposide*, *5-fluorouracil* and *temozolomide*. To clarify the mechanism of synergism between **C2** and the clinically used drugs, more advanced studies are needed.

In turn, in microbiological tests, the investigated 1-(isoquinolin-3-yl)heteroalkyl-2-one ligands **L2–4** and their copper(II) complexes **C2–4** were inactive, except for a 1-(isoquinolin-3-yl)-3-methylimidazolidin-2-one ligand (**L3**), which exhibited moderate anti-*Candida* activity.

The antioxidant activity results suggest that in the DPPH test, the coordination of ligands **L1–4** with a copper(II) metal center is beneficial for antiradical potency. No direct correlation was found between antiproliferative and antioxidant effects; the exception to this was dichloro[1-ethyl-3-(isoquinolin-3-yl)imidazolidin-2-one]copper(II) (**C4**), which demonstrated remarkable growth-inhibitory properties against cancer cells and the strongest antioxidant activity in both the DPPH and ABTS assays within the tested compounds. On the other hand, the copper(II) compound **C4**, which had the highest potency against the tested tumor cell lines, exhibited moderate antiradical properties.

Generally, the prediction of ADME/drug-likeness properties revealed that the tested copper(II) complexes may be considered as drug-likeness molecules.

In summary, the results obtained may be useful as a starting point for the development of novel copper-based antitumor agents.

**Supplementary Materials:** The following supporting information can be downloaded at https://www.mdpi.com/article/10.3390/ijms25010008/s1. CCDC 2307086 contains the supplementary crystallographic data for this paper. The data are provided free of charge by the Cambridge Crystallographic Data Centre via https://www.ccdc.cam.ac.uk/structures (accessed on 10 November 2023).

**Author Contributions:** Conceptualization, A.K.; methodology, Ł.B., T.P., I.K.-G., A.H., M.S., P.J.B. and A.K.; software, Ł.B., T.P, I.K.-G., A.H., M.S., A.O., J.K. and A.K.; validation, Ł.B., T.P., I.K.-G., A.H. and M.S.; formal analysis, Ł.B., T.P., I.K.-G., A.H., M.S., A.O. and A.K.; investigation, Ł.B., T.P., I.K.-G., A.H., M.S., P.J.B., J.K. and A.K.; data curation, A.K.; writing—original draft preparation, A.K., Ł.B., T.P., I.K.-G., A.H., M.S. and A.O.; writing—review and editing, A.K., T.P., I.K.-G., A.H., M.S. and P.J.B.; visualization; Ł.B., T.P., I.K.-G., M.S., A.O., A.H., J.K. and A.K.; supervision, A.K., I.K.-G. and M.S.; project administration, A.K.; funding acquisition, A.K. All authors have read and agreed to the published version of the manuscript.

**Funding:** This research was supported by the "MUGs' Experienced Researcher Program" (grant no. 01-0514/08/513). The APC was funded by the Medical University of Gdansk under the "Excellence Initiative—Research University" program and the Statutory Activity of the Medical University of Gdansk (ST 01-50023/0004931/513/513/0/2023). Tomasz Plech wishes to acknowledge the funding support received from the statutory funds of the Medical University of Lublin (DS. 544).

**Institutional Review Board Statement:** Not applicable.

**Informed Consent Statement:** Not applicable.

**Data Availability Statement:** The data are contained within the article and Supplementary Materials.

**Acknowledgments:** Łukasz Balewski would like to thank the Medical University of Gdansk for providing financial support for his research internship (maintenance and accommodation). Łukasz Balewski is grateful to Patrick J. Bednarski for making it possible for him to undergo an internship at the Department of Pharmaceutical and Medicinal Chemistry, Institute of Pharmacy, University of Greifswald, Germany.

**Conflicts of Interest:** The authors declare no conflict of interest.

## References

1. Gaynor, D.; Griffith, D.M. The prevalence of metal-based drugs as therapeutic or diagnostic agents: Beyond platinum. *Dalton Trans.* **2012**, *41*, 13239. [CrossRef] [PubMed]
2. Tesauro, D. Metal complexes in diagnosis and therapy. *Int. J. Mol. Sci.* **2022**, *23*, 4377. [CrossRef] [PubMed]
3. Johnstone, T.C.; Suntharalingam, K. Lippard. The next generation of platinum drugs: Targeted Pt(II) agents, nanoparticle delivery, and Pt(IV) prodrugs. *Chem. Rev.* **2016**, *116*, 3436–3486. [CrossRef] [PubMed]
4. Zhong, T.; Yu, J.; Pan, Y.; Zhang, N.; Qi, Y.; Huang, Y. Recent advances of platinum-based anticancer complexes in combinational multimodal therapy. *Adv. Healthcare Mater.* **2023**, *12*, 2300253. [CrossRef] [PubMed]
5. Oun, R.; Moussa, Y.E.; Wheate, N.J. The side effects of platinum-based chemotherapy drugs: A review for chemists. *Dalton Trans.* **2018**, *47*, 6645–6653. [CrossRef] [PubMed]
6. Lugones, Y.; Loren, P.; Salazar, L.A. Cisplatin resistance: Genetic and epigenetic factors involved. *Biomolecules* **2022**, *12*, 1365. [CrossRef] [PubMed]
7. Zhou, J.; Kang, Y.; Chen, L.; Wang, H.; Liu, J.; Zeng, S.; Yu, L. The drug-resistance mechanisms of five platinum-based antitumor agents. *Front. Pharmacol.* **2020**, *11*, 343. [CrossRef]
8. Waters, J.E.; Stevens-Cullinane, L.; Siebenmann, L.; Hess, J. Recent advances in the development of metal complexes as antibacterial agents with metal-specific modes of action. *Curr. Opin. Microbiol.* **2023**, *75*, 102347. [CrossRef]
9. Sharma, B.; Shukla, S.; Rattan, R.; Fatima, M.; Goel, M.; Bhat, M.; Dutta, S.; Ranjan, R.K.; Sharma, M. Antimicrobial agents based on metal complexes: Present situation and future prospects. *Int. J. Biomater.* **2022**, *2022*, 6819080. [CrossRef]
10. Lucaciu, R.L.; Hangan, A.C.; Sevastre, B.; Oprean, L.S. Metallo-drugs in cancer therapy: Past, present and future. *Molecules* **2022**, *27*, 6485. [CrossRef]
11. Ndagi, U.; Mhlongo, N.; Soliman, M.E. Metal complexes in cancer therapy—An update from drug design perspective. *Drug Des. Devel. Ther.* **2017**, *11*, 599–616. [CrossRef] [PubMed]

12. Xu, X.; Dai, F.; Mao, Y.; Zhang, K.; Qin, Y.; Zheng, J. Metallodrugs in the battle against non-small cell lung cancer: Unlocking the potential for improved therapeutic outcomes. *Front. Pharmacol.* **2023**, *14*, 1242488. [CrossRef] [PubMed]
13. Muhammad, N.; Hanif, M.; Yang, P. Beyond cisplatin: New frontiers in metallodrugs for hard-to-treat triple negative breast cancer. *Coord. Chem. Rev.* **2024**, *499*, 215507. [CrossRef]
14. Gaál, A.; Orgován, G.; Mihucz, V.G.; Pape, I.; Ingerle, D.; Streli, C.; Szoboszlai, N.J. Metal transport capabilities of anticancer copper chelators. *Trace Elem. Med. Biol.* **2018**, *47*, 79–88. [CrossRef] [PubMed]
15. Mahendiran, D.; Amuthakala, S.; Bhuvanesh, N.S.P.; Kumar, R.S.; Rahiman, A.K. Copper complexes as prospective anticancer agents: In vitro and in vivo evaluation, selective targeting of cancer cells by DNA damage and S phase arrest. *RSC Adv.* **2018**, *8* 16973–16990. [CrossRef] [PubMed]
16. Lelièvre, P.; Sancey, L.; Coll, J.-L.; Deniaud, A.; Busser, B. The multifaceted roles of copper in cancer: A trace metal element with dysregulated metabolism, but also a target or a bullet for therapy. *Cancers* **2020**, *12*, 3594. [CrossRef] [PubMed]
17. Krasnovskaya, O.; Naumov, A.; Dmitry Guk, D.; Gorelkin, P.; Erofeev, A.; Beloglazkina, E.; Majouga, A. Copper coordination compounds as biologically active agents. *Int. J. Mol. Sci.* **2020**, *21*, 3965. [CrossRef]
18. Ji, P.; Wang, P.; Chen, H.; Xu, Y.; Ge, J.; Tian, Z.; Yan, Z. Potential of copper and copper compounds for anticancer applications *Pharmaceuticals* **2023**, *16*, 234. [CrossRef]
19. Jiang, Y.; Huo, Z.; Qi, X.; Zuo, T.; Wu, Z. Copper-induced tumor cell death mechanisms and antitumor theragnostic applications of copper complexes. *Nanomedicine* **2022**, *17*, 303–324. [CrossRef]
20. Carcelli, M.; Tegoni, M.; Bartoli, J.; Marzano, C.; Pelosi, G.; Salvalaio, M.; Rogolino, D.; Gandin, V. In vitro and in vivo anticancer activity of tridentate thiosemicarbazone copper complexes: Unravelling an unexplored pharmacological target. *Eur. J. Med. Chem.* **2020**, *194*, 112266. [CrossRef]
21. Khan, R.A.; Usman, M.; Dhivya, R.; Balaji, P.; Alsalme, A.; AlLohedan, H.; Arjmand, F.; AlFarhan, K.; Akbarsha, M.A.; Marchetti, F.; et al. Heteroleptic copper(I) complexes of "scorpionate" bis-pyrazolyl carboxylate ligand with auxiliary phosphine as potential anticancer agents: An insight into cytotoxic mode. *Sci. Rep.* **2017**, *7*, 45229–45246. [CrossRef] [PubMed]
22. Molinaro, C.; Martoriati, A.; Pelinski, L.; Cailliau, K. Cooper complexes as anticancer agents targeting topoisomerases I and II *Cancers* **2020**, *12*, 2863. [CrossRef] [PubMed]
23. Qi, J.; Zheng, Y.; Li, B.; Ai, Y.; Chen, M.; Zheng, X. Pyridoxal hydrochloride thiosemicarbazones with copper ions inhibit cell division via Topo-I and Topo-IIα. *J. Inorg. Biochem.* **2022**, *232*, 111816. [CrossRef] [PubMed]
24. Qin, Q.-P.; Meng, T.; Tan, M.-X.; Liu, Y.-C.; Luo, X.-J.; Zou, B.-Q.; Liang, H. Synthesis, crystal structure and biological evaluation of a new dasatinib copper(II) complex as telomerase inhibitor. *Eur. J. Med. Chem.* **2018**, *143*, 1597–1603. [CrossRef] [PubMed]
25. Xin, C.; Xiao, Z.; Jinghong, C.; Qianqi, Y.; Li, Y. Hinokitiol copper complex inhibits proteasomal deubiquitination and induces paraptosis-like cell death in human cancer cells. *Eur. J. Pharmacol.* **2017**, *815*, 147–155. [CrossRef]
26. Zhang, Z.; Wang, H.; Yan, M.; Wang, H.; Zhang, C. Novel copper complexes as potential proteasome inhibitors for cancer treatment. *Mol. Med. Rep.* **2017**, *15*, 3–11. [CrossRef] [PubMed]
27. Konarikova, K.; Frivaldska, J.; Gbelcova, H.; Sveda, M.; Ruml, T.; Janubova, M.; Zitnanova, I. Schiff base Cu(II) complexes as inhibitors of proteasome in human cancer cells. *Bratisl. Lek Listy* **2019**, *120*, 646–649. [CrossRef] [PubMed]
28. Anu, D.; Naveen, P.; Vijaya, P.B.; Frampton, C.S.; Kaveri, M.V. An unexpected mixed valence tetranuclear copper (I/II) complex: Synthesis, structural characterization, DNA/protein binding, antioxidant and anticancer properties. *Polyhedron* **2019**, *167*, 137–150. [CrossRef]
29. Bollua, V.S.; Bathinib, T.; Baruia, A.K.; Roya, A.; Ragic, N.C.; Maloth, S.; Sripadic, P.; Sreedharb, B.; Nagababub, P.; Patra, C.R. Design of DNA-intercalators based copper(II) complexes, investigation of their potential anti-cancer activity and sub-chronic toxicity. *Mater. Sci. Eng. C* **2019**, *105*, 110079. [CrossRef]
30. Movahedi, E.; Razmazma, H.; Rezvani, A.; Nowroozi, A.; Ebrahimi, A.; Eigner, V.; Dusek, M.; Arjmand, F. A novel Cu(II)-based DNA-intercalating agent: Structural and biological insights using biophysical and *in silico* techniques. *Spectrochim. Acta A Mol. Biomol. Spectrosc.* **2023**, *293*, 122438. [CrossRef]
31. Lesiów, M.K.; Bieńko, A.; Sobańska, K.; Kowalik-Jankowska, T.; Rolka, K.; Łęgowska, A.; Ptaszyńska, N. Cu(II) complexes with peptides from FomA protein containing –His-Xaa-Yaa-Zaa-His and –His-His-motifs. ROS generation and DNA degradation. *J. Inorg. Biochem.* **2020**, *212*, 111250. [CrossRef] [PubMed]
32. Rodríguez, M.R.; Lavecchia, M.J.; Parajón-Costa, B.S.; González-Baró, A.C.; González-Baró, M.R.; Cattáneo, E.R. DNA cleavage mechanism by metal complexes of Cu(II), Zn(II) and VO(IV) with a schiff-base ligand. *Biochemie* **2021**, *186*, 43–50. [CrossRef] [PubMed]
33. Shao, J.; Li, M.; Guo, Z.; Jin, C.; Zhang, F.; Ou, C.; Xie, Y.; Tan, S.; Wang, Z.; Zheng, S.; et al. TPP-related mitochondrial targeting copper(II) complex induces p53-dependent apoptosis in hepatoma cells through ROS-mediated activation of Drp1. *Cell Commun. Signal.* **2019**, *17*, 149. [CrossRef] [PubMed]
34. Done, G.; Ari, F.; Akgun, O.; Akgun, H.; Cevatemre, B.; Gençkal, H.M. The mechanism for anticancer and apoptosis-inducing properties of Cu(II) complex with quercetin and 1,10-phenanthroline. *ChemistrySelect* **2022**, *7*, e202203242. [CrossRef]
35. Martinez-Bulit, P.; Garza-Ortíz, A.; Mijangos, E.; Barrón-Sosa, L.; Sánchez-Bartéz, F.; Gracia-Mora, I.; Flores-Parra, A.; Contreras, R.; Reedijk, J.; Barba-Behrens, N. 2,6-Bis(2,6-diethylphenyliminomethyl)pyridine coordination compounds with cobalt(II), nickel(II), copper(II), and zinc(II): Synthesis, spectroscopic characterization, X-ray study and in vitro cytotoxicity. *J. Inorg. Biochem.* **2015**, *142*, 1–7. [CrossRef] [PubMed]

36. Sîrbu, A.; Palamarciuc, O.; Babak, M.V.; Lim, J.M.; Ohui, K.; Enyedy, E.A.; Shova, S.; Darvasiová, D.; Rapta, P.; Ang, W.H.; et al. Copper(II) thiosemicarbazone complexes induce marked ROS accumulation and promote nrf2-mediated antioxidant response in highly resistant breast cancer cells. *Dalton Trans.* **2017**, *46*, 3833–3847. [CrossRef] [PubMed]
37. Peña, Q.; Lorenzo, J.; Sciortino, G.; Rodríguez-Calado, S.; Maréchal, J.-D.; Bayón, P.; Simaan, A.J.; Iranzo, O.; Capdevila, M.; Palacios, O. Studying the reactivity of "old" Cu(II) complexes for "novel" anticancer purpose. *J. Inorg. Biochem.* **2019**, *195*, 51–60. [CrossRef]
38. Zehra, S.; Tabassum, S.; Arjmand, F. Biochemical pathways of copper complexes: Progress over the past 5 years. *Drug Discov. Today* **2021**, *26*, 1086–1096. [CrossRef]
39. Meraz-Torres, F.; Plöger, S.; Garbe, C.; Niessner, H.; Sinnberg, T. Disulfiram as a therapeutic agent for metastatic malignant melanoma-old myth or new logos? *Cancers* **2020**, *12*, 3538. [CrossRef]
40. Xu, B.; Shi, P.; Fombon, I.S.; Zhang, Y.; Huang, F.; Wang, W.; Zhou, S. Disulfiram/copper complex activated JNK/c-jun pathway and sensitized cytotoxicity of doxorubicin in doxorubicin resistant leukemia HL60 cells. *Blood Cells Mol. Dis.* **2011**, *47*, 264–269. [CrossRef]
41. Climova, A.; Pivovarova, E.; Szczesio, M.; Gobis, K.; Ziembicka, D.; Korga-Plewko, A.; Kubik, J.; Iwan, M.; Antos-Bielska, M.; Krzyżowska, M.; Czylkowska, A. Anticancer and antimicrobial activity of new copper (II) complexes. *J Inorg Biochem.* **2023**, *240*, 112108. [CrossRef] [PubMed]
42. Gul, N.S.; Khan, T.M.; Chen, M.; Huang, K.B.; Hou, C.; Choudhary, M.I.; Liang, H.; Chen, Z.F. New copper complexes inducing bimodal death through apoptosis and autophagy in A549 cancer cells. *J. Inorg. Biochem.* **2020**, *213*, 111260. [CrossRef] [PubMed]
43. Heffeter, P.; Jungwirth, U.; Jakupec, M.; Hartinger, C.; Galanski, M.; Elbling, L.; Micksche, M.; Keppler, B. Resistance against novel anticancer metal compounds: Differences and similarities. *Drug Resist. Updates* **2008**, *11*, 1–16. [CrossRef] [PubMed]
44. Chen, S.-Y.; Chang, Y.-L.; Liu, S.-T.; Chen, G.-S.; Lee, S.-P.; Huang, S.-M. Differential cytotoxicity mechanisms of copper complexed with disulfiram in oral cancer cells. *Int. J. Mol. Sci.* **2021**, *22*, 3711. [CrossRef] [PubMed]
45. da Silva, D.A.; De Luca, A.; Squitti, R.; Rongioletti, M.; Rossi, L.; Machado, C.M.L.; Cerchiaro, G. Copper in tumors and the use of copper-based compounds in cancer treatment. *J. Inorg. Biochem.* **2022**, *226*, 111634. [CrossRef]
46. Hussain, A.; AlAjmi, M.F.; Rehman, M.T.; Amir, S.; Husain, F.M.; Alsalme, A.; Siddiqui, M.A.; AlKhedhairy, A.A.; Khan, R.A. Copper(II) complexes as potential anticancer and nonsteroidal anti-inflammatory agents: In vitro and in vivo studies. *Sci. Rep.* **2019**, *9*, 5237. [CrossRef]
47. Malis, G.; Geromichalou, E.; Geromichalos, G.D.; Hatzidimitriou, A.G.; Psomas, G. Copper(II) complexes with non-steroidal anti-inflammatory drugs: Structural characterization, in vitro and in silico biological profile. *J. Inorg. Biochem.* **2021**, *224*, 111563. [CrossRef]
48. Bellia, F.; Lanza, V.; Naletova, I.; Tomasello, B.; Ciaffaglione, V.; Greco, V.; Sciuto, S.; Amico, P.; Inturri, R.; Vaccaro, S.; et al. Copper(II) complexes with carnosine conjugates of hyaluronic acids at different dipeptide loading percentages behave as multiple SOD mimics and stimulate Nrf2 translocation and antioxidant response in in vitro inflammatory model. *Antioxidants* **2023**, *12*, 1632. [CrossRef]
49. Gordon, N.A.; McGuire, K.L.; Wallentine, S.K.; Mohl, G.A.; Lynch, J.D.; Harrison, R.G.; Busath, D.D. Divalent copper complexes as influenza A M2 inhibitors. *Antivir. Res.* **2017**, *147*, 100–106. [CrossRef]
50. Kowalczyk, M.; Golonko, A.; Świsłocka, R.; Kalinowska, M.; Parcheta, M.; Swiergiel, A.; Lewandowski, W. Drug design strategies for the treatment of viral disease. Plant phenolic compounds and their derivatives. *Front. Pharmacol.* **2021**, *12*, 709104. [CrossRef]
51. Salah, I.; Parkin, I.P.; Allan, E. Copper as an antimicrobial agent: Recent advances. *RSC Adv.* **2021**, *11*, 18179. [CrossRef] [PubMed]
52. Pereira, A.L.; Vasconcelos, M.A.; Andrade, A.L.; Martins, I.M.; Holanda, A.K.M.; Gondim, A.C.S.; Penha, D.P.S.; Bruno, K.L.; Silva, F.O.N.; Teixeira, E.H. Antimicrobial and antibiofilm activity of copper-based metallic compounds against bacteria related with healthcare-associated infections. *Curr. Microbiol.* **2023**, *80*, 133. [CrossRef] [PubMed]
53. Iranshahy, M.; Quinnb, R.J.; Iranshahi, M. Biologically active isoquinoline alkaloids with drug-like properties from the genus Corydalis. *RSC Adv.* **2014**, *4*, 15900–15913. [CrossRef]
54. Yuan, H.L.; Zhao, Y.L.; Qin, X.J.; Liu, Y.P.; Yang, X.W.; Luo, X.D. Diverse isoquinolines with anti-inflammatory and analgesic bioactivities from *Hypecoum erectum*. *J. Ethnopharmacol.* **2021**, *270*, 113811. [CrossRef] [PubMed]
55. Alagumuthu, M.; Sathiyanarayanan, K.I.; Arumugam, S. Molecular docking, design, synthesis, in vitro antioxidant and anti-inflammatory evaluations of new isoquinoline derivatives. *Int. J. Pharm. Pharm. Sci.* **2015**, *7*, 200–208.
56. Zajdel, P.; Marciniec, K.; Maślankiewicz, A.; Grychowska, K.; Satała, G.; Duszyńska, B.; Lenda, T.; Siwek, A.; Nowak, G.; Partyka, A.; et al. Antidepressant and antipsychotic activity of new quinoline- and isoquinoline-sulfonamide analogs of aripiprazole targeting serotonin 5-HT1A/5-HT2A/5-HT7 and dopamine D2/D3 receptors. *Eur. J. Med. Chem.* **2013**, *60*, 42–50. [CrossRef] [PubMed]
57. Luo, C.; Ampomah-Wireko, M.; Wang, H.; Wu, C.; Wang, Q.; Zhang, H.; Cao, Y. Isoquinolines: Important cores in many marketed and clinical drugs. *Anticancer Agents Med. Chem.* **2021**, *21*, 811–824. [CrossRef] [PubMed]
58. Faheem; Kumar, B.K.; Chandra Sekhar, K.V.G.; Chander, S.; Kunjiappan, S.; Murugesan, S. Medicinal chemistry perspectives of 1,2,3,4-tetrahydroisoquinoline analogs—Biological activities and SAR studies. *RSC Adv.* **2021**, *11*, 12254. [CrossRef]
59. Theeramunkong, S.; Thiengsusuk, A.; Vajragupta, O.; Muhamad, P. Synthesis, characterization and antimalarial activity of isoquinoline derivatives. *Med. Chem. Res.* **2021**, *30*, 109–119. [CrossRef]

60. Galán, A.; Moreno, L.; Párraga, J.; Serrano, Á.; Sanz, M.J.; Cortes, D.; Cabedo, N. Novel isoquinoline derivatives as antimicrobial agents. *Bioorg. Med. Chem.* **2013**, *21*, 3221–3230. [CrossRef]
61. Karanja, C.W.; Naganna, N.; Abutaleb, N.S.; Dayal, N.; Onyedibe, K.I.; Aryal, U.; Seleem, M.N.; Sintim, H.O. Isoquinoline antimicrobial agent: Activity against intracellular bacteria and effect on global bacterial proteome. *Molecules* **2022**, *27*, 5085. [CrossRef] [PubMed]
62. Mao, Y.; Soni, K.; Sangani, C.; Yao, Y. An overview of privileged scaffold: Quinolines and isoquinolines in medicinal chemistry as anticancer agents. *Curr. Top. Med. Chem.* **2020**, *20*, 2599–2633. [CrossRef] [PubMed]
63. Yadav, P.; Kumar, A.; Althagafi, I.; Nemaysh, V.; Rai, R.; Pratap, R. The recent development of tetrahydro-quinoline/isoquinoline based compounds as anticancer agents. *Curr. Top. Med. Chem.* **2021**, *21*, 1587–1622. [CrossRef] [PubMed]
64. Balewski, Ł.; Sączewski, F.; Gdaniec, M.; Kornicka, A.; Cicha, K.; Jalińska, A. Synthesis and fluorescent properties of novel isoquinoline derivatives. *Molecules* **2019**, *24*, 4070. [CrossRef] [PubMed]
65. Sączewski, F.; Dziemidowicz-Borys, E.; Bednarski, P.J.; Grünert, R.; Gdaniec, M.; Tabin, P. Synthesis, crystal structure and biological activities of copper(II) complexes with chelating bidentate 2-substituted benzimidazole ligands. *J. Inorg. Biochem.* **2006**, *100*, 1389–1398. [CrossRef] [PubMed]
66. Sączewski, F.; Dziemidowicz-Borys, E.; Bednarski, P.J.; Gdaniec, M. Synthesis, crystal structure, cytotoxic and superoxide dismutase activities of copper(II) complexes of *N*-(4,5-dihydroimidazol-2-yl)azoles. *Arch. Pharm.* **2007**, *340*, 333–338. [CrossRef] [PubMed]
67. Balewski, Ł.; Sączewski, F.; Bednarski, P.J.; Gdaniec, M.; Borys, E.; Makowska, A. Structural diversity of copper(II) complexes with *N*-(2-pyridyl)imidazolidin-2-ones(thiones) and their in vitro antitumor activity. *Molecules* **2014**, *19*, 17026–17051. [CrossRef]
68. Makowska, A.; Sączewski, F.; Bednarski, P.J.; Gdaniec, M.; Balewski, Ł.; Warmbier, M.; Kornicka, A. Synthesis, structure and cytotoxic properties of copper(II) complexes of 2-iminocoumarins bearing a 1,3,5-triazine or benzoxazole/benzothiazole moiety. *Molecules* **2022**, *27*, 7155. [CrossRef]
69. Sączewski, F.; Bułakowska, A.; Gdaniec, M. 2-Chloro-4,5-dihydroimidazole, Part X. Revisiting route to *N*-heteroarylimidazolidin-2-ones. *J. Heterocycl. Chem.* **2002**, *39*, 911–915. [CrossRef]
70. Balewski, Ł.; Sączewski, F.; Gdaniec, M.; Bednarski, P.J.; Jara, I. Synthesis of *N*-(2-pyridyl)imidazolidin-2-ones and 1-(2-pyridyl)-2,3,7,8-tetrahydro-1*H*-imidazo[2,1-*b*][1,3,5]-triazepin-5(6*H*)-ones with potential biological activity. *Heterocycl. Commun.* **2013**, *19*, 331–341. [CrossRef]
71. Becke, A.D. Density-functional thermochemistry. I. The effect of the exchange-only gradient correction. *J. Chem. Phys.* **1992**, *96*, 2155–2160. [CrossRef]
72. Becke, A.D. Density-functional thermochemistry. II. The effect of the Perdew-Wang generalized-gradient correlation correction. *J. Chem. Phys.* **1992**, *97*, 9173–9177. [CrossRef]
73. Becke, A.D. Density-functional thermochemistry. III. The role of exact exchange. *J. Chem. Phys.* **1993**, *98*, 5648–5652. [CrossRef]
74. Ali, A.A.S.; Khan, D.; Naqvi, A.; Al-blewi, F.F.; Rezki, N.; Aouad, M.R.; Hagar, M. Design, synthesis, molecular modeling, anticancer studies, and density functional theory calculations of 4-(1,2,4-triazol-3-ylsulfanylmethyl)-1,2,3-triazole derivatives. *ACS Omega* **2021**, *6*, 301–316. [CrossRef]
75. Zhang, Y.L.; Deng, C.X.; Zhou, W.F.; Zhou, L.Y.; Cao, Q.Q.; Shen, W.Y.; Liang, H.; Chen, Z.F. Synthesis and *in vitro* antitumor activity evaluation of copper(II) complexes with 5-pyridin-2-yl-[1,3]dioxolo[4,5-*g*]isoquinoline derivatives. *J. Inorg. Biochem.* **2019**, *201*, 110820. [CrossRef] [PubMed]
76. Le, T.T.; Wu, M.; Lee, J.H.; Bhatt, N.; Berger, J.M.; Wang, M.D. Etoposide promotes DNA loop trapping and barrier formation by topoisomerase II. *Nat. Chem. Biol.* **2023**, *19*, 641–650. [CrossRef] [PubMed]
77. Dasari, S.; Tchounwou, P.B. Cisplatin in cancer therapy: Molecular mechanisms of action. *Eur. J. Pharmacol.* **2014**, *740*, 364–378. [CrossRef]
78. Strobel, H.; Baisch, T.; Fitzel, R.; Schilberg, K.; Siegelin, M.D.; Karpel-Massler, G.; Debatin, K.M.; Westhoff, M.A. Temozolomide and other alkylating agents in glioblastoma therapy. *Biomedicines* **2019**, *7*, 69. [CrossRef]
79. Longley, D.; Harkin, D.; Johnston, P. 5-Fluorouracil: Mechanisms of action and clinical strategies. *Nat. Rev. Cancer* **2003**, *3*, 330–338. [CrossRef]
80. Khan, T.M.; Gul, N.S.; Lu, X.; Wei, J.H.; Liu, Y.C.; Sun, H.; Liang, H.; Orvig, C.; Chen, Z.F. In vitro and in vivo anti-tumor activity of two gold(III) complexes with isoquinoline derivatives as ligands. *Eur. J. Med. Chem.* **2019**, *163*, 333–343. [CrossRef]
81. Huang, K.B.; Mo, H.Y.; Chen, Z.F.; Wei, J.H.; Liu, Y.C.; Liang, H. Isoquinoline derivatives Zn(II)/Ni(II) complexes: Crystal structures, cytotoxicity, and their action mechanism. *Eur. J. Med. Chem.* **2015**, *100*, 68–76. [CrossRef] [PubMed]
82. Wang, F.Y.; Xi, Q.Y.; Huang, K.B.; Tang, X.M.; Chen, Z.F.; Liu, Y.C.; Liang, H. Crystal structure, cytotoxicity and action mechanism of Zn(II)/Mn(II) complexes with isoquinoline ligands. *J. Inorg. Biochem.* **2017**, *169*, 23–31. [CrossRef] [PubMed]
83. Huang, K.B.; Chen, Z.F.; Liu, Y.C.; Wang, M.; Wei, J.H.; Xie, X.L.; Zhang, J.L.; Hu, K.; Liang, H. Copper(II/I) complexes of 5-pyridin-2-yl-[1,3]dioxolo[4,5-g]isoquinoline: Synthesis, crystal structure, antitumor activity and DNA interaction. *Eur. J. Med. Chem.* **2013**, *70*, 640–648. [CrossRef] [PubMed]
84. Gomaa, E.Z. Human gut microbiota/microbiome in health and diseases: A review. *Antonie Van Leeuwenhoek* **2020**, *113*, 2019–2040. [CrossRef] [PubMed]
85. George, S.; Abrahamse, H. Redox potential of antioxidants in cancer progression and prevention. *Antioxidants* **2020**, *9*, 1156. [CrossRef] [PubMed]

86. Fox, J.T.; Sakamuru, S.; Huang, R.L.; Teneva, N.; Simmons, S.O.; Xia, M.H.; Tice, R.R.; Austin, C.P.; Myung, K. High-throughput genotoxicity assay identifies antioxidants as inducers of DNA damage response and cell death. *Proc. Natl. Acad. Sci. USA* **2012**, *109*, 5423–5428. [CrossRef] [PubMed]
87. Veiga, N.; Alvarez, N.; Castellano, E.E.; Ellena, J.; Facchin, G.; Torre, M.H. Comparative study of antioxidant and pro-oxidant properties of homoleptic and heteroleptic copper complexes with amino acids, dipeptides and 1,10-phenanthroline: The quest for antitumor compounds. *Molecules* **2021**, *26*, 6520. [CrossRef]
88. Pinheiro, A.C.; Nunes, I.J.; Ferreira, W.V.; Tomasini, P.P.; Trindade, C.; Martins, C.C.; Wilhelm, E.A.; Oliboni, R.D.S.; Netz, P.A.; Stieler, R.; et al. Antioxidant and anticancer potential of the new Cu(II) complexes bearing imine-phenolate ligands with pendant amine N-donor groups. *Pharmaceutics* **2023**, *15*, 376. [CrossRef]
89. Yusuf, T.L.; Waziri, I.; Olofinsan, K.A.; Akintemi, E.O.; Hosten, E.C.; Muller, A.J. Evaluating the in vitro antidiabetic, antibacterial and antioxidant properties of copper(II) Schiff base complexes: An experimental and computational studies. *J. Mol. Liq.* **2023**, *389*, 122845. [CrossRef]
90. Gurgul, I.; Hricovíniovà, J.; Mazuryk, O.; Hricovíniovà, Z.; Brindell, M. Enhancement of the cytotoxicity of quinazolinone Schiff base derivatives with copper coordination. *Inorganics* **2023**, *11*, 391. [CrossRef]
91. Hazra, M.; Dolai, T.; Pandey, A.; Dey, S.K.; Patra, A. Synthesis and characterisation of copper(II) complexes with tridentate NNO functionalized ligand: Density Function Theory study, DNA binding mechanism, optical properties, and biological application. *Bioinorg. Chem. Appl.* **2014**, *2014*, 104046. [CrossRef] [PubMed]
92. Daina, A.; Michielin, O.; Zoete, V. SwissADME: A free web tool to evaluate pharmacokinetics, drug-likeness and medicinal chemistry friendliness of small molecules. *Sci. Rep.* **2017**, *7*, 42717. [CrossRef] [PubMed]
93. Shityakov, S. Analysing molecular polar surface descriptors to predict blood-brain barrier permeation. *Int. J. Comput. Biol. Drug Des.* **2013**, *6*, 146–156. [CrossRef] [PubMed]
94. Trani, A.; Bellasio, E. Synthesis of 2-chloro-2-imidazoline and its reactivity with aromatic amines, phenols, and thiophenols. *J. Heterocycl. Chem.* **1974**, *11*, 257–262. [CrossRef]
95. Sheldrick, G.M. SHELXT—Integrated space-group and crystal-structure determination. *Acta Crystallogr. Sect. Found. Adv.* **2015**, *71*, 3–8. [CrossRef]
96. Sheldrick, G.M. A short history of SHELX. *Acta Crystallogr. A* **2008**, *64*, 112–122. [CrossRef]
97. Hübschle, C.B.; Sheldrick, G.M.; Dittrich, B. ShelXle: A Qt graphical user interface for SHELXL. *J. Appl. Cryst.* **2011**, *44*, 1281–1284. [CrossRef]
98. Macrae, C.F.; Sovago, I.; Cottrell, S.J.; Galek, P.T.A.; McCabe, P.; Pidcock, E.; Platings, M.; Shields, G.P.; Stevens, J.S.; Towler, M.; et al. Mercury 4.0: From visualization to analysis, design and prediction. *J. Appl. Cryst.* **2020**, *53*, 226–235. [CrossRef]
99. Dolomanov, O.V.; Bourhis, L.J.; Gildea, R.J.; Howard, J.A.K.; Puschmann, H. OLEX2: A complete structure solution, refinement and analysis program. *J. Appl. Crystallogr.* **2009**, *42*, 339–341. [CrossRef]
100. Westrip, S.P. publCIF: Software for editing, validating and formatting crystallographic information files. *J. Apply. Cryst.* **2010**, *43*, 920–925. [CrossRef]
101. $IC_{50}$ Tool Kit: A Free Web Server. Available online: http://www.ic50.tk (accessed on 20 September 2023).
102. Motyka, S.; Kusznierewicz, B.; Ekiert, H.; Korona-Głowniak, I.; Szopa, A. Comparative analysis of metabolic variations, antioxidant profiles and antimicrobial activity of *Salvia hispanica* (*Chia*) seed, sprout, leaf, flower, root and herb extracts. *Molecules* **2023**, *28*, 2728. [CrossRef] [PubMed]
103. SwissADME: A Free Web Tool to Compute Physicochemical Descriptors as Well as to Predict ADME Parameters, Pharmacokinetic Properties, Druglike Nature and Medicinal Chemistry Friendliness of One or Multiple Small Molecules to Support Drug Discovery. Available online: http://www.swissadme.ch (accessed on 10 November 2023).

International Journal of
*Molecular Sciences*

MDPI

*Article*

# Decavanadate-Bearing Guanidine Derivatives Developed as Antimicrobial and Antitumor Species

**Andreea Dumitrescu [1], Catalin Maxim [1], Mihaela Badea [1], Arpad Mihai Rostas [2], Alexandra Ciorîță [3,*], Alina Tirsoaga [4] and Rodica Olar [1,*]**

[1] Department of Inorganic and Organic Chemistry, Biochemistry and Catalysis, Faculty of Chemistry, University of Bucharest, 90-92 Panduri Str., District 5, 050663 Bucharest, Romania; andreea.dumitrescu@s.unibuc.ro (A.D.); catalin.maxim@chimie.unibuc.ro (C.M.); mihaela.badea@chimie.unibuc.ro (M.B.)

[2] Department of Physics of Nanostructured Systems, National Institute for Research and Development of Isotopic and Molecular Technologies, 67-103 Donat Str., 400293 Cluj-Napoca, Romania; arpad.rostas@itim-cj.ro

[3] Department of Molecular Biology and Biotechnology, Faculty of Biology and Geology, Babeș-Bolyai University, 5-7 Clinicilor St., 400001 Cluj-Napoca, Romania

[4] Department of Analytical and Physical Chemistry, Faculty of Chemistry, University of Bucharest, 4-12 Regina Elisabeta Av., District 3, 030018 Bucharest, Romania; alina.jurca@unibuc.ro

* Correspondence: alexandra.ciorita@ubbcluj.ro (A.C.); rodica.olar@chimie.unibuc.ro (R.O.)

**Abstract:** To obtain biologically active species, a series of decavanadates $(Hpbg)_4[H_2V_{10}O_{28}]\cdot 6H_2O$ (**1**) $(Htbg)_4[H_2V_{10}O_{28}]\cdot 6H_2O$; (**2**) $(Hgnd)_2(Hgnu)_4[V_{10}O_{28}]$; (**3**) $(Hgnu)_6[V_{10}O_{28}]\cdot 2H_2O$; and (**4**) (pbg = 1-phenyl biguanide, tbg = 1-(*o*-tolyl)biguanide, gnd = guanidine, and gnu = guanylurea) were synthesized and characterized by several spectroscopic techniques (IR, UV-Vis, and EPR) as well as by single crystal X-ray diffraction. Compound (**1**) crystallizes in space group *P*-1 while (**3**) and (**4**) adopt the same centrosymmetric space group *P*21/n. The unusual signal identified by EPR spectroscopy was assigned to a charge-transfer $\pi(O)\rightarrow d(V)$ process. Both stability in solution and reactivity towards reactive oxygen species ($O_2^-$ and $OH\cdot$) were screened through EPR signal modification. All compounds inhibited the development of *Escherichia coli*, *Pseudomonas aeruginosa*, *Staphylococcus aureus*, and *Enterococcus faecalis* bacterial strains in a planktonic state at a micromolar level, the most active being compound (**3**). However, the experiments conducted at a minimal inhibitory concentration (MIC) indicated that the compounds do not disrupt the biofilm produced by these bacterial strains. The cytotoxicity assayed against A375 human melanoma cells and BJ human fibroblasts by testing the viability, lactate dehydrogenase, and nitric oxide levels indicated compound (**1**) as the most active in tumor cells.

**Keywords:** antimicrobial; cytotoxicity; decavanadate; DFT calculations; guanidine; guanylurea

## 1. Introduction

The disease evolution in the modern epoch requests the diversification of treatment approaches. For instance, bacterial infections are increasingly resistant to treatment with antibiotics [1–3] or hard to treat due to complications with biofilms, both mono- and poly-microbial ones [4–6]. On the other hand, both organic and inorganic cytostatics used in the current treatment of cancer are expensive and generate resistance. Moreover, these have limited effectiveness, especially for recurrent and metastatic cancers, and are highly toxic due to their unspecific mechanisms of action [7–9]. As a result, new approaches are required for these diseases' treatment, one involving inorganic–organic hybrid species such as polyoxometalates (POMs) [10–12].

Among POMs, decavanadates ($[H_nV_{10}O_{28}]^{(n-6-)}$, (n = 0–4)) (DVs) are oxo-clusters that involve ten distorted $VO_6$ octahedra sharing edges and vertices [13]. These species exhibit in structure V(V) ions linked by oxo anions that act as a bridge between two, three, or four metallic ions. The structure is completed by some anions involved in double bonds as terminal ligands. Because of the high anionic charge, DV can be mainly stabilized in a solid state by counterions through electrostatic interactions. As a result, DV can be combined with various cations, including proton, representative and transition metal ions, ammonium, cationic complexes, alkylammonium, phosphonium, and organic or organometallic species [10–13].

Moreover, DV in combination with different inorganic and organic cations or a mixture of them generates species with a wide range of biological activities, such as hypoglycemic, antimicrobial, anticancer, or antiviral.

Like other V(V) derivatives, DVs were first studied as insulin-mimetic species [10]. Hence, ammonium DV (ADV) exhibits insulin-like activity and enhances glucose uptake in the presence of insulin [14]. Moreover, the activity was improved for benzyl ammonium [15] or *N,N′*-dimethylbiguanidinium DVs [16].

On the other hand, the DVs combined with ammonium [17], ammonium and $[M(OH_2)_6]^{2+}$ (M = Ni [18,19], Co [19], Mg [20]) or hexamethylenetetramine [21], and 3/4-pyridinium carboxamide [22] cations inhibit the growth of some Gram-negative bacteria like *Escherichia coli* [17–22] or *Pseudomonas aeruginosa* [19,20] in a micromolar range. For other species with protonated 4-picolinic acid [23], sodium [24], or calcium [25] cations, the inhibition spectrum was extended against Gram-positive bacteria, such as *Bacillus ciroflagellosus* [23], *Mycobacterium smegmatis* [24], *Mycobacterium tuberculosis* [24], and *Staphylococcus aureus* strains [25]. These studies evidenced that DVs are generally more effective than mono-, di- and tetra vanadate, which rapidly decompose in biological media [10,11]. The enhanced activity can be related to many V(V) centers and general properties such as total net charge, size, and redox activity.

Concerning the antitumor activity, the DVs bearing Co(II) and sodium cations inhibit cell proliferation in both human liver (SMMC-7721) and ovary (SKOV-3) cancer cell lines, a better activity compared to the approved drug fluorouracil [26]. At the same time, species with ammonium and lithium demonstrated dose-dependent antiproliferative activity on glioblastoma (U87), breast (MDA-MB-231), and melanoma (IGR39) human invasive cell lines [27]. The same cancer cell lines were inhibited in a dose-dependent manner by DV with mixed $[Mg(H_2O)_6]^{2+}$ and 2-methylimidazolium species [28].

Using the organic cations *N,N,N′,N′*-tetramethylethylenediamine, and ethylenediamine, some DVs exhibiting very good activity against human lung carcinoma (A549) and murine leukemia (P388) cells were obtained. It is worth mentioning that against A549 cells, both DVs exhibit inhibition values lower in comparison with cisplatin, but they have cytotoxicity on normal human hepatocytes (LO2) [29]. Unlike these, species with both sodium and carnitine do not affect normal cells, but the cytotoxicity against A549 is lower compared to the cisplatin one. In addition, this compound is 400 times more active against the MDA-MB-231 cell line than cisplatin [30]. The exposure of human breast epithelial (MCF-7) and A549 cancer cells to trimethyl ammonium acetate DV reduced the cell viability in a concentration-dependent fashion [31], while ADV and *N,N′*-dimethylbiguanidine DV exert an antiproliferative effect on the human melanoma cell line (UACC-62) at lower concentrations than ammonium metavanadate [32]. Moreover, the tetra-(benzyl ammonium) dihydrogen DV inhibited the proliferation and migration of MDA-MB231 at a micromolar range [33].

Moreover, the ability of ADV to associate with the receptor-binding domain of the SARS-CoV-2 spike protein and disrupt the protein's binding to its host cell surface receptor was recently demonstrated. Unfortunately, in vitro studies on SARS-CoV-2 infected cells identify enhanced ADV cytotoxicity [34].

All these results demonstrated both the potential of DVs for medical applications and the fact that the biological activity can be fine-tuned through single- or mixed-type counter-cations.

Considering the known biological activity of guanidine derivatives [35] and the proven antitumor activity of *N,N′*-dimethylbiguanidinium DV [32], the biological potential of some new DV-bearing guanidine cations was studied to obtain species with good antimicrobial or antitumor activities. Moreover, the anti-biofilm potential of such compounds is first reported here. DVs bearing guanidium derivatives were fully characterized based on data provided by single X-ray diffraction.

## 2. Results and Discussion

The DVs were prepared in a two-step procedure. First, the ADV was prepared by adding HCl in an aqueous solution of ammonium metavanadate up to pH 5, and a biguanide in a proper ratio was added (Scheme 1). For 1-phenyl biguanide (pbg) and 1-(*o*-tolyl)biguanide (tbg), the isolated species correspond to $(Hpbg)_4[H_2V_{10}O_{28}]\cdot 6H_2O$ (**1**) and $(Htbg)_4[H_2V_{10}O_{28}]\cdot 6H_2O$ (**2**). By adding hydrogen peroxide to the system containing tbg, an interesting compound bearing both cations of guanidine (gnd) and guanylurea (gnu), $(Hgnd)_2(Hgnu)_4[V_{10}O_{28}]$ (**3**) was isolated. These cations were generated by the oxidative cleavage of tbg and water addition to one of the resulting species. In an attempt to obtain the DV with 1-cyanoguanidine, the compound $(Hgnu)_6[V_{10}O_{28}]\cdot 2H_2O$ (**4**) was isolated, containing the same cation as guanylurea, this time by water addition to the guanidine derivative.

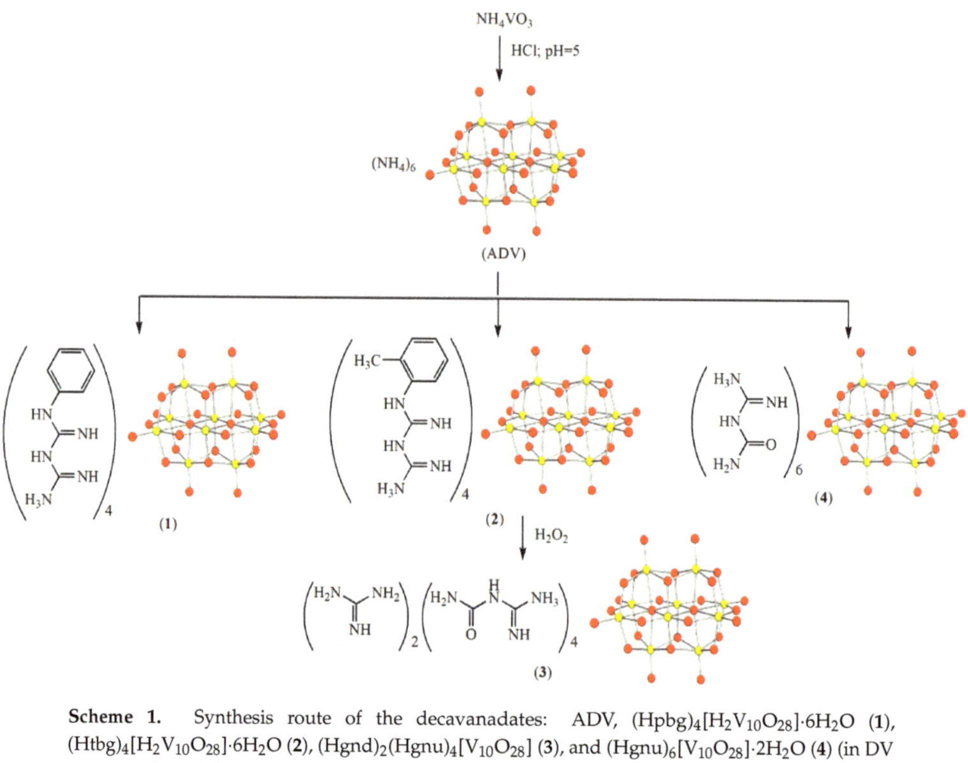


**Scheme 1.** Synthesis route of the decavanadates: ADV, $(Hpbg)_4[H_2V_{10}O_{28}]\cdot 6H_2O$ (**1**), $(Htbg)_4[H_2V_{10}O_{28}]\cdot 6H_2O$ (**2**), $(Hgnd)_2(Hgnu)_4[V_{10}O_{28}]$ (**3**), and $(Hgnu)_6[V_{10}O_{28}]\cdot 2H_2O$ (**4**) (in DV structure the red dots represent vanadium atoms and yellow dots oxygen atoms).

The compounds were characterized through a plethora of physicochemical methods, both in solid state and in solution. Compounds (**1**), (**3**), and (**4**) were fully characterized through single crystal X-ray diffraction. Moreover, all species were screened comparatively with ADV for biological activity. All these findings are detailed below.

*2.1. Description of the Crystal Structure of Compounds*

For compounds, (**1**), (**3**), and (**4**) single crystals suitable for X-ray analysis were obtained. This method provided information regarding the molecular structure and non-covalent interactions that appear between component moieties.

The crystallographic data and experimental details associated with the newly synthesized compounds are detailed in Supplementary Table S1. The selected bond distances and angles for all species are shown in Supplementary Tables S2–S7.

For (**1**) the $(H_2V_{10}O_{28})^{4-}$ anion is situated on an inversion center in space group *P*-1 and approaches the expected $D_{2h}$ symmetry, which has already been reported. The asymmetric unit for (**1**), presented in Figure 1, also contains two crystallographically independent phenyl biguanide cations and one water molecule.

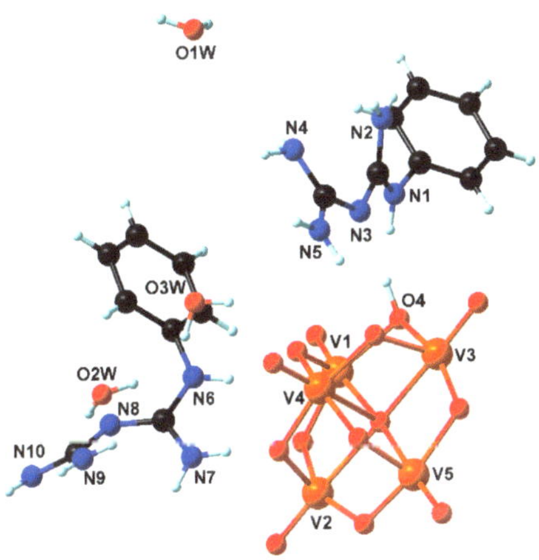


**Figure 1.** The asymmetric unit with the atom-labelling scheme of (**1**).

The Hpbg$^+$ monocation has its charge located mainly on the triangles N3N4N5 and N8N9N10. Furthermore, this cation is characterized by a dihedral angle of 73.02° and 22.5° between these planes and the phenyl ring.

The decavanadate anion, the cations, and the crystal water molecules engage in an extensive network of hydrogen bonds (Figure 2a). All N–H and O–H groups present in the asymmetric unit serve as donor groups. The two strongest hydrogen bonds are formed between the anion $V_{10}$-$\mu_3$OH and one nitrogen from the cation (N3–O6 = 2.743). A packing diagram showing the π–π interactions between the phenyl rings of the biguanide cations is presented in Figure 2b. The distances between aromatic moieties are 3.68 Å.

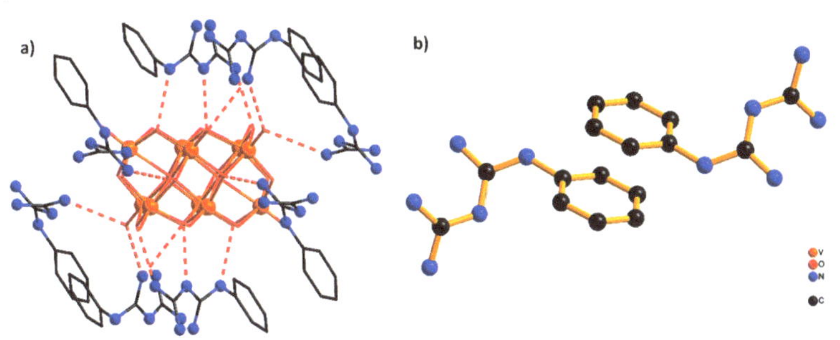


**Figure 2.** Perspective view of the crystal structure of (**1**) showing (**a**) hydrogen bond interactions between anionic and cationic units and (**b**) noncovalent π–π interactions.

By replacing the pbg with 1-(*o*-tolyl)biguanide, in different reaction conditions, two compounds $(Hgnd)_2(Hgnu)_4[V_{10}O_{28}]$ (**3**) and $(Hgnu)_6[V_{10}O_{28}]\cdot 2H_2O$ (**4**) have been crystallized. These compounds crystallize in the same centrosymmetric space group $P2_1/n$, with the $[V_{10}O_{28}]^{6-}$ anion lying in an inversion center. The difference in the asymmetric units (Figure 3) of these compounds is the presence of different cations and solvent molecules.

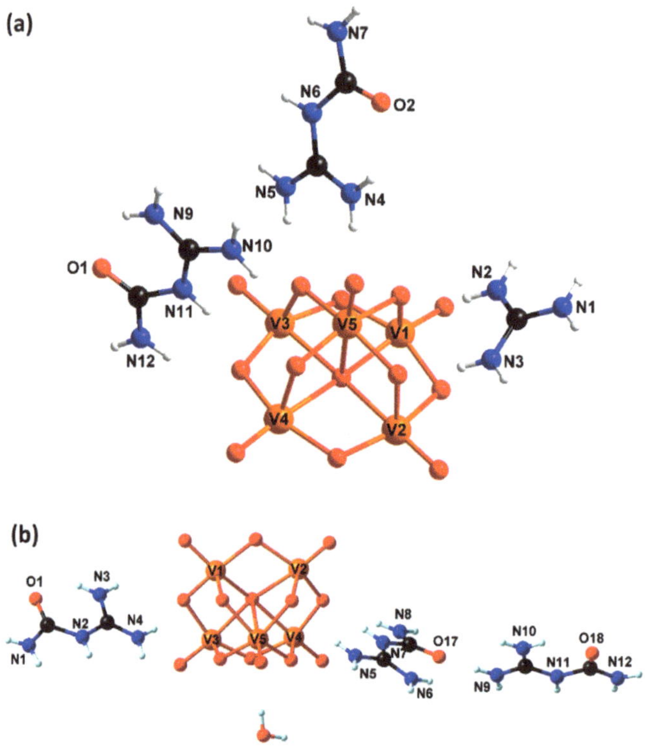


**Figure 3.** The asymmetric unit with the atom-labelling scheme of (**3**) (**a**) and (**4**) (**b**).

For (**3**), the apparition of guanylurea and guanidinium cations was observed, while for (**4**), only guanylurea and crystallization water molecules were observed.

In the two compounds, the $V_{10}$ is built with three types of vanadium atoms with different environments as follows: (i) two V atoms in the center ($V_1$), each binding to six bridging O atoms, including both of the $\mu_6$-O atoms of the complex; (ii) four V atoms ($V_{2,3}$), each corresponding to one $\mu_6$-O atom, four $\mu_2$-O atoms, and a terminal O atom; and (iii) four V atoms ($V_{4,5}$), each with five bridging O atoms and one terminal O atom.

The resulting 3D hydrogen network (Figure 4) is complicated because of the presence of different cations and crystallizations of water molecules. Practically all the amines, guanidinium, water molecules, and oxo donor groups are engaged in hydrogen bonds. In particular, the guanidinium cations interact with decavanadate anions in (**3**). Thus, the presence of planar cations within the lattice results in several NH donors with different orientations compared to those of the donors found in the phenyl-guanidine cation, and this supplies these V(V) clusters with more unpredictability in the composition and nature of the hydrogen-bond interactions.

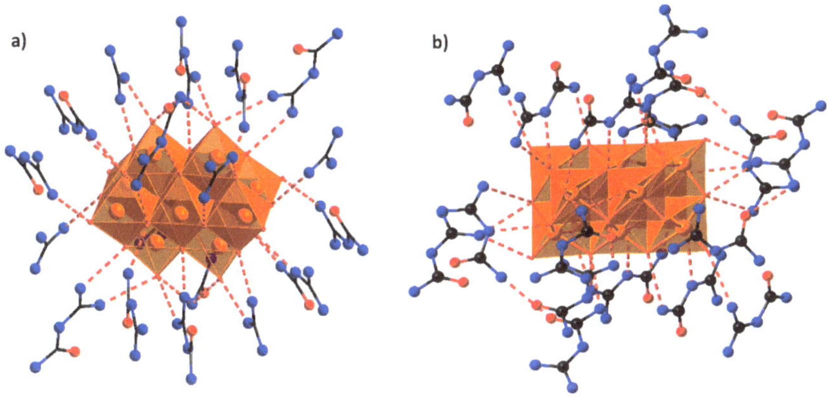


**Figure 4.** Extended hydrogen bond networks in compounds (**3**) (**a**) and (**4**) (**b**).

### 2.2. Physicochemical Characterization of Decavanadates

#### 2.2.1. FT-IR Spectra

The IR spectra of DVs were studied to obtain information concerning the characteristic bands associated with the main functional groups of guanidine derivatives and DV core, respectively.

These spectra for ADV and the (**1–4**) species are depicted in Figure 5.

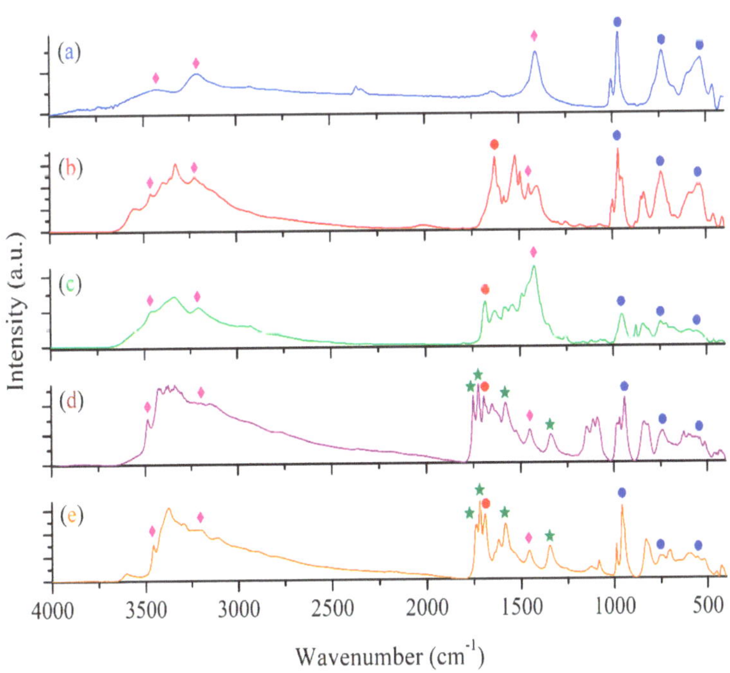


**Figure 5.** IR spectra of compounds ADV (**a**), (**1**) (**b**), (**2**) (**c**), (**3**) (**d**), and (**4**) (**e**) registered in KBr pellets. The characteristic bands of the DV core were highlighted with blue circles, the ammonium group with magenta rhomb, the C=N group of the guanidine moiety with red circles, and the amide group of the urea moiety with green stars.

All spectra show characteristic vibration bands of all building units, DV, and the corresponding cations. The strong IR band at around 950 $cm^{-1}$ is assigned to the stretching vibration of the V=O group. The asymmetric and symmetric vibrations of the bridge V–O–V groups appear in the 730–880 $cm^{-1}$ range, while the vibration of the V–O group is around 515 $cm^{-1}$ [18–33]. The spectra exhibit the characteristic bands of ammonium cations in the 3200–3460 and 1410–1450 $cm^{-1}$ ranges [21]. The band assigned to stretching vibration of the C=N bond from the guanidine moiety appears in the 1620–1690 $cm^{-1}$ range. For complexes (**3**) and (**4**), the spectra are more complex since they contain supplementary bands, two in the 1710–1755 $cm^{-1}$ range and one around 1580 and 1340 $cm^{-1}$ that arise from combined vibration modes for the amide group [36].

### 2.2.2. UV-Vis Spectra

For compounds bearing ions with a $3d^0$ configuration, such as V(V), UV-Vis spectra provide information concerning the presence of charge transfer and intraligand transitions. On the other hand, the absence of absorptions at wavelengths higher than 500 nm is a confirmation of the preservation of the oxidation state of the vanadium ion.

The UV–Vis absorption spectrum of ADV displays two well-defined absorption bands at 280 and 350 nm corresponding to the ligand-to-metal charge transfer (LMCT) transitions of the $O^{2-} \rightarrow V^{5+}$ type, as reported in the literature [37]. In the spectra of the (**1–4**) DVs (Supplementary Figure S1), both bands are shifted to higher wavenumbers (Table 1). They are accompanied in the UV region by a band assigned to the $\pi \rightarrow \pi^*$ transition for guanidinium units. In the spectra of compound (**3**), two well-defined bands appear at 215 and 280 nm due to the two guanidinium derivatives.

**Table 1.** Absorption maxima for ADV and compounds (**1–4**) in UV-Vis spectra.

| Compound | Absorption Maxima (nm) | |
|---|---|---|
| | $\pi \rightarrow \pi^*$ | LMCT |
| ADV | - | 280, 350 |
| (1) | 260 | 390, 450 |
| (2) | 265 | 395, 425 |
| (3) | 215, 280 | 335, 395 |
| (4) | 265 | 365, 450 |

### 2.2.3. Solid-State EPR Spectroscopy

EPR spectroscopy is a highly sensitive technique useful for paramagnetic species with unpaired electron characterization.

The solid-state EPR spectra of ADV and the (**1–4**) compounds are depicted in Figure 6, showing a resonance with a g-value of 1.97 and a complex hyperfine splitting. Similar EPR spectra are reported in the literature for vanadium-based compounds with a four-oxidation state, which exhibits a characteristic g-factor of 1.968 [38]. Since the V(IV) ion has an electronic spin S = 1/2 and the nuclear spin for the $^{51}V$ isotope (natural abundance 99.5%) is I = 7/2, the isotropic EPR spectrum of isolated V(IV) species consists of a set of eight hyperfine lines due to the dipole–dipole interaction between the magnetic moment of the $^{51}V$ nucleus and the electronic moment of the unpaired V(IV) electron [39]. The EPR spectrum of compound (**1**) presents only one broad EPR line due to strong spin–spin interactions between the paramagnetic centers. The presence of the V(IV) centers is correlated with the charge-transfer process ($\pi(O) \rightarrow d(V)$ transition probability) that can occur in the decavanadate units, as observed in the UV-Vis spectra. All these signals disappear through compound solubilization in DMSO.

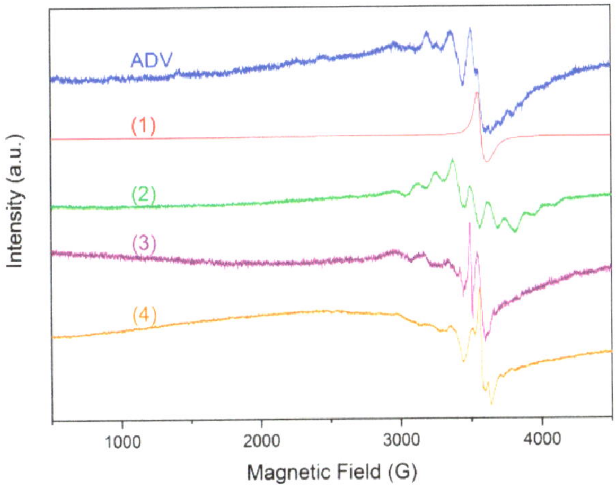


**Figure 6.** X-band powder EPR spectra of the ADV and (**1–4**) compounds.

### 2.3. *Antibacterial and Antitumor Activities*

#### 2.3.1. Antibacterial Activity

Based on the literature data, some Gram-negative (*Escherichia coli* and *Pseudomonas aeruginosa*) and Gram-positive (*Staphylococcus aureus* and *Enterococcus faecalis*) bacterial strains with ATCC provenience were chosen for biological assay. The compounds generally inhibited the development of the bacterial strains (Figure 7). *E. coli* and *P. aeruginosa* were most sensitive to ADV and complex (**3**) (the recorded minimal inhibitory concentration (MIC) was 7.9 μM for ADV and 31.7 μM for complex (**3**)). Except for *Enterococcus faecalis* (31.7 μM for (**3**)), all other compounds showed inhibitory abilities at the highest concentration used in this experiment (63.5 μM).

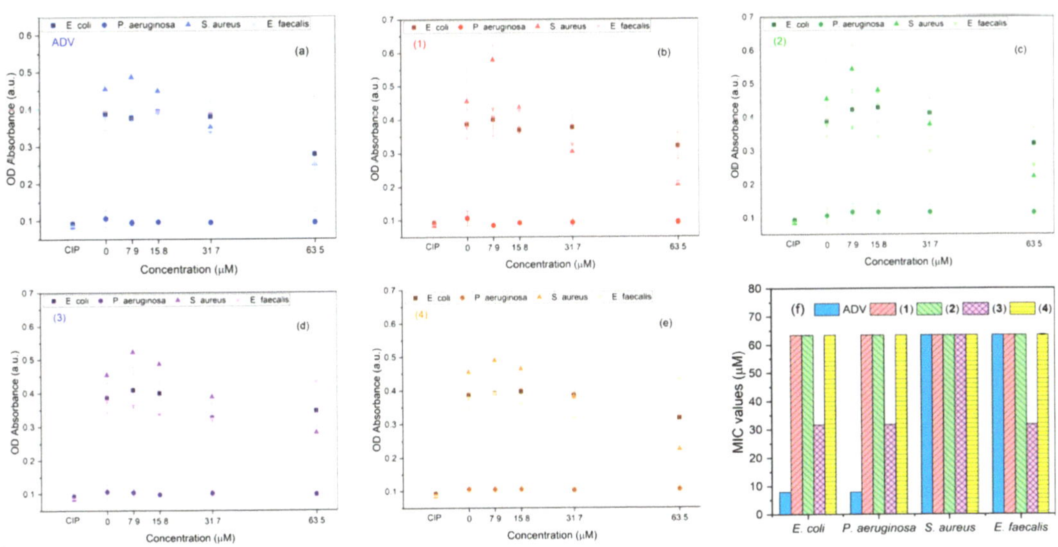


**Figure 7.** Selected bacteria inhibition by the complexes ADV (**a**), (**1**) (**b**), (**2**) (**c**), (**3**) (**d**), and (**4**) (**e**) at different concentrations, as well as the MIC values (**f**) for the selected bacterial strains.

Based on these results, the MIC values were chosen to determine the inhibitory capacity of the compounds in biofilm formation (Supplementary Figure S2). Except for *P. aeruginosa*, all samples exhibited a proliferative response, with registered values over those of the untreated control. This might indicate that either the compounds are not able to disrupt the biofilm in three hours or that the compounds favor biofilm formation if they are left in contact with the bacterial strains for less than 24 h.

Other studies showed that Gram-positive bacteria are more susceptible to ADV than Gram-negative strains [23]. The bacterial wall–ADV interaction is possible due to the membrane potential and charged components. The mechanism of action is explained by several factors: (i) the nutrient uptake from the media is inhibited by the blockage of membrane proteins by ADV [22,25]; (ii) the ADV enters the bacteria and forms intralysosomal decavanadate that could later lead to cellular damage and eventually death [24]; and (iii) decavanadate disrupts the function of some proteins (i.e., ATPases), which later leads to cell death [17,22]. Although probable, these modes of action do not explain completely how ADV inhibits bacterial growth, and further investigations are still required.

### 2.3.2. Antitumor Activity

The cytotoxicity was assayed against A 375 human melanoma cells and BJ human fibroblasts by testing the viability by the 3-(4,5-dimethylthiazol-2-yl)-2,5-diphenyltetrazolium bromide (MTT) colorimetric method, lactate dehydrogenase (LDH), and nitric oxide (NO) release. The concentration at which 50% of the cells are affected ($IC_{50}$) was calculated based on the obtained results. All compounds exhibited high cytotoxicity against both melanoma and fibroblast cells (Figure 8a,c), with the best results observed for compound **(1)**, followed by **(3)**, **(2)**, **ADV**, and **(4)** (Table 2). According to the statistical analyses conducted, compounds **(3)** and **(4)** (at 250 μM concentration) had similar toxicity levels as the control treated with Tween 20 against melanoma.

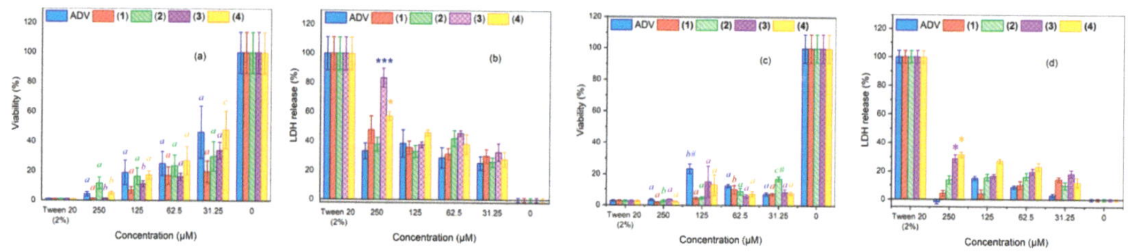


**Figure 8.** Cytotoxicity assay of the obtained compounds against human normal and cancerous cells; the MTT viability assay (**a**) and LDH release (**b**) of A375 cells; the MTT viability assay (**c**) and LDH release (**d**) of BJ cells; the bars marked with different letters are significantly different from the others between concentrations of the same compound, and those marked with a letter and '#' are significantly different within the same concentration as well ($p < 0.05$ according to the Tukey test); bars marked with '*' are significantly different from the rest according to one way ANOVA (*** $p < 0.0001$, and * $p < 0.01$).

**Table 2.** The $IC_{50}$ values registered for the melanoma and fibroblast cells treated with the vanadium compounds.

| Cell | ADV | (1) | (2) | (3) | (4) |
|---|---|---|---|---|---|
| | | | $IC_{50}$ (μM) | | |
| A375 | 57.5 | 17.6 | 39.01 | 34.6 | 59.6 |
| BJ | 3.63 | 8.04 | 4.43 | 7.21 | 6.56 |

LDH measurements indicate the degree of cellular damage. Thus, the LDH release was observed for all compounds at all tested concentrations (Figure 8b,d), with the highest value registered for compound (**3**) at 250 μM for melanoma and compound (**4**) for fibroblasts.

The tested fibroblasts showed a significant decrease in viability at all concentrations (Figure 8c), while the $IC_{50}$ values determined were similar for all compounds. However, the LDH release registered indicates that the compounds might not affect the external membrane of the BJ cells to the same extent as for A375 cells (Figure 8d).

The NO release in the media was determined to further investigate the degree of induced cytotoxicity. All tested compounds generated NO for at least one concentration of nitrates/nitrites, which could indicate that the vanadium compounds might induce oxidative stress in cancerous cells. However, no changes in the NO level were observed for the BJ cells.

In the case of A375 cells, for every concentration of vanadium compounds at which a low viability was registered, a high LDH concentration was observed in the media. Moreover, NO absorbance indicated nitrites/nitrates in the media at the same concentrations where cell toxicity was detected. Following what was observed herein, vanadium compounds were previously reported to have cytotoxicity in vitro [26–33]. These results indicate that the cells might reach apoptosis through necrosis [40], making the compounds good candidates for potential cancer theranostics. The BJ fibroblasts registered significantly low viability, but as indicated by the other two assays, these values could be associated with either a lower number of cells or a lower number of mitochondria in the cells [41]. Hence, the vanadium compounds might not affect the overall metabolism of normal cells as they do for cancerous cells.

The mechanism of action is dependent on the membrane potential, elasticity, and surface charge, as in prokaryotic cells, which leads to the activation of apoptotic pathways [28]. Since the membrane of tumor cells is more permeable, it can facilitate a higher intake of ADV compared to normal cells [41], similar to other findings [27,29,32].

2.3.3. Solution EPR Spectroscopy

The ability of the proposed DV-based compounds to trap or scavenge reactive oxygen species was tested via EPR spectroscopy, for which $KO_2$ and $H_2O_2$ were added as $O_2^-$ and $OH^-$ radical donors. The results are presented in Figure 9, showing that compound (**1**) can trap both types of radicals, and thus, the V(IV) characteristic EPR signal appears after adding the radical donors. The same is valid also for compound (**2**), but only for $O_2^-$ radicals. The other compounds, ADV, (**3**), and (**4**), show no activity at all, indicating that these compounds are not as sensitive in the presence of ROS as (**1**) and (**2**).

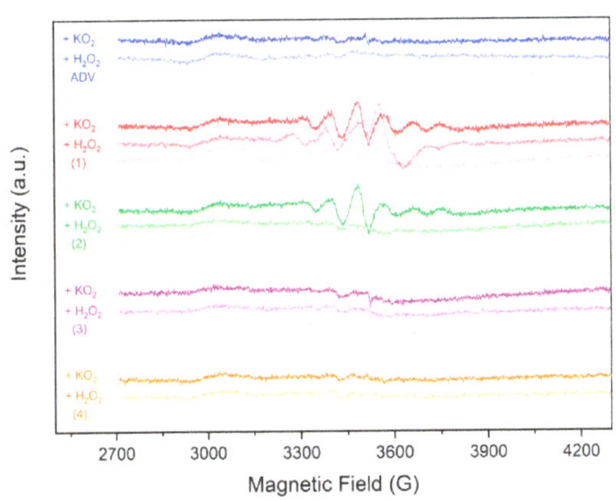


**Figure 9.** EPR spectra of the ADV and (**1–4**) compounds in a 50 mM DMSO solution where $KO_2$ and $H_2O_2$ were added as $O_2^-$ and $OH^-$ radical donors.

### 2.4. Molecular Docking Studies

The cytochrome $bc_1$ complex (complex III) is an essential multi-functional protein located in the inner mitochondrial membrane of eukaryotes or the cytoplasmatic membrane of prokaryotic organisms, being an integral part of the cellular respiratory chain, considering its contribution to the generation of both electrochemical potential and reactive oxygen species [42]. Alteration of the $bc_1$ mechanism by chemical or genetic factors leads to harmful consequences for tissue integrity.

Briefly, the cytochrome $bc_1$ complex is involved in the translocation of protons across the inner mitochondrial membrane due to a joint action of two redox processes between ubiquinol ($QH_2$) and ubiquinone (Q), which take place at the distinct catalytic sites located on the opposite side of the lipidic membrane (a quinone reduction site ($Q_i$) and a quinol oxidation site ($Q_0$)), as well as a third redox process of cytochrome c located in the water phase (intermembrane space or periplasm). An interesting feature of *b*-type cytochrome is that the catalytic activity of the two sites ($Q_i$ and $Q_0$) is closely related, i.e., the product of one catalytic site becomes the substrate for the second one, and vice versa [43]. When the $bc_1$-complex is blocked by specific respiratory inhibitors, the electron transfer between cytochrome b and cytochrome c diminishes drastically, and the ROS production is increased, resulting in these species being released into the inter-membrane space or the mitochondrial matrix [44]. Such oxidative stress contributes to the pathogenesis of many diseases.

For this reason, developing drugs (antibiotics, anti-fungal agents, or anti-parasite agents) acting as specific $bc_1$ inhibitors is of commercial interest due to their use in medicine or agriculture to control diseases or cure severe infections. According to the state-of-the-art literature, specific $bc_1$ inhibitors are divided into three classes based on their points of action [45], such as (i) complex III $Q_i$ inhibitors (antimycin A, amisulbrom, cyazofamid); (ii) complex III $Q_0$ inhibitors (stigmatellin, ametoctradin, azoxystrobin); and (iii) dual site inhibitors (ascochlorin). Moreover, based on in vivo studies, Aureliano et al. [46] pointed out that mitochondria might be a target for DV units that could affect its bioenergetics through different toxicity mechanisms by blocking the respiratory electron transport chain and acting as an antimycin A-like inhibitor.

Our docking study aims to investigate the binding affinities of DV units to the $Q_i$ and $Q_0$ sites of complex III to predict their activity as mitochondrial inhibitors. It is noteworthy that the structure of cytochrome $bc_1$ presents a dimeric association structure between two identical units, each featuring both $Q_i$ and $Q_0$ sites. Due to their identical environment, the docking study has been performed on a single set of $Q_i$ and $Q_0$ sites, as found in chains C or P of the cytochrome $bc_1$-associated structure.

Regarding the $Q_i$ binding pocket located in the C (or P) chain of the enzyme sequence, some key residues were mentioned to be involved in antimycin H-bond interactions (*via* water molecules), namely Asp228, Lys 227, Ser35, His201, and Ser205, in close vicinity of the heme $b_H$, as reported by Huang et al. [47].

The results of docking calculations indicate that all four DV units were conveniently accommodated inside the $Q_i$ site, i.e., in the proximity of the heme $b_H$ with a negative binding energy of −6.29 kcal/mol for the reference ADV compound and for (**1**), (**3**), and (**4**) slightly smaller values (−6.37, −6.31, and −6.3 kcal/mol, respectively), indicating a viable possibility of binding in each case.

The $Q_0$ binding site is located close to both the heme $b_L$ and the Rieske cluster (2Fe-2S cluster, Figure 10) in such an orientation that stigmatellin is positioned between the two prosthetic residues [48] (i.e., it binds to cytochrome *b* at a distal domain from the heme $b_L$) and interacts with the Rieske cluster through the formation of H bonds with His161, while the Pro270 residue plays an important role in the aromatic $\pi$–$\pi$ interaction [47].

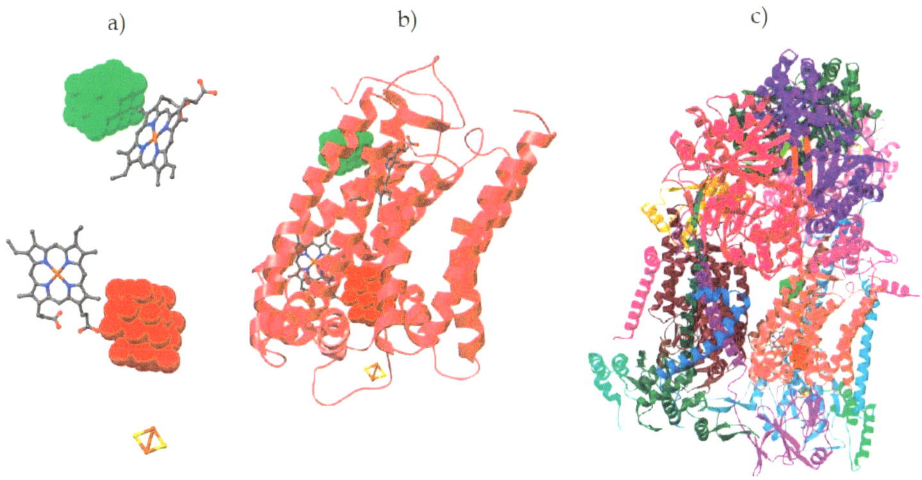


**Figure 10.** Docking poses of the DV unit in the cytochrome $bc_1$ structure as simulated with the AutoDock-Vina algorithm: successful docking of the DV unit (green) in the $Q_i$ site; the forced pose of the DV unit (red) in the $Q_0$ site (energetically disfavored docking, see text). Both poses represent a guide for the eye to locate $Q_i$ and $Q_0$ sites (**a**); a representation of chain C of the enzyme (**b**); and a full representation of the dimeric enzyme structure (**c**). More details can be found in Supplementary Figure S3.

Attempts to dock any DV units into the $Q_0$ site failed: negative binding energy was obtained (around −7 kcal/mol) only if the DV unit left the $Q_0$ site space, whereas forcing the DV unit to pose in the $Q_0$ site near the heme $b_L$ structure leads to unrealistically high (positive) values of the binding energy.

## 3. Materials and Methods

### 3.1. Materials and Physical Measurements

Chemicals were purchased from Sigma-Aldrich (Darmstadt, Germany) (ammonium vanadate, 1-phenyl biguanide 98%, 1-(*o*-tolyl)biguanide 98%, and 1-cyanoguanidine 99%) in reagent grade. All these were used as received without further purification.

Fourier Transform Infrared spectroscopy (FT-IR) spectra were recorded in KBr pellets with a Tensor 37 spectrometer (Bruker, Billerica, MA, USA) in the 400–4000 $cm^{-1}$ range. UV-Vis spectra in solid-state were recorded using a Jasco V 670 spectrophotometer (Jasco, Easton, MD, USA) with Spectralon as a standard in the 200–1500 nm range. Quartz cells measuring 1 cm were used for all measurements. X-band EPR measurements were carried out with a continuous-wave dual-band E500 ELEXSYS EPR spectrometer (Bruker, Karlsruhe, Germany). The room temperature measurements in the X-band were carried out at a microwave frequency of 9.879 GHz. The free radical scavenging ability of the complexes was tested through EPR spectroscopy using $KO_2$ and $H_2O_2$ as sources for $O_2^-$ and $OH^-$. $KO_2$ was dissolved in DMSO by complexation with dibenzo-18-crown-6-ether. The final concentrations used were 500 μM $KO_2$ and 3% $H_2O_2$. X-ray diffraction data were collected at 293 K on a Rigaku XtaLAB Synergy-S diffractometer operating with a Mo-Kα (λ = 0.71073 Å) micro-focus sealed X-ray tube. The structure was solved by direct methods and refined using full-matrix least-squares techniques based on F2. The non-H atoms were refined using anisotropic displacement parameters. The calculations were performed using the crystallographic software package SHELX-2018 [49]. The crystallographic data (excluding structure factors) have been deposited with the Cambridge Crystallographic Data Centre with CCDC reference numbers 2303445–2303447. These data can be obtained free of charge from http://www.ccdc.cam.ac.uk/conts/retrieving.html, accessed on 20 October

2023, or from the Cambridge Crystallographic Data Centre, 12 Union Road, Cambridge CB2 1EZ, UK; fax: (+44) 1223-336-033; or e-mail: deposit@ccdc.cam.ac.uk. X-ray powder diffraction measurements (XRPD) were carried out on a D8 Advance Bruker diffractometer (CuK$\alpha$ radiation) equipped with a linear Vantec Super Speed detector and a TTK-450 temperature chamber. Additional XRPD patterns were recorded on a PROTO AXRD Benchtop Powder Diffractometer (CuK$\alpha$ radiation).

### 3.2. Synthesis of Compounds

The compounds $(Hpbg)_4[H_2V_{10}O_{28}]\cdot 6H_2O$ (**1**) and $(Htbg)_4[H_2V_{10}O_{28}]\cdot 6H_2O$ (**2**) were prepared as follows: to a solution containing ammonium vanadate (0.585 g, 5 mmol) in 100 mL of water, a few drops of 37% HCl (pH = 5) were added, and then a solution containing pbg/tbg (0.531/0.573 g, 3 mmol) was dissolved in 50 mL of water. The reaction mixture was magnetically stirred at 50 °C for 1 h and left at room temperature for slow evaporation. Orange crystals were obtained after the solution stood at room temperature for several days, was filtered off, washed several times with cold ethanol, and dried in the air. Yield 86/87%. Anal. Calcd. for $C_{32}H_{62}N_{20}O_{34}V_{10}$ (**1**) (%): C, 21.59; H, 3.51; N, 15.73; found (%): C, 21.67; H, 3.38; N, 15.81; calcd. for $C_{36}H_{70}N_{20}O_{34}V_{10}$ (**2**) (%): C, 23.65; H, 3.42; N, 15.32; and found (%): C, 23.67; H, 3.32; N, 15.86. The crystals, suitable for X-ray diffraction, were obtained by slow diffusion of an aqueous solution of fbg in an aqueous solution of ADV.

The same solution as for (**2**), completed with 5 mL of hydrogen peroxide at 30%, afforded $(Hgnd)_2(Hgnu)_4[V_{10}O_{28}]$ (**3**) with a yield of 81%. Calcd. for $C_{10}H_{40}N_{22}O_{32}V_{10}$ (%): C, 8.06; H, 2.71; N, 20.68; found (%): C, 8.07; H, 2.68; N, 20.71. Crystals suitable for X-ray diffraction were obtained by slow diffusion of an aqueous solution of tbg and $H_2O_2$ 30% in an aqueous solution of ADV.

The compound $(Hgdu)_6[V_{10}O_{28}]\cdot 2H_2O$ (**4**) was synthesized by mixing a 100 mL solution containing 0.585 g (5 mmol) of ammonium vanadate in water with a few drops of HCl (37% up to pH 5) and a 50 mL solution with 0.252 g (3 mmol) of cyanoguanidine. The reaction mixture was magnetically stirred at 50 °C for 1 h and left at room temperature for slow evaporation. The orange crystals obtained after the solution stood at room temperature for several days were filtered off, washed several times with cold ethanol, and dried in the air. Yield 89%. Anal. Calcd. for $C_{12}H_{46}N_{24}O_{36}V_{10}$ (%): C, 8.94; H, 2.88; N, 20.85; and found (%): C, 8.97; H, 2.85; N, 20.91. X-ray diffraction-suitable crystals were obtained by slow diffusion of an aqueous solution of 1-cyanoguanidine in an aqueous solution of ADV.

All compounds were obtained as crystalline materials. The powder XRD analysis of polycrystalline samples is presented in Supplementary Figure S4. For compound (**2**), several attempts to obtain good diffraction single crystals were unsuccessful.

### 3.3. Antibacterial Assay

The antibacterial properties of the components were assayed against *E. coli* (ATCC 25922), *P. aeruginosa* (ATCC 27853), *S. aureus* (ATCC 25923), and *E. faecalis* (ATCC: 29212) through the microdilution method [50]. Four concentrations were used (7.9, 15.8, 31.7, and 63.5 μM), and the first that registered bacterial inhibition below untreated control or similar to antibiotic control was chosen as the minimal inhibitory concentration (MIC). This concentration was further used to determine the inhibitory capacity of the component on biofilm formation, according to the literature data [51]. Briefly, the samples were incubated with the bacteria (24 h fresh cultures at 1:40 dilution with nutrient media) for 3 h at 37 °C. The solution was removed after this interval and replaced with MTT (1.25 mg/mL final concentration in the well) for 2 h. Later, the MTT solution was removed and replaced with isopropanol for 10 min to dissolve the formazan crystals. The absorbance was read at 550 and 630 nm (Epoch plate reader; BioTek, Bad Friedrichshal, Germany), and the percentage of inhibition was calculated according to the following Equation (1).

$$\%\text{inhibition} = (\text{Mean of registered absorbance})/(\text{Mean of untreated control}) \times 100. \quad (1)$$

All experiments had four replicates, and the mean was calculated for at least two independent values. One plate containing the components and the media, without bacteria, was also used as a control. A student *t*-test and one-way ANOVA were used to determine the significance level between the compounds applied to one strain.

### 3.4. In Vitro Cytotoxicity Assay

The cytotoxicity was assayed against human melanoma cells (ATCC CRL-1619, Wesel, Germany) and human fibroblasts (BJ, CRL-2522, Wesel, Germany), as previously reported [52]. First, the viability was determined through the MTT test, and the $IC_{50}$ values were calculated. Briefly, the cells inoculated at $12 \times 10^3$ cells/well were incubated with the solutions for 24 h at 37 °C. After this, the media was removed and tested for LDH and NO. The wells were incubated with 3-(4,5-dimethylthiazol-2-yl)-2,5-diphenyltetrazolium bromide (MTT, Sigma Aldrich, Merck KGaA, Darmstadt, Germany) for 1.5 h at 37 °C, and the formazan crystals were dissolved with isopropanol. Next, the LDH and NO releases in the media were determined. For LDH, 50 µL lithium lactate solution of 50 mM, 50 µL tris solution of 200 mM, and 50 µL of NAD solution (a mixture of ionitrotetrazolium violet, phenazine metosulfate, and nicotinamide dinucleotide; Sigma Aldrich, Merck KGaA, Darmstadt, Germany) were mixed with 50 µL of media. NO sulphanilamide and N-(-1-napthyl)-ethylenediamine (Sigma Aldrich, Merck KGaA, Darmstadt, Germany) were mixed with 50 µL of media. The plates were read using a BioTek Epoch plate reader (BioTek, Bad Friedrichshal, Germany) and Gen5 software (version 1.09). The concentrations used ranged from 250 to 31.25 µM for all compounds. One-way ANOVA and Tukey statistical analyses were used to determine the significance level within the concentrations and between the compounds.

### 3.5. Computational Strategy

Docking Studies

The X-ray diffraction structure of mitochondrial complex III (PDB entry: 1PPJ, deposition on 2003; latest revision in 2017, accessed on 12 March 2023) was obtained from the RCSB Protein Data Bank (https://www.rcsb.org/structure/1PPJ, accessed on 12 March 2023). This crystal structure of bovine heart mitochondrial complex III was chosen for its best resolution (2.1 Å) compared to similar structures reported. Moreover, the co-crystallized ligands (both antimycin A and stigmatellin that also served to identify the location of $Q_i$ and $Q_0$ sites correctly) were afterward removed (together with all non-proteic residues) from the enzyme structure before initiating the docking procedure. Theoretical calculations were performed using the following software suites: AutoDock-Vina v1.2.3-52 source [53,54] was compiled statically with GCC 8.5.0 using system default boost libs v1.66 and Python Molecule Viewer v1.5.7p1 as implemented in the MGL Tools v.1.5.7 tarball installer [55]. The geometries of the investigated DV units were taken from experimental single crystal measurements without further optimization. Molecular docking calculations were performed with the AutoDock-Vina engine [53,54] using box sizes up to $24 \times 24 \times 26$ Å$^3$ and a spacer of 0.375 Å (exhaustiveness parameter 32).

## 4. Conclusions

Several decavanadates bearing guanide moieties were synthesized and characterized. For three species, the structure was determined through single crystal X-ray diffraction. These contain the DV units in interaction with guanidium cations and/or water molecules, generating a supramolecular arrangement through a complex hydrogen bond network. Complex (**3**) was the most active on all planktonic bacterial strains, while compound (**1**) exhibited the highest cytotoxicity on A375 human melanoma cells, this being lower on BJ healthy ones. Both lactate dehydrogenase and nitric oxide-enhanced levels in tumor cells accompanied this activity. Molecular docking calculations predict a specific interaction of the investigated DV units with the $Q_i$ site of the cytochrome bc1 complex with very close binding energies, in concordance with the similar order of magnitude of the inhibitory capacity for each bacterial strain case.

**Supplementary Materials:** The following supporting information can be downloaded at: https://www.mdpi.com/article/10.3390/ijms242417137/s1.

**Author Contributions:** Conceptualization, R.O., M.B. and A.T.; methodology, R.O., A.M.R. and A.C.; formal analysis, C.M., M.B., A.M.R., A.T. and R.O.; investigation, A.D., C.M., M.B., A.M.R., A.C., A.T. and R.O.; data curation, A.D., C.M., M.B., A.M.R., A.C., A.T. and R.O.; writing—original draft preparation, A.D., C.M., M.B., A.M.R., A.C., A.T. and R.O.; writing—review and editing, A.M.R., A.C., A.T. and R.O.; supervision, R.O. All authors have read and agreed to the published version of the manuscript.

**Funding:** This work was supported by a grant from the Ministry of Research, Innovation, and Digitalization, CNCS-UEFISCDI, project number PN-III-P1-1.1-PD-2021-0024, within PNCDI III.

**Institutional Review Board Statement:** Not applicable.

**Informed Consent Statement:** Not applicable.

**Data Availability Statement:** Data are contained within the article and Supplementary Materials.

**Acknowledgments:** The authors thank the student Sălăgeanu Andreea for his help in synthesizing and obtaining single crystals for compound (**4**).

**Conflicts of Interest:** The authors declare no conflict of interest.

## References

1. Schrader, S.M.; Botella, H.; Vaubourgeix, J. Reframing antimicrobial resistance as a continuous spectrum of manifestations. *Curr. Opin. Microbiol.* **2023**, *72*, 102259. [CrossRef]
2. Rogers, P.D.; Lee, R.E. Editorial overview: Recent advances in antimicrobial drug discovery and resistance. *Curr. Opin. Microbiol.* **2023**, *71*, 102242. [CrossRef]
3. Ye, J.; Chen, X. Current Promising Strategies against Antibiotic-Resistant Bacterial Infections. *Antibiotics* **2023**, *12*, 67. [CrossRef]
4. Olar, R.; Badea, M.; Chifiriuc, M.C. Metal Complexes—A Promising Approach to Target Biofilm Associated Infections. *Molecules* **2022**, *27*, 758. [CrossRef]
5. Mishra, S.; Gupta, A.; Upadhye, V.; Singh, S.C.; Sinha, R.P.; Häder, D.-P. Therapeutic Strategies against Biofilm Infections. *Life* **2023**, *13*, 172. [CrossRef]
6. Diban, F.; Di Lodovico, S.; Di Fermo, P.; D'Ercole, S.; D'Arcangelo, S.; Di Giulio, M.; Cellini, L. Biofilms in Chronic Wound Infections: Innovative Antimicrobial Approaches Using the In Vitro Lubbock Chronic Wound Biofilm Model. *Int. J. Mol. Sci.* **2023**, *24*, 1004. [CrossRef]
7. Paprocka, R.; Wiese-Szadkowska, M.; Janciauskiene, S.; Kosmalski, T.; Kulik, M.; Helmin-Basa, A. Latest developments in metal complexes as anticancer agents. *Coord. Chem. Rev.* **2022**, *452*, 214307. [CrossRef]
8. Lucaciu, R.L.; Hangan, A.C.; Sevastre, B.; Oprean, L.S. Metallo-Drugs in Cancer Therapy: Past, Present and Future. *Molecules* **2022**, *27*, 6485. [CrossRef] [PubMed]
9. Tirsoaga, A.; Cojocaru, V.; Badea, M.; Badea, I.A.; Rostas, A.M.; Stoica, R.; Bacalum, M.; Chifiriuc, M.C.; Olar, R. Copper (II) Species with Improved Anti-Melanoma and Antibacterial Activity by Inclusion in Cyclodextrin. *Int. J. Mol. Sci.* **2023**, *24*, 2688. [CrossRef] [PubMed]
10. Bijelic, A.; Aureliano, M.; Rompel, A. The antibacterial activity of polyoxometalates: Structures, antibiotic effects and future perspectives. *Chem. Commun.* **2018**, *54*, 1153–1169. [CrossRef]
11. Aureliano, M.; Gumerova, N.I.; Sciortino, G.; Garribba, E.; Rompel, A.; Crans, D.C. Polyoxovanadates with emerging biomedical activities. *Coord. Chem. Rev.* **2021**, *447*, 214143. [CrossRef]
12. Wang, X.; Wei, S.; Zhao, C.; Li, X.; Jin, J.; Shi, X.; Su, Z.; Li, J.; Wang, J. Promising application of polyoxometalates in the treatment of cancer, infectious diseases and Alzheimer's disease. *J. Biol. Inorg. Chem.* **2022**, *27*, 405–419. [CrossRef]
13. Bošnjaković-Pavlović, N.; Prévost, J.; Spasojević-de Biré, A. Crystallographic statistical study of decavanadate anion based-structures: Toward a prediction of noncovalent interactions. *Cryst. Growth Des.* **2011**, *11*, 3778–3789. [CrossRef]
14. Pereira, M.J.; Carvalho, E.; Eriksson, J.W.; Crans, D.C.; Aureliano, M. Effects of decavanadate and insulin enhancing vanadium compounds on glucose uptake in isolated rat adipocytes. *J. Inorg. Biochem.* **2009**, *103*, 1687–1692. [CrossRef] [PubMed]
15. García-Vicente, S.; Yraola, F.; Marti, L.; González-Muñoz, E.; García-Barrado, M.J.; Cantó, C.; Abella, A.; Bour, S.; Artuch, R.; Sierra, C.; et al. Oral insulin-mimetic compounds that act independently of insulin. *Diabetes* **2007**, *56*, 486–493. [CrossRef] [PubMed]
16. Treviño, S.; González-Vergara, E. Metformin-decavanadate treatment ameliorates hyperglycemia and redox balance of the liver and muscle in a rat model of alloxan-induced diabetes. *New J. Chem.* **2019**, *43*, 17850–17862. [CrossRef]

17. Marques-da-Silva, D.; Fraqueza, G.; Lagoa, R.; Vannathan, A.A.; Mal, S.S.; Aureliano, M. Polyoxovanadate inhibition of Escherichia coli growth shows a reverse correlation with $Ca^{2+}$-ATPase inhibition. *New J. Chem.* **2019**, *43*, 17577–17587. [CrossRef]
18. Abd-Elmonsef Mahmoud, G.; Ibrahim, A.B.M.; Mayer, P. $(NH_4)_2[Ni(H_2O)_6]_2V_{10}O_{28}\cdot4H_2O$; Structural Analysis and Bactericidal Activity against Pathogenic Gram Negative Bacteria. *Chem. Select.* **2021**, *6*, 3782–3787. [CrossRef]
19. Mamdouh, A.-A.; Ibrahim, A.B.M.; Reyad, N.E.-H.A.; Elsayed, T.R.; Santos, I.C.; Paulo, A.; Mahfouz, R.M. $(NH_4)_2[Co(H_2O)_6]_2V_{10}O_{28}\cdot4H_2O$ Vs. $(NH_4)_2[Ni(H_2O)_6]_2V_{10}O_{28}\cdot4H_2O$: Structural, Spectral and Thermal Analyses and Evaluation of Their Antibacterial Activities. *J. Clust. Sci.* **2022**, *34*, 1535–1546. [CrossRef]
20. Mamdouh, A.-A.; Ibrahim, A.B.M.; Reyad, N.E.-H.A.; Elsayed, T.R.; Cordeiro Santos, I.; Paulo, A.; Mahfouz, R.M. Monoclinic- vs. triclinic-$(NH_4)_2[Mg(H_2O)_6]2V_{10}O_{28}\bullet4H_2O$: Structural studies and variation in antibacterial activities with the polymorph type. *J. Mol. Struct.* **2022**, *1253*, 132247. [CrossRef]
21. Jammazi, D.; Ratel-Ramond, N.; Rzaigui, M.; Akriche, S. Trapped mixed $[(water)_4$–$(ammonium)_4{}^{4+}$ octamer in a 3D-binodal *(4,8)*-connected decavanadate core with hexamethylenetetramine: Synthesis, structure, photophysical and antimicrobial properties. *Polyhedron* **2019**, *168*, 146–154. [CrossRef]
22. Missina, M.; Gavinho, B.; Postal, K.; Santana, F.S.; Valdameri, G.; de Souza, E.M.; Hughes, D.L.; Ramirez, M.I.; Soares, J.F.; Nunes, G.G. Effects of decavanadate salts with organic and inorganic cations on Escherichia Coli, Giardia Intestinalis, and vero cells. *Inorg. Chem.* **2018**, *57*, 11930–11941. [CrossRef] [PubMed]
23. Shahid, M.; Sharma, P.K.; Anjuli; Chibber, S.; Siddiqi, Z.A. Isolation of a Decavanadate Cluster $[H_2V_{10}O_{28}][4\text{-picH}]_4\cdot2H_2O$ (4-pic = 4-picoline): Crystal Structure, Electrochemical Characterization, Genotoxic and Antimicrobial Studies. *J. Clust. Sci.* **2014**, *25*, 1435–1447. [CrossRef]
24. Samart, N.; Arhouma, Z.; Kumar, S.; Murakami, H.A.; Crick, D.C.; Crans, D.C. Decavanadate inhibits microbacterial growth more potently than other oxovanadates. *Front. Chem.* **2018**, *6*, 519. [CrossRef] [PubMed]
25. Chen, S.; Wu, G.; Long, D.; Liu, Y. Preparation, characterization and antibacterial activity of chitosan–Ca3V10O28 complex membrane. *Carbohydr. Polym.* **2006**, *64*, 92–97. [CrossRef]
26. Zhai, F.; Wang, X.; Li, D.; Zhang, H.; Li, R.; Song, L. Synthesis and biological evaluation of decavanadate $Na_4Co(H_2O)_6V_{10}O_{28}\cdot18H_2O$. *Biomed. Pharmacother.* **2009**, *63*, 51–55. [CrossRef]
27. Ksiksi, R.; Abdelkafi-Koubaa, Z.; Mlayah-Bellalouna, S.; Aissaoui, D.; Marrakchi, N.; Srairi-Abid, N.; Zid, F.M.; Graia, M. Synthesis, structural characterization and antitumoral activity of $(NH_4)_4Li_2V_{10}O_{28}.10H_2O$ compound. *J. Mol. Struct.* **2021**, *1229*, 129492. [CrossRef]
28. Louati, M.; Ksiksi, R.; Elbini-Dhouib, I.; Mlayah-Bellalouna, S.; Doghri, R.; Srairi-Abid, N.; Zid, M.-F. Synthesis, structure and characterization of a novel decavanadate, $Mg(H_2O)_6(C_4N_2H_7)4V_{10}O_{28}\cdot4H_2O$, with a potential antitumor activity. *J. Mol. Struct.* **2021**, *1242*, 130711. [CrossRef]
29. Li, Y.-T.; Zhu, C.-Y.; Wu, Z.-Y.; Jiang, M.; Yan, C.-W. Synthesis, crystal structures and anticancer activities of two decavanadate compounds. *Transit. Met. Chem.* **2010**, *35*, 597–603. [CrossRef]
30. Galani, A.; Tsitsias, V.; Stellas, D.; Psycharis, V.; Raptopoulou, C.P.; Karaliota, A. Two novel compounds of vanadium and molybdenum with carnitine exhibiting potential pharmacological use. *J. Inorg. Biochem.* **2015**, *142*, 109–117. [CrossRef]
31. Kioseoglou, E.; Gabriel, C.; Petanidis, S.; Psycharis, V.; Raptopoulou, C.P.; Terzis, A.; Salifoglou, A. Binary decavanadate-betaine composite materials of potential anticarcinogenic activity. *Z. Anorg. Allg. Chem.* **2013**, *639*, 1407–1416. [CrossRef]
32. De Sousa-Coelho, A.L.; Aureliano, M.; Fraqueza, G.; Serrão, G.; Gonçalves, J.; Sánchez Lombardo, I.; Link, W.; Ferreira, B.I. Decavanadate and metformin-decavanadate effects in human melanoma cells. *J. Inorg. Biochem.* **2022**, *235*, 111915. [CrossRef] [PubMed]
33. Ksiksi, R.; Essid, A.; Kouka, S.; Boujelbane, F.; Daoudi, M.; Srairi-Abid, N.; Zid, M.F. Synthesis and characterization of a tetra-(benzylammonium) dihydrogen decavanadate dihydrate compound inhibiting MDA-MB-231 human breast cancer cells proliferation and migration. *J. Mol. Struct.* **2022**, *1250*, 131929. [CrossRef]
34. Favre, D.; Harmon, J.F.; Zhang, A.; Miller, M.S.; Kaltashov, I.A. Decavanadate interactions with the elements of the SARS-CoV-2 spike protein highlight the potential role of electrostatics in disrupting the infectivity cycle. *J. Inorg. Biochem.* **2022**, *234*, 111899. [CrossRef]
35. Saczewski, F.; Balewski, Ł. Biological activities of guanidine compounds. *Expert Opin. Ther. Pat.* **2009**, *19*, 1417–1448. [CrossRef]
36. Ji, Y.; Yang, X.; Ji, Z.; Zhu, L.; Ma, N.; Chen, D.; Jia, X.; Tang, J.; Cao, Y. DFT-Calculated IR Spectrum Amide I, II, and III Band Contributions of N-Methylacetamide Fine Components. *ACS Omega* **2020**, *5*, 8572–8578. [CrossRef]
37. Omri, I.; Mhiri, T.; Graia, M. Novel decavanadate cluster complex $(HImz)_{12}(V_{10}O_{28})_2\cdot3H_2O$: Synthesis, characterization, crystal structure, optical and thermal properties. *J. Mol. Struct.* **2015**, *1098*, 324–331. [CrossRef]
38. Magon, C.J.; Lima, J.F.; Donoso, J.P.; Lavayen, V.; Benavente, E.; Navas, D.; Gonzalez, G. Deconvolution of the EPR spectra of vanadium oxide nanotubes. *J. Magn. Reson.* **2012**, *222*, 26–33. [CrossRef]
39. Brückner, A. In situ electron paramagnetic resonance: A unique tool for analyzing structure–reactivity relationships in heterogeneous catalysis. *Chem. Soc. Rev.* **2010**, *39*, 4673–4684. [CrossRef]
40. Suyama, K.; Kesamaru, H.; Okubo, T.; Kasatani, K.; Tomohara, K.; Matsushima, A.; Nose, T. High cytotoxicity of a degraded TBBPA, dibromobisphenol A, through apoptotic and necrosis pathways. *Heliyon* **2023**, *16*, e13003. [CrossRef]
41. Ciorîță, A.; Suciu, M.; Macavei, S.; Kacso, I.; Lung, I.; Soran, M.-L.; Pârvu, M. Green Synthesis of Ag-$MnO_2$ Nanoparticles using Chelidonium majus and Vinca minor Extracts and Their In Vitro Cytotoxicity. *Molecules* **2020**, *25*, 819. [CrossRef] [PubMed]

42. Xia, D.; Esser, L.; Tang, W.-K.; Zhou, F.; Zhou, Y.; Yu, L.; Yu, C.-A. Structural analysis of cytochrome bc1 complexes: Implications to the mechanism of function. *Biochimica Biophysica Acta* **2013**, *1827*, 1278–1294. [CrossRef] [PubMed]
43. Sarewicz, M.; Osyczka, A. Electronic connection between the quinone and cytochrome c redox pools and its role in regulation of mitochondrial electron transport and redox signaling. *Physiol. Rev.* **2015**, *95*, 219–243. [CrossRef] [PubMed]
44. Chen, Q.; Vazquez, E.J.; Moghaddas, S.; Hoppel, C.L.; Lesnefsky, E.J. Production of reactive oxygen species by mitochondria: Central role of complex III. *J. Biol. Chem.* **2003**, *278*, 36027–36031. [CrossRef] [PubMed]
45. Rosell-Hidalgo, A.; Moore, L.A.; Ghafourian, T. Prediction of grug-induced mitochondiral dysfunction using succinate-cytochrome c reductase activity, QSAR and molecular docking. *Toxicology* **2023**, *485*, 153412. [CrossRef]
46. Soares, S.S.; Gutiérrez-Merino, C.; Aureliano, M. Decavanadate induces mitochondrial membrane depolarization and inhibits oxygen consumption. *J. Inorg. Biochem.* **2007**, *101*, 789–796. [CrossRef]
47. Huang, L.S.; Cobessi, D.; Tung, E.; Berry, E.A. Binding of the respiratory chain inhibitor antimycin to the mitochondrial bc1 complex: A new crystal structure reveals an altered intramolecular hydrogen-bonding pattern. *J. Mol. Biol.* **2005**, *351*, 573–597. [CrossRef]
48. Kim, H.J.; Khalimonchuk, O.; Smith, P.M.; Winge, D.R. Structure, function and assembly of heme centers in mitochondrial respiratory complexes. *Biochimia Biophysica Acta Mol. Cell Res.* **2012**, *1823*, 1604–1616. [CrossRef]
49. Sheldrick, G.M. *SHELXL-97, Program for Crystal Structure Refinement*; University of Göttingen: Göttingen, Germany, 1998.
50. Nekvapil, F.; Ganea, I.-V.; Ciorîță, A.; Hirian, R.; Ogresta, L.; Glamuzina, B.; Roba, C.; Cintă Pinzaru, S. Wasted Biomaterials from Crustaceans as a Compliant Natural Product Regarding Microbiological, Antibacterial Properties and Heavy Metal Content for Reuse in Blue Bioeconomy: A Preliminary Study. *Materials* **2021**, *14*, 4558. [CrossRef]
51. Walencka, E.; Różalska, S.; Sadowska, B.; Różalska, B. The influence of Lactobacillus acidophilus-derived surfactants on staphylococcal adhesion and biofilm formation. *Folia Microbiol.* **2008**, *53*, 61–66. [CrossRef]
52. Ciorîță, A.; Zăgrean-Tuza, C.; Mot, A.C.; Carpa, R.; Pârvu, M. The Phytochemical Analysis of Vinca L. Species Leaf Extracts Is Correlated with the Antioxidant, Antibacterial, and Antitumor Effects. *Molecules* **2021**, *26*, 3040. [CrossRef] [PubMed]
53. Trott, O.; Olson, A.J. AutoDock Vina: Improving the speed and accuracy of docking with a new scoring function, efficient optimization and multithreading. *J. Comp. Chem.* **2010**, *31*, 455–461. [CrossRef] [PubMed]
54. Eberhardt, J.; Santos Martins, D.; Tillack, A.F.; Forli, S. AutoDock Vina 1.2.0: New Docking Methods, Expanded Force Field and Python Bindings. *J. Chem. Inf. Model.* **2021**, *61*, 3891–3898. [CrossRef] [PubMed]
55. Sanner, M.F. Python: A programming language for software integration and development. *J. Mol. Graphics Mod.* **1999**, *17*, 57–61. [CrossRef]

International Journal of
*Molecular Sciences*

MDPI

*Article*

# Monolacunary Wells-Dawson Polyoxometalate as a Novel Contrast Agent for Computed Tomography: A Comprehensive Study on In Vivo Toxicity and Biodistribution

**Marko Stojanović [1,†], Mirjana B. Čolović [2,†], Jovana Lalatović [3], Aleksandra Milosavljević [4], Nada D. Savić [5], Kilian Declerck [5], Branimir Radosavljević [6], Mila Ćetković [4], Tamara Kravić-Stevović [4], Tatjana N. Parac-Vogt [5,*] and Danijela Krstić [6,*]**

1 Department of Pharmacology, Clinical Pharmacology and Toxicology, Faculty of Medicine, University of Belgrade, 11000 Belgrade, Serbia; marko.stojanovic@med.bg.ac.rs
2 "Vinča" Institute of Nuclear Sciences—National Institute of the Republic of Serbia, University of Belgrade, 11351 Belgrade, Serbia; colovicm@vin.bg.ac.rs
3 Department of Radiology, University Hospital Medical Center Bežanijska Kosa, 11080 Belgrade, Serbia; lalatovic.jovana@bkosa.edu.rs
4 Institute of Histology and Embryology, Faculty of Medicine, University of Belgrade, 11000 Belgrade, Serbia; aleksandra.milosavljevic@med.bg.ac.rs (A.M.); mila.cetkovic-milisavljevic@med.bg.ac.rs (M.Ć.); tamara.kravic-stevovic@med.bg.ac.rs (T.K.-S.)
5 Department of Chemistry, KU Leuven, Celestijnenlaan 200F, 3001 Leuven, Belgium; nada.savic@kuleuven.be (N.D.S.); kilian.declerck@kuleuven.be (K.D.)
6 Institute of Medical Chemistry, Faculty of Medicine, University of Belgrade, 11000 Belgrade, Serbia; branimir.radosavljevic@med.bg.ac.rs
* Correspondence: tatjana.vogt@kuleuven.be (T.N.P.-V.); danijela.krstic@med.bg.ac.rs (D.K.)
† These authors contributed equally to this work.

**Abstract:** Polyoxotungstate nanoclusters have recently emerged as promising contrast agents for computed tomography (CT). In order to evaluate their clinical potential, in this study, we evaluated the in vitro CT imaging properties, potential toxic effects in vivo, and tissue distribution of monolacunary Wells–Dawson polyoxometalate, $\alpha_2$-$K_{10}P_2W_{17}O_{61}\cdot 20H_2O$ (mono-WD POM). Mono-WD POM showed superior X-ray attenuation compared to other tungsten-containing nanoclusters (its parent WD-POM and Keggin POM) and the standard iodine-based contrast agent (iohexol). The calculated X-ray attenuation linear slope for mono-WD POM was significantly higher compared to parent WD-POM, Keggin POM, and iohexol (5.97 ± 0.14 vs. 4.84 ± 0.05, 4.55 ± 0.16, and 4.30 ± 0.09, respectively). Acute oral (maximum-administered dose (MAD) = 960 mg/kg) and intravenous administration (1/10, 1/5, and 1/3 MAD) of mono-WD POM did not induce unexpected changes in rats' general habits or mortality. Results of blood gas analysis, CO-oximetry status, and the levels of electrolytes, glucose, lactate, creatinine, and BUN demonstrated a dose-dependent tendency 14 days after intravenous administration of mono-WD POM. The most significant differences compared to the control were observed for 1/3 MAD, being approximately seventy times higher than the typically used dose (0.015 mmol W/kg) of tungsten-based contrast agents. The highest tungsten deposition was found in the kidney (1/3 MAD—0.67 ± 0.12; 1/5 MAD—0.59 ± 0.07; 1/10 MAD—0.54 ± 0.05), which corresponded to detected morphological irregularities, electrolyte imbalance, and increased BUN levels.

**Keywords:** blood gas analysis; CO-oximetry status; biochemical parameters; histological analysis; in vitro computed tomography imaging; in vivo toxicity; monolacunary Wells-Dawson polyoxotungstate; tissue distribution; X-ray attenuation

## 1. Introduction

Computed Tomography (CT) has been employed as a pivotal imaging technique in both clinical and research domains, providing 3D images with high spatial and temporal resolution. The constant evolution of CT technology, characterized by innovations in image reconstruction algorithms and reduced radiation doses, is largely able to address its limitations. In the field of diagnostic medicine, this non-invasive technique, combined with contrast media, provides detailed 3-dimensional anatomical information of blood vessels, internal organs, and pathological lesions [1–3].

However, compared to other clinical imaging techniques such as Magnetic Resonance Imaging, the lack of sensitivity to contrast agents is the main limitation of CT. Furthermore, the conventionally used intravenous contrast agents for CT, iodinated small molecules (derivatives of a 2,4,6-triiodinated benzene ring), have several drawbacks, including adverse reactions [4–6] and a heightened risk for patients with specific conditions [7]. In addition, these contrast agents are not specific for any pathology and have limited blood circulation time. Thus, in the realm of medical imaging, the quest for innovative CT contrast agents to enhance diagnostic accuracy and imaging capabilities remains an ongoing endeavor. Recent studies highlight the potential of metallic nanoparticles and polyoxometalate (POM) nanoclusters as promising candidates for developing new-generation CT contrast agents [8–12]. These metal-based compounds possess extended circulation times and favorable X-ray attenuation, whereas their shape, composition, size, and surface characteristics are tunable for desired biomedical applications [13,14].

In this study, in vitro CT imaging properties and potential toxic effects of monolacunary Wells–Dawson POM, $\alpha_2$-$K_{10}P_2W_{17}O_{61}\cdot 20H_2O$ (mono-WD POM), after intravenous administration in rats were evaluated to achieve a careful balance between imaging performance and potential side effects in the development of new-generation POM-based contrast agents. In addition, ex vivo biodistribution of mono-WD POM was evaluated two weeks after treatment with 1/10, 1/5, and 1/3 of the maximum-administered dose (MAD) by measuring tungsten concentrations in different tissue samples, using Inductively Coupled Plasma-Optical Emission Spectroscopy (ICP-OES). Mono-WD POM (Scheme 1) is a derivative of parent WD-POM ($\alpha$-/$\beta$-$K_6P_2W_{18}O_{62}$), which was studied as a potential CT contrast in our previous research [9]. Parent WD-POM is partially converted into mono-WD POM at physiological pH by spontaneous removal of one -WO unit [12]. Moreover, de Bournonville et al. reported mono-WD POM as the most suitable non-destructive staining agent for contrast-enhanced microfocus CT of bone and kidney [12]. Generally, the results of this study contribute novel insights into the relationship between structure and toxic properties of POMs, which have been widely explored as promising diagnostic tools and therapeutics [10–12,15–17].

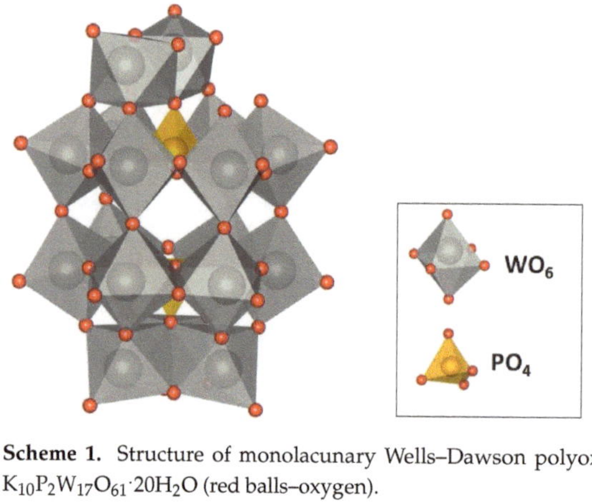


**Scheme 1.** Structure of monolacunary Wells–Dawson polyoxotungstate (mono-WD POM), $\alpha_2$-$K_{10}P_2W_{17}O_{61}\cdot 20H_2O$ (red balls–oxygen).

## 2. Results

### 2.1. In Vitro CT Performances

X-ray attenuation (X-ray source voltage of 80 kV) was measured in vitro in the presence of increasing concentrations of tungsten (3.125–100 mM), one Keggin structure POM (12-tungstosilicic acid, WSiA), and two analogous WD POMs: parent WD [9] and its monolacunary derivative, mono-WD POM (Scheme 1). The corresponding in vitro CT phantom images of mono-WD and WSiA are presented in Figure 1a. In order to compare the contrast performances of these tungsten-containing compounds with a commercially available CT contrast agent that has been employed in clinical practice, X-ray attenuation was also presented for the same iodine concentrations of iodine-based iohexol solution [9]. The results are expressed in Hounsfield units (HU) and presented as a function of W/I concentration in Figure 1b. The obtained experimental results for all tested compounds showed an excellent fit to a linear function ($R^2 > 0.99$). The highest HU values for particular concentrations were obtained for mono-WD POM compared to those for both its parent WD compound and WSiA, a POM with Keggin structure. Accordingly, the calculated value of the slope for mono-WD POM was significantly higher compared to parent WD and WSiA ($5.97 \pm 0.14$ vs. $4.84 \pm 0.05$ and $4.55 \pm 0.16$, respectively). The obtained values of X-ray attenuation for all studied concentrations were evidently higher for all three tungsten-containing POMs in comparison with the standard iodine-based CT contrast agent, iohexol, for which the slope of the linear function had a value of $4.30 \pm 0.09$ (Figure 1b).

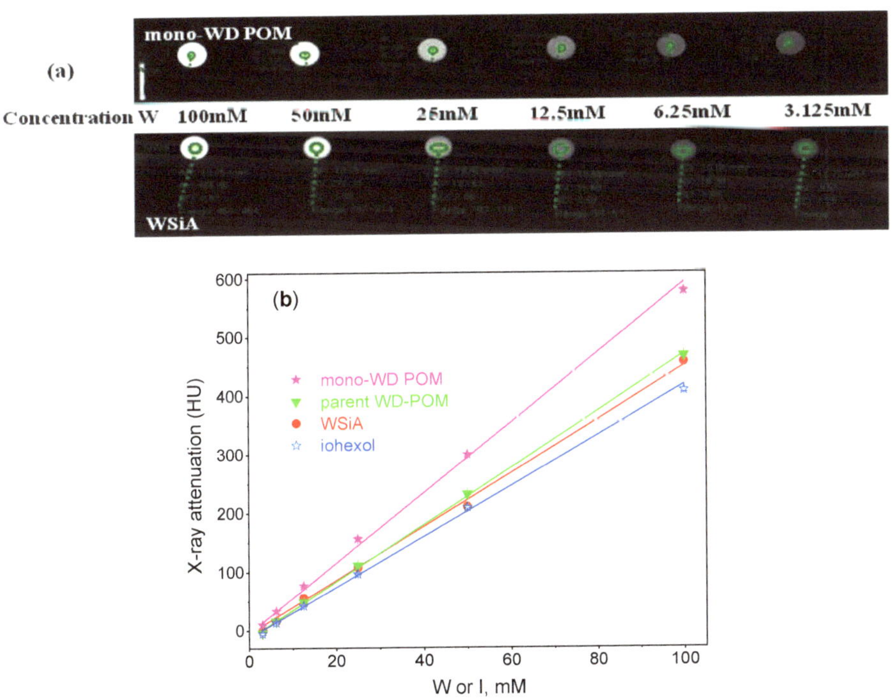


**Figure 1.** (**a**) Images generated from a clinical CT scanner phantom scan of mono-WD POM and WSiA, ranging from 3.125 to 100 mM W; (**b**) the linear dependence of X-ray attenuation (expressed in Hounsfield units (HU)) on W/I concentration for the tungsten-containing compounds of various structures: Keggine structure POM (12-tungstosilicic acid, WSiA) (red circle), parent WD-POM (green triangle) [9] and its monolacunary derivative (mono-WD POM) (pink solid asterisk), and iodine-containing omnipaque (blue open asterisk) [9].

### 2.2. *General Habit of Animals*

After the acute oral application of MAD (960 mg/kg mono-WD POM), any unexpected changes in the appearance of skin and visible mucosa, sleeping, feeding, and body weight, as well as mortality, were not observed in the treated Wistar rats. This indicates that the orally applied MAD does not produce severe toxic effects or death in animals.

During a 14-day follow-up period after the intravenous administration of mono-WD POM, a mortality rate was investigated for four experimental groups (five rats per one group): (1) control (saline), (2) 1/3 MAD (320 mg/kg), (3) 1/5 MAD (192 mg/kg), and (4) 1/10 MAD (96 mg/kg), and was 0% in all cases. Indeed, all observed experimental rats (both the treated and control groups) survived (a 100% survival rate) 14 days post-mono-WD POM application intravenously. Furthermore, during the 14 days after the intravenous application of saline/mono-WD POM, the rats were observed in the morning and evening [18,19]. No significant changes in behavior (sleeping and feeding) and body weight (not following the rat age) were detected throughout the follow-up period. The inspection of skin and visible mucosa did not reveal any suspected changes in appearance in all experimental groups (both control and POM-administered animals).

### 2.3. *Laboratory Blood Analysis*

On the 14th day post-treatment, arterial blood gas analysis, CO-oxymetry status, electrolyte levels, glucose level, lactate level, creatine level, and BUN values were measured from arterial blood.

#### 2.3.1. Arterial Blood Gas Analysis

Arterial blood gas analysis (Table 1) demonstrates low pH values that were under the normal range for all groups, both control and three treated groups, without statistical differences between them ($p$ = 0.0545). $pCO_2$ levels were above the upper limit of the normal range for all treated groups. A statistically significant difference was noted only between the control and 1/3 MAD group ($p$ = 0.0007). $HCO_3^-$ levels detected in the treated groups (1/5 and 1/3 MAD) were above the upper limit of the normal range. A statistically significant difference in $HCO_3^-$ level was detected between all treated groups and the control group ($p < 0.0001$). The base excess of blood values was significantly higher in the treated groups ($p$ = 0.0008) relative to control. The detected control group values were below the lower limit of the normal range. The anion gap values were below the lower limit of the normal range in the 1/3 MAD group, whereas these values were within the normal range for the other groups. A statistically significant difference was obtained between the control and 1/3 MAD, as well as the 1/5 MAD group ($p < 0.0001$).

**Table 1.** The results of arterial blood gas analysis for the control group and animals treated with 1/10 MAD, 1/5 MAD, and 1/3 MAD of mono-WD POM.

| Analyte | Units | Control (n = 5) | 1/10 MAD (n = 5) | 1/5 MAD (n = 5) | 1/3 MAD (n = 5) | ANOVA *p*-Value | Normal Range |
|---|---|---|---|---|---|---|---|
| pH | / | 7.16 ± 0.03 | 7.16 ± 0.03 | 7.28 ± 0.17 | 7.07 ± 0.12 | 0.0545 | 7.3–7.4 |
| $pCO_2$ | kPa | 4.97 ± 1.02 | 9.84 ± 2.54 | 9.92 ± 4.50 | 15.78 ± 3.57 | 0.0007 *** | 4.66–5.99 |
| $HCO_3^-$ | mmol/L | 17.52 ± 3.48 | 26.43 ± 5.57 | 29.79 ± 2.46 | 34.41 ± 1.77 | <0.0001 ***†††‡‡ | 21–28 |
| Base excess of blood | mmol/L | (−) 6.64 ± 1.54 | (−) 2.85 ± 3.47 | (−) 0.7 ± 1.79 | 1.56 ± 2.86 | 0.0008 ***† | −2 to +3 |
| Anion gap | mmol/L | 15.94 ± 1.88 | 13.60 ± 2.57 | 9.01 ± 3.53 | 5.11 ± 1.45 | <0.0001 ***†† | 8.00–16.00 |

For determining the difference in values of arterial blood gas analysis between control and three treated groups, one-way ANOVA was used, followed by the Tukey post hoc test. After Tukey post hoc analysis, the difference between the control and treated groups was presented as: * for 1/3 MAD group, † for 1/5 MAD group, and ‡ for 1/10 MAD group. The significant values: *** $p < 0.001$, ††† $p < 0.001$, †† $p < 0.01$, † $p < 0.05$, ‡‡ $p < 0.01$.

#### 2.3.2. Arterial Blood CO-Oxymetry Status

The results of CO-oximetry analysis for the control and all three treated groups are shown in Table 2. It was found that the values of total hemoglobin, carboxyhemoglobin, and the oxygen capacity of hemoglobin were within the normal range for all studied groups, and there was no statistically significant difference among groups regarding these parameters ($p > 0.05$). The oxyhemoglobin values were below the lower limit of the normal range in the treated groups. These values were statistically significantly lower in the treated groups compared to the control group ($p = 0.0001$). Obtained methemoglobin values were in the normal range in both control and treated groups. However, values for the treated groups were significantly lower ($p = 0.0008$) relative to the control values. Then, the values of deoxyhemoglobin in the treated groups were highly above the upper limit of the normal range and were statistically significantly higher than the values measured in the control group ($p < 0.0001$). The oxygen saturation was below the normal range in the treated groups, and these values were statistically significantly lower in the 1/5 MAD and 1/10 MAD groups when compared to the control group ($p = 0.004$). Furthermore, the oxygen content of hemoglobin was below the normal range in all treated groups, whereas a statistically significant difference was observed when the control and 1/5 MAD group were compared ($p = 0.0202$).

**Table 2.** The results of CO-oximetry analysis obtained for the control group and animals treated with 1/10 MAD, 1/5 MAD, and 1/3 MAD of mono-WD POM.

| Analyte | Units | Control (n = 5) | 1/10 MAD (n = 5) | 1/5 MAD (n = 5) | 1/3 MAD (n = 5) | ANOVA *p*-Value | Normal Range |
|---|---|---|---|---|---|---|---|
| Total hemoglobin | g/L | 164.50 ± 7.02 | 173 ± 10.97 | 173.20 ± 13.88 | 156.40 ± 8.96 | 0.0669 | 140–178 |
| Oxyhemoglobin | % | 97.87 ± 0.52 | 67.36 ± 12.78 | 60.04 ± 7.88 | 73.61 ± 13.69 | 0.0001 **†††‡‡ | 94–97 |
| Carboxyhemoglobin | % | 0.45 ± 0.04 | 0.43 ± 0.03 | 0.7 ± 0.46 | 0.61 ± 0.31 | 0.3869 | 0.0–1.5 |
| Methemoglobin | % | 0.82 ± 0.29 | 0.39 ± 0.04 | 0.4 ± 0.07 | 0.40 ± 0.06 | 0.0008 **††‡‡ | 0.0–1.5 |
| Deoxyhemoglobin | % | 0.55 ± 0.07 | 30.91 ± 11.68 | 36.92 ± 4.22 | 23.93 ± 10.96 | <0.0001 **†††‡‡‡ | 0.0–5.0 |
| Oxygen saturation | % | 97.8 ± 3.82 | 69.50 ± 14.16 | 60 ± 8.68 | 74.63 ± 15.67 | 0.004 ††‡ | 95–98 |
| Oxygen content of hemoglobin | vol % | 21.97 ± 3.46 | 14.93 ± 5.04 | 13.38 ± 3.50 | 16.18 ± 3.61 | 0.0202 † | 18–24 |
| Oxygen capacity of hemoglobin | vol % | 23.42 ± 1.85 | 23.89 ± 1.56 | 24.04 ± 1.96 | 21.24 ± 1.21 | 0.0994 | 18–25 |

For determining the difference in CO-oximetry values between control and three treated groups, one-way ANOVA was used, followed by the Tukey *post hoc test. After* Tukey post hoc analysis, the difference between control and treated groups was presented as: * for 1/3 MAD group, † for 1/5 MAD group, and ‡ for 1/10 MAD group. The significant values: ** $p < 0.01$, ††† $p < 0.001$, †† $p < 0.01$, † $p < 0.05$, ‡‡‡ $p < 0.001$, ‡‡ $p < 0.01$, ‡ $p < 0.05$.

### 2.3.3. Arterial Blood Electrolyte Concentrations

The obtained blood values of different electrolytes ($Na^+$, $K^+$, $Cl^-$, $Ca^{2+}$, $Mg^{2+}$, $Ca^{2+}$ normalized to pH 7.4 and $Mg^{2+}$ normalized to pH 7.4) are presented in Table 3. When comparing the values of $Na^+$, $K^+$, and $Cl^-$ between the control and treated groups, there is no statistically significant difference, nor do the obtained values deviate significantly from the normal range. $Ca^{2+}$ values were within the normal range in all groups except for the 1/3 MAD group, whose values were above the upper limit of the normal range. Statistically significantly higher values were detected in the 1/3 MAD group compared to the control group ($p = 0.0058$). The $Mg^{2+}$ values were above the upper limit in all three treated groups, whereas a statistically significant difference was detected between the obtained values for the control and 1/3 MAD group ($p = 0.0001$). The values of $Ca^{2+}$ normalized to pH 7.4 were in the normal range in both control and treated groups, and a statistically significant difference was observed only when the control group was compared to the 1/5 MAD group ($p = 0.0151$). Finally, the values of $Mg^{2+}$ normalized to pH 7.4 were above the upper limit for all three treated groups. A statistically significant difference was obtained between the control group and 1/3 MAD group, as well as the 1/5 MAD group ($p = 0.002$).

**Table 3.** Electrolyte concentrations in arterial blood obtained for the control group and treated animals (1/10, 1/5, and 1/3 MAD of mono-WD POM).

| Analyte | Units | Control (n = 5) | 1/10 MAD (n = 5) | 1/5 MAD (n = 5) | 1/3 MAD (n = 5) | ANOVA *p* Value | Normal Range |
|---|---|---|---|---|---|---|---|
| $Na^+$ | mmol/L | 135.8 ± 1.52 | 135.9 ± 3.02 | 137.7 ± 2.78 | 135.4 ± 4.12 | 0.6297 | 136–146 |
| $K^+$ | mmol/L | 5.42 ± 0.46 | 5.59 ± 0.55 | 5.22 ± 0.49 | 5.52 ± 0.26 | 0.6168 | 3.5–5.2 |
| $Cl^-$ | mmol/L | 103.70 ± 2.39 | 102.0 ± 1.39 | 104.10 ± 2.63 | 103.8 ± 3.33 | 0.554 | 98–106 |
| $Ca^{2+}$ | mmol/L | 1.23 ± 0.04 | 1.27± 0.04 | 1.29 ± 0.06 | 1.39 ± 0.08 | 0.0058 ** | 1.09–1.30 |
| $Mg^{2+}$ | mmol/L | 0.66 ± 0.10 | 0.96 ± 0.13 | 0.95 ± 0.16 | 1.32 ± 0.23 | 0.0001 *** | 0.45–0.65 |
| Ionized $Ca^{2+}$ normalized to pH 7.4 (nCa) | mmol/L | 1.12 ± 0.04 | 1.12 ± 0.03 | 1.22 ± 0.08 | 1.19 ± 0.07 | 0.0151 † | 1.09–1.30 |
| Ionized $Mg^{2+}$ normalized to pH 7.4 (nMg) | mmol/L | 0.61 ± 0.04 | 0.83 ± 0.10 | 0.89 ± 0.23 | 1.04 ± 0.13 | 0.002 **† | 0.45–0.6 |

For determining the difference in electrolyte concentrations between the control and three treated groups, one-way ANOVA was used, followed by the Tukey post hoc test. After Tukey post hoc analysis, the difference between the control and treated groups was presented as: * for the 1/3 MAD group and † for the 1/5 MAD group. The significant values: *** $p < 0.001$, ** $p < 0.01$, † $p < 0.05$.

### 2.3.4. Arterial Blood Glucose Level

The glucose levels (Figure 2) in the control group (9.22 ± 0.91 mmol/L), 1/10 MAD group (11.58 ± 1.53 mmol/L), 1/5 MAD group (12.40 ± 2.05 mmol/L), and 1/3 MAD group (14.56 ± 3.28 mmol/L) were above the upper limit of the normal range (2.8–7.5 mmol/L). A statistically significant difference in glucose level was detected only between the control and 1/3 MAD groups ($p = 0.0095$).

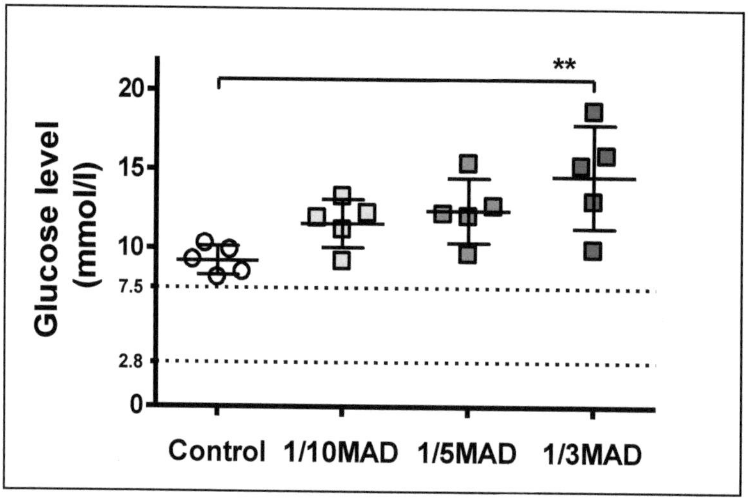


**Figure 2.** Arterial blood glucose levels in the control group and animals treated with 1/10 MAD, 1/5 MAD, and 1/3 MAD of mono-WD POM. The results are expressed as mean value ± SD. The significant values: ** $p < 0.01$.

### 2.3.5. Arterial Blood Lactate Level

Lactic acid levels (Figure 3) were above the upper limit of the normal range (0.7–2.5 mmol/L) in the control group (7.21 ± 1.65 mmol/L), 1/10 MAD group (7.68 ± 1.19 mmol/L)), 1/5 MAD group (5.98 ± 0.89 mmol/L), and 1/3 MAD group (6.05 ± 2.60 mmol/L). No statistically significant differences were detected between the control and all treated groups.

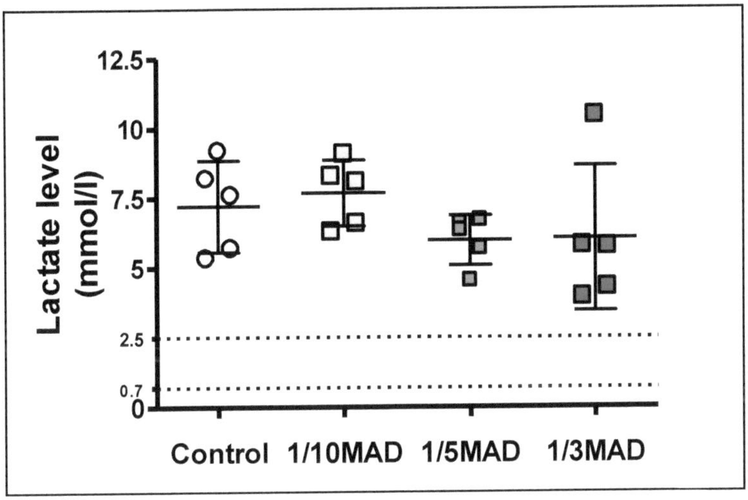


**Figure 3.** Arterial blood lactate levels in the control group and animals treated with 1/10 MAD, 1/5 MAD, and 1/3 MAD of mono-WD POM. The results are expressed as mean value ± SD.

2.3.6. Arterial Blood Creatinine Concentrations

The creatinine values measured in arterial blood (Figure 4) were within the limits of the normal range (35–75 µmol/L) in the control (71.50 ± 13.42 µmol/L) and 1/5 MAD group (42.38 ± 9.39 µmol/L), just below the upper limit of the normal range in the 1/10 MAD group (74.60 ± 15.06 µmol/L), and above the upper limit of the normal range in the 1/3 MAD group (88.62 ± 14.38 µmol/L), creating a U-shape curve. There were no statistically significant differences between the control group and any of the treated groups ($p = 0.0507$).

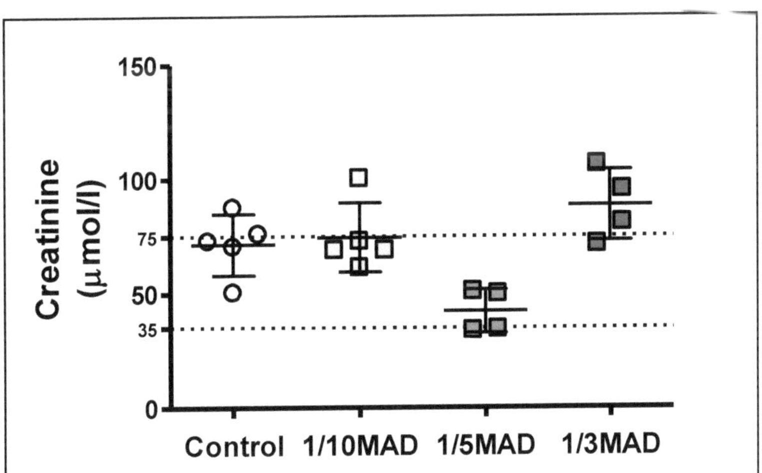


**Figure 4.** Arterial blood creatinine concentrations in the control group and animals treated with 1/10 MAD, 1/5 MAD, and 1/3 MAD of mono-WD POM. The results are expressed as mean value ± SD.

#### 2.3.7. Arterial Blood BUN Concentrations

In the control group, the detected BUN values (7.35 ± 1.27 µmol/L) were within the normal range (5.36–7.86 µmol/L). However, BUN values (µmol/L) were above the upper limit of the normal range in the 1/10 MAD (8.53 ± 0.62), 1/5 MAD (10.39 ± 1.22), and 1/3 MAD (11.36 ± 1.16) groups (Figure 5). A statistically significant difference was obtained between the control group and the 1/5 MAD as well as the 1/3 MAD groups ($p = 0.0001$).

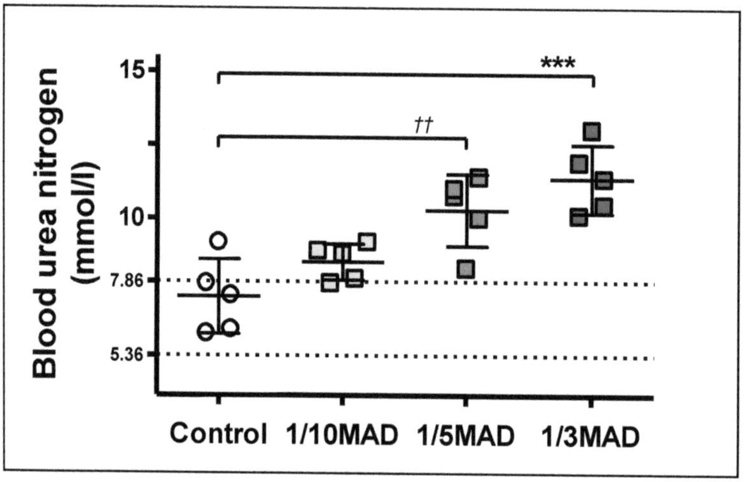


**Figure 5.** Arterial blood urea nitrogen (BUN) concentrations in the control group and animals treated with 1/10 MAD, 1/5 MAD, and 1/3 MAD of mono-WD POM. The results are expressed as mean value ± S.D. The significant values: *** $p < 0.001$, †† $p < 0.01$.

### 2.4. Histological Analysis

#### 2.4.1. Histopathological Evaluation of Mono-WD POM—Induced Renal Toxicity

The results of histopathological analysis, performed using light microscopy and transmission electronic microscopy (TEM) techniques, of rat kidney tissues after 14-day treatment with 1/10, 1/5, and 1/3 MAD of mono-WD POM are presented in Figure 6. The photographs of the HE-stained kidney tissues show that there is no difference between the kidney tissues from the untreated group (control) (a,b) and rats treated with 1/10 MAD (c,d) and 1/5 MAD (e,f), whereas a necrosis of tubular cells (arrowhead) and glomerular sclerosis (arrows) were observed in the 1/3 MAD group (g,h). TEM analysis confirmed the necrosis of tubular cells after 1/3 MAD administration (4) and additionally revealed a similar effect in 1/5 MAD (3) as well as an apoptosis of tubular cells in the 1/10 MAD group (2).

#### 2.4.2. Histopathological Evaluation of Mono-WD POM—Induced Hepatotoxicity

Photographs of rat liver sections excised from the untreated (control) and treated rats (1/10, 1/5, and 1/3 MAD of mono-WD POM), captured under a light microscope and TEM, are shown in Figure 7. Photomicrographs of HE-stained liver tissues demonstrated that the administration of 1/5 (e,f) and 1/3 MAD (g,h) induced discrete necrosis. Moreover, a necrosis of hepatocytes was also visible under TEM (3,4) for these experimental groups. On the other hand, normal tissue morphology (as in the control group (a,b)) was seen in the liver tissues of 1/10 MAD-treated rats (c,d) using light microscopy, whereas apoptosis of endothelial cells was detected in TEM micrographs obtained for this experimental group (2).

2.4.3. Histopathological Evaluation of Mono-WD POM—Induced Lung Toxicity

TEM micrographs and photographs of HE-stained lung tissues analyzed after rat treatment with 1/10, 1/5, and 1/3 MAD of mono-WD POM are presented in Figure 8. The results of both light microscopy and TEM analysis showed that 1/10 treatment did not induce any histological changes in the lung sections (c,d,2) compared with the control (untreated) group (a,b,1). However, a thickening of the interstitial spaces was observed in HE-stained lung sections excised from 1/3 MAD-treated rats (g,h), whereas TEM micrographs detected a fibrosis and a prominent fibrosis of lung interstitium in both 1/5 MAD (3) and 1/3 MAD (4) experimental groups.

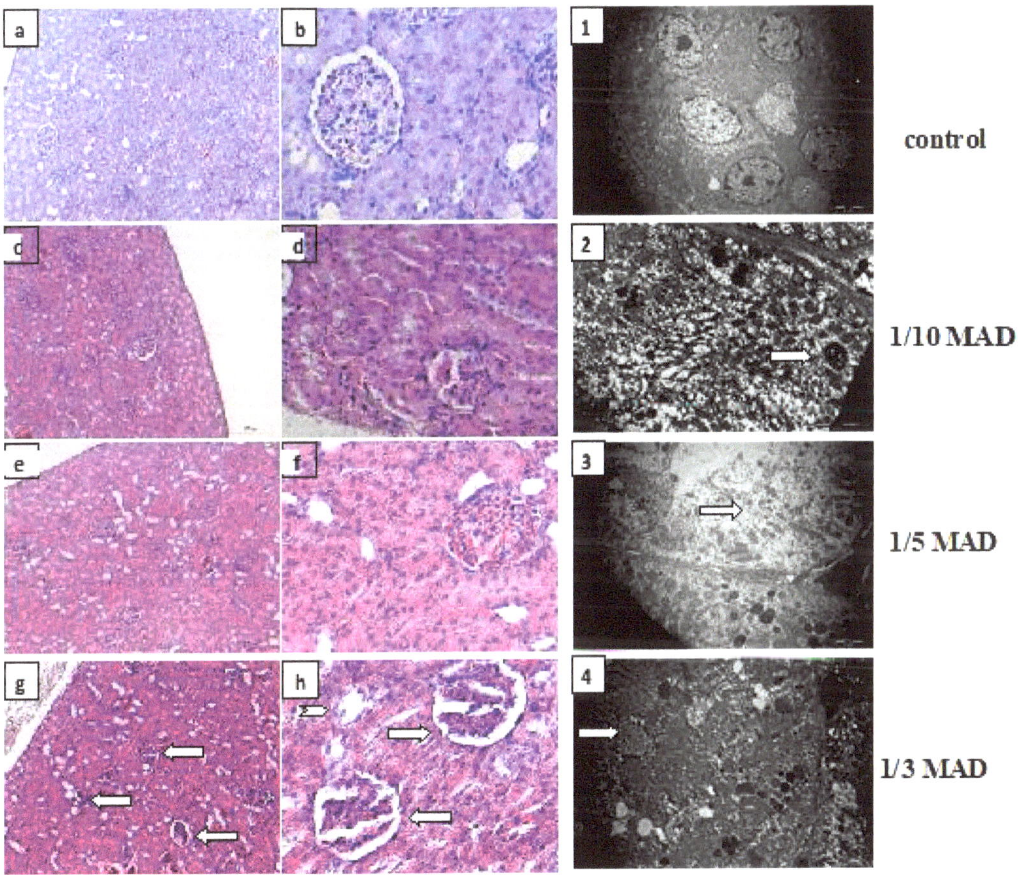


**Figure 6.** Histopathological evaluation of mono-WD POM-induced renal toxicity. Photographs of HE-stained rat kidney sections (**a–h**); image was captured under a light microscope with ×10 (**a**,**c**,**e**,**g**) and ×40 (**b**,**d**,**f**,**h**) magnifications. No differences were observed between the kidney tissues from the control group (**a**,**b**) and rats treated with 1/10 MAD (**c**,**d**) and 1/5 MAD (**e**,**f**), whereas tissue from rats treated with 1/3 MAD (**g**,**h**) showed necrosis of tubular cells (arrowhead) and glomerular sclerosis (arrows). TEM micrographs of kidney sections (**1–4**) show: (**1**) (×2800) control rat; (**2**) rat treated with 1/10 MAD showing (×4400) apoptosis of tubular cells (arrows); and (**3**) (×2800), (**4**) (×3500)) rats treated with 1/5 MAD and 1/3 MAD, respectively, showing necrosis of tubular cells (arrows).

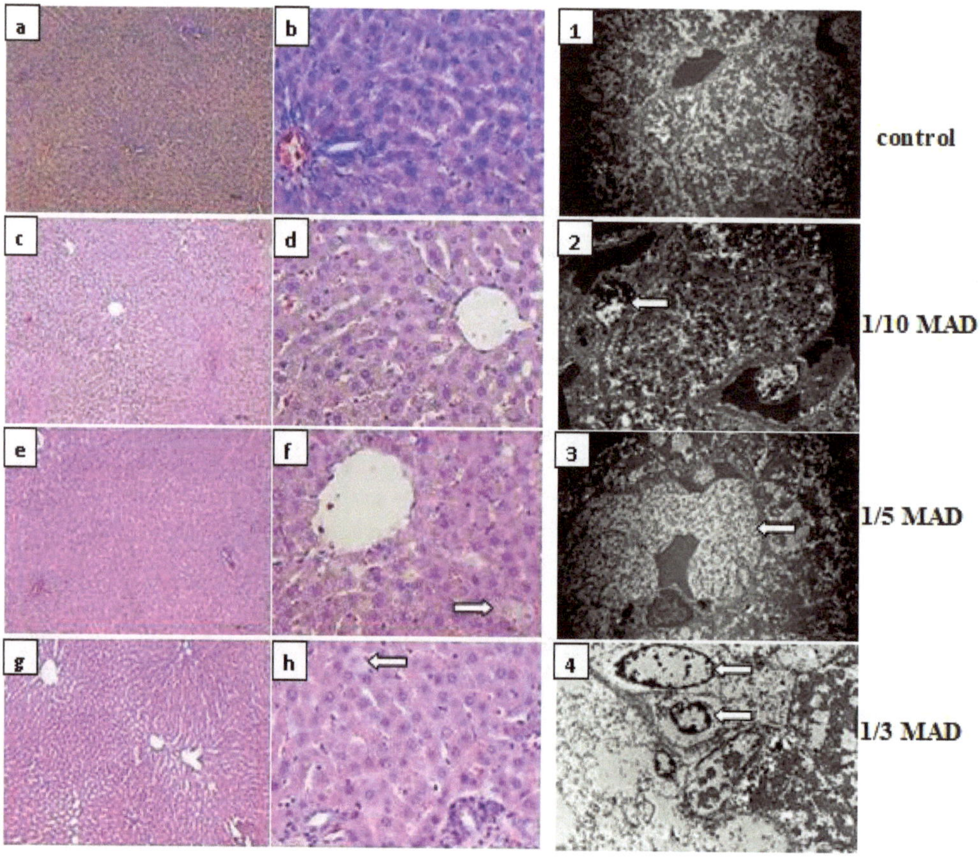


**Figure 7.** Histopathological evaluation of mono-WD POM-induced hepatotoxicity. Photographs of HE-stained rat liver sections (**a–h**); image was captured under a light microscope with ×10 (**a**,**c**,**e**,**g**) and ×40 (**b**,**d**,**f**,**h**) magnifications. No differences were observed between the liver tissue from the control group (**a**,**b**) and rat treated with 1/10 MAD (**c**,**d**), whereas tissue from rats treated with 1/5 MAD (**e**,**f**) showed discrete necrosis (arrow), and 1/3 MAD (**g**,**h**) showed dilatation of sinusoidal spaces and discrete necrosis (arrow). TEM micrographs of liver sections (**1–4**) show: (**1**) (×2200) control rat; (**2**) (×2800) rat treated with 1/10 MAD showing apoptosis of endothelial cells (arrow); and (**3**) (×2200), (**4**) (×3500)) rats treated with 1/5 MAD and 1/3 MAD, respectively, showing necrosis of hepatocytes (arrows).

### 2.4.4. Histopathological Evaluation of Mono-WD POM—Induced Cardiotoxicity

The results of light microscopy and TEM analysis obtained for rat heart tissues from the control and treated groups (the application of 1/10, 1/5, and 1/3 MAD of mono-WD POM) are shown in Figure 9. No difference was obtained between the heart tissues from the control group (a,b,1) and rats treated with 1/10 MAD (c,d,2), indicating that this lowest administrated dose did not induce cardiotoxicity in the experimental rats. On the contrary, photomicrographs of HE-stained heart sections revealed a loss of striation in cardiac muscle cells (arrows) in rats treated with 1/5 MAD (e,f) and 1/3 MAD (g,h). Moreover, interstitial bleeding (arrow) was observed in TEM micrographs of heart sections in rats treated with 1/5 MAD (3), whereas necrosis of cardiac muscle and endothelial cells (arrow) was detected in 1/3 MAD-treated experimental animals (4).

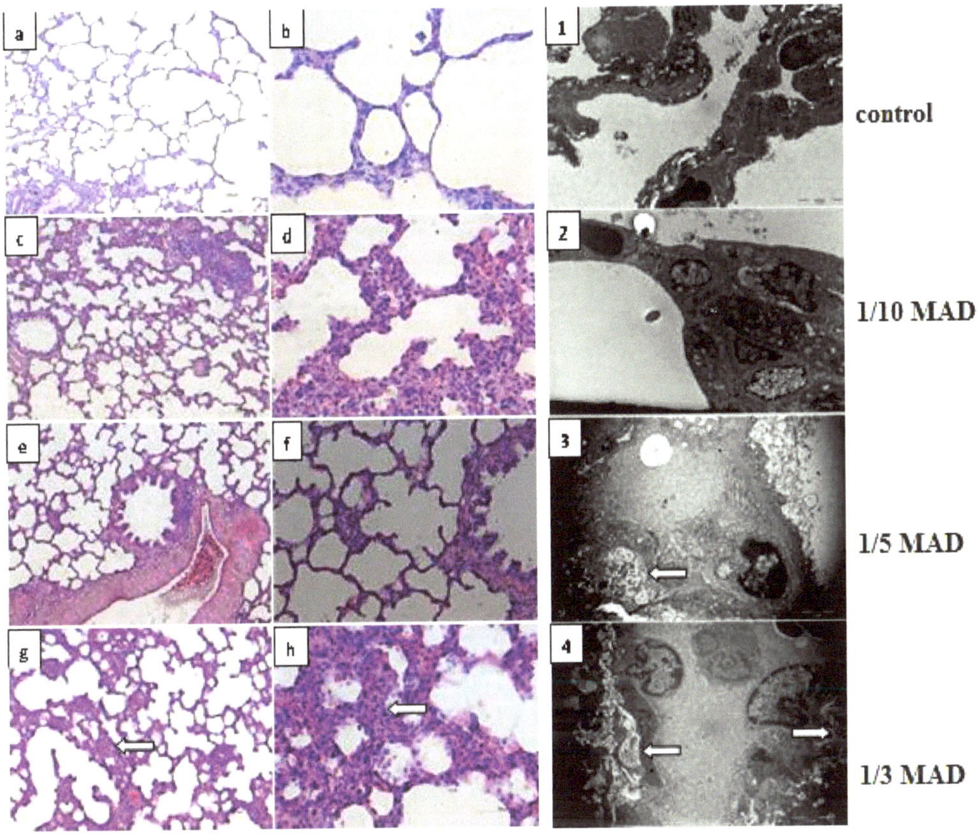


**Figure 8.** Histopathological evaluation of mono-WD POM-induced lung toxicity. Photographs of HE-stained rat lung sections (**a–h**); image was captured under a light microscope with ×10 (**a**,**c**,**e**,**g**), ×20 (**f**), and ×40 (**b**,**d**,**h**) magnifications. No differences were observed between the lung tissues from the control group (**a**,**b**) and rats treated with 1/10 MAD (**c**,**d**) and 1/5 MAD (**e**,**f**), whereas tissue from rats treated with 1/3 MAD (g, h) showed a thickening of interstitial spaces (arrows). TEM micrographs of lung sections (**1–4**) show: (**1**) (×2800) control rat; (**2**) (×3500) rat treated with 1/10 MAD showing normal lung tissue; and (**3**) (×8900), (**4**) (×4400)) rats treated with 1/5 MAD and 1/3 MAD, showing fibrosis and prominent fibrosis of lung interstitium (arrows), respectively.

### *2.5. Biodistribution Study In Vivo*

To investigate the in vivo biodistribution of mono-WD POM in animals subjected to 1/10, 1/5, and 1/3 of MAD (960 mg/kg), the tungsten content in kidney, liver, lung, heart, and femur tissues was measured using ICP-OES, two weeks post-intravenous administration (Figure 10). As expected, a gradual increase in the given dose of mono-WD POM resulted in a corresponding increase in tungsten concentration in all investigated tissues. The highest tungsten concentrations (ppm) were detected in the kidney samples (1/3 MAD—0.67 ± 0.12; 1/5 MAD—0.59 ± 0.07; 1/10 MAD—0.54 ± 0.05). Approximately three times lower tungsten concentrations were obtained in the liver tissue (1/3 MAD—0.22 ± 0.03; 1/5 MAD—0.16 ± 0.01; 1/10 MAD—0.18 ± 0.01). Then, the lungs were the third most deposited organ with tungsten (1/3 MAD—0.16 ± 0.04; 1/5 MAD—0.14 ± 0.02; 1/10 MAD—0.13 ± 0.02). Finally, similar tungsten concentrations were detected in the heart (1/3 MAD—0.11 ± 0.01; 1/5 MAD—0.8 ± 0.01;

1/10 MAD—0.11 ± 0.00) and femur (1/3 MAD—0.14 ± 0.03; 1/5 MAD—0.12 ± 0.01; 1/10 MAD—0.12 ± 0.01).

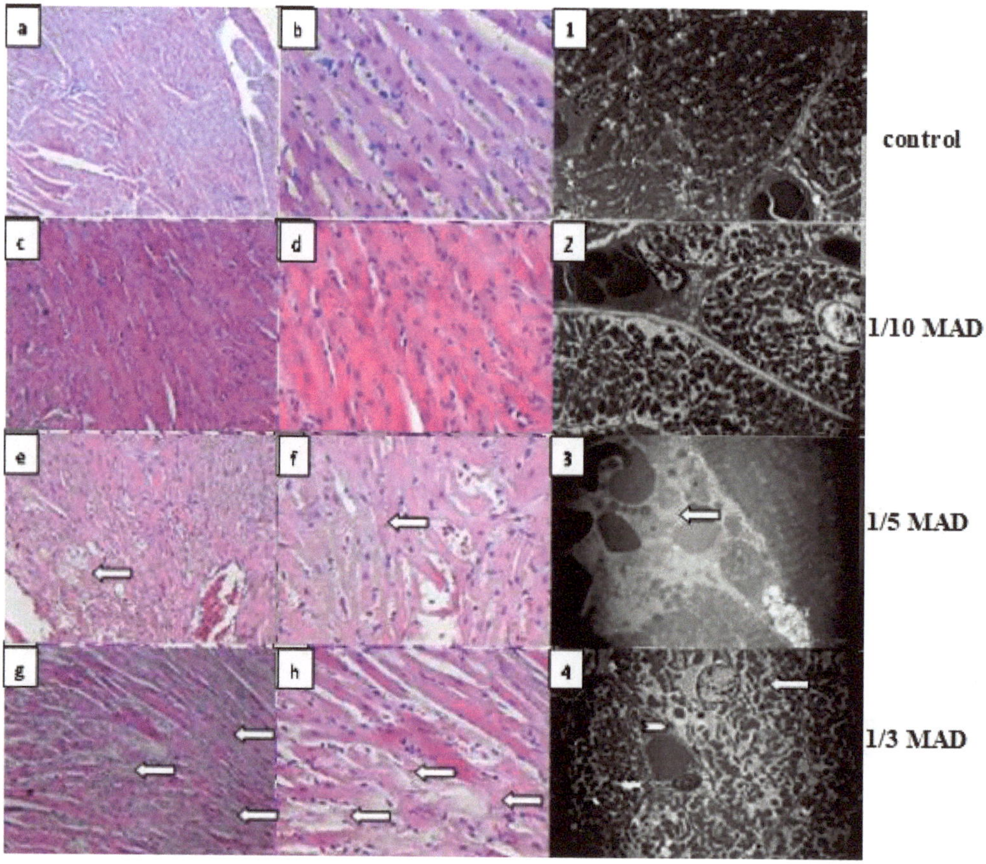


**Figure 9.** Histopathological evaluation of mono-WD POM-induced cardiotoxicity. Photographs of HE-stained rat heart sections (**a–h**); image was captured under a light microscope with ×10 (**a**), ×20 (**c**,**e**,**g**), and ×40 (**b**,**d**,**f**,**h**) magnifications. No differences were observed between the heart tissues from the control group (a, b) and rats treated with 1/10 MAD (**c**,**d**), whereas tissues from rat treated with 1/5 MAD (**e**,**f**) and 1/3 MAD (**g**,**h**) showed a loss of striation in cardiac muscle cells (arrows). TEM micrographs of heart sections (**1–4**) show: (**1**) (×2800) control rat; (**2**) (×2800) rat treated with 1/10 MAD showing normal heart tissue; and (**3**) (×3300) rats treated with 1/5 MAD showing an interstitial bleeding (arrow); (**4**) (×2800) rats treated with 1/3 MAD showing necrosis of cardiac muscle cell (arrow) and necrosis of endothelial cell (arrowhead).

### *2.6. Biodistribution Study In Vivo*

To investigate the in vivo biodistribution of mono-WD POM in animals subjected to 1/10, 1/5, and 1/3 of MAD (960 mg/kg), the tungsten content in kidney, liver, lung, heart, and femur tissues was measured using ICP-OES, two weeks post-intravenous administration (Figure 10). As expected, a gradual increase in the given dose of mono-WD POM resulted in a corresponding increase in tungsten concentration in all investigated tissues. The highest tungsten concentrations (ppm) were detected in the kidney samples (1/3 MAD—0.67 ± 0.12; 1/5 MAD—0.59 ± 0.07; 1/10 MAD—0.54 ± 0.05). Approximately three times lower tungsten concentrations were obtained in the liver tissue (1/3 MAD—0.22 ± 0.03; 1/5 MAD—0.16 ± 0.01; 1/10 MAD—0.18 ± 0.01). Then, the lungs were the third most deposited organ with tungsten (1/3 MAD—0.16 ± 0.04; 1/5 MAD—0.14 ± 0.02; 1/10 MAD—0.13 ± 0.02). Finally, similar tungsten concentrations were detected in the heart (1/3 MAD—0.11 ± 0.01; 1/5 MAD—0.8 ± 0.01; 1/10 MAD—0.11 ± 0.00) and femur (1/3 MAD—0.14 ± 0.03; 1/5 MAD—0.12 ± 0.01; 1/10 MAD—0.12 ± 0.01).

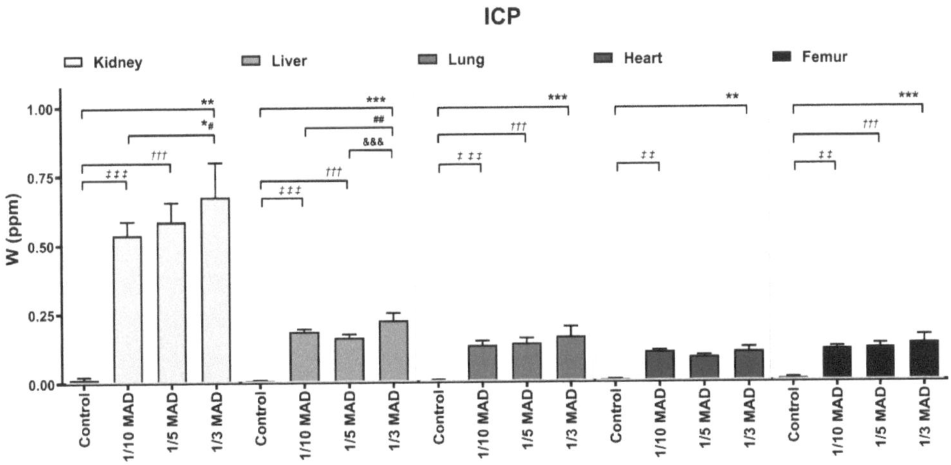


**Figure 10.** Tissue distribution of mono-WD POM in *Wistar* rats two weeks post-intravenously applied 1/10, 1/5, and 1/3 MAD, expressed as tungsten concentration (mean ± SD) in kidney, liver, lung, heart, and femur tissues. The difference between control and treated groups was presented as: * for 1/3 MAD group, † for 1/5 MAD group, and ‡ for 1/10 MAD group. The difference between 1/3 MAD and 1/5 MAD was presented as &, and between 1/3 MAD and 1/10 MAD as #. The significant values: *** $p < 0.001$, ** $p < 0.01$, ††† $p < 0.001$, ‡‡‡ $p < 0.001$, ‡‡ $p < 0.01$, &&& $p < 0.001$, ## $p < 0.01$, # $p < 0.05$.

## 3. Discussion

This study builds on our previous research, exploring the WD polyoxotungstate nanocluster family as effective CT contrast agents with the aim to identify the compound with the most favorable toxicological profile [9].

In the current work, in vitro CT studies (Figure 1) revealed significantly better contrast performances of mono-WD POM compared to the Keggine structure WSiA, differing in shape, size, charge density, and chemical composition. Thus, the ability of a molecule to attenuate X-rays could be associated with these parameters as well, not only with the presence of tungsten atoms. Moreover, higher X-ray attenuation values were obtained for mono-WD POM respective to its structural analog WD-POM [9], which is formed by the removal of one WO unit from its parent compound. Since mono-WD POM contains fewer W atoms per nanocluster (17 W and 18 W for mono-WD POM and WD-POM, respectively), the same W concentration is reached with a higher number of mono-WD POM molecules, which might be the reason for the observed stronger X-ray attenuation. Furthermore, the studied W-containing mono-WD POM showed superior contrast properties in vitro to the standard CT contrast, I-based iohexol (Omnipaque$^{TM}$), containing three I atoms, which is in line with our previous study on parent WD-POM [9] and the literature regarding the comparison of tungsten-based nanoparticles and commercial iodine contrast agents [20,21]. From a toxicity point of view, the key advantage of mono-WD POM is that it contains a higher quantity of W atoms (17 W) per molecule compared to the quantity of I atoms per iohexol molecule (3 I) and thus can induce a higher X-ray attenuation and better contrast in CT imaging with a lower number of moles per kg body weight.

In this study, administered mono-WD POM doses (320, 192, and 96 mg/kg) corresponding to W doses of 1.107, 0.664, and 0.332 (in mmol W/kg) were significantly higher (up to 70 times) than the dose of 0.015 mmol W/kg that is usually administered for in vivo CT imaging [20,21]. Therefore, more severe toxic effects than those observed in this study are not expected if mono-WD POM is applied in effective doses for clinical CT imaging.

During the study, no death of the experimental animals was registered. This finding is different from the results of our previous study in which parent Wells–Dawson polyoxotungstate (WD-POM) showed an unexpectedly high mortality rate after intravenous administration [9]. Comparing the applied doses of the mono-WD POM studied here with its parent analog [9], it is obvious that the maximum dose of mono-WD POM applied intravenously, 320 mg/kg (1/3 MAD), did not induce the death of any treated rat (100% survival rate, 5/5), whereas for the lower dose (1.6 times) of parent WD-POM (200 mg/kg, 1/10 MTD), the survival rate was 80% (4/5). Moreover, the higher doses of parent WD-POM, 400 and 666.7 mg/kg, resulted in a significant increase in mortality rate (the survival rates of 40 and 0%, respectively). These results suggest that mono-WD POM has a better safety profile and is safer for intravenous use. Furthermore, lack of mortality and no observed physical changes after the oral application of MAD (960 mg/kg) might be associated with a low absorption rate of mono-WD POM due to its large molecular weight. Even if mono-WD POM could be absorbed, the efflux pumps or the first-pass metabolism effect could decrease its bioavailability [22]. The low bioavailability of POM-based compounds was previously shown in a study that investigated the pharmacokinetic profile of $Cs_2K_4Na[SiW_9Nb_3O_{40}]$(POM-based compound) in rats [23]. Additionally, in our previous studies that examined the oral toxicity of polyoxotungstate-based compounds, only mild toxic effects were found, additionally confirming that these compounds probably have low bioavailability [16,19]. The low POM bioavailability was additionally corroborated by the high value of MTD (2000 mg/kg) obtained for parent WD-POM [9], which is a structural analogue of the studied mono-WD POM. The low bioavailability of POM-based compounds and consequently low toxicity after *per os* application suggest that they could be considered promising candidates for the development of a new oral CT contrast agent. Moreover, the excellent survival rate found after intravenous application indicates that mono-WD POM could also be further developed as a new intravenous CT agent.

The results of blood pH values, which were below the normal range, indicate that animals in both control and all three treated groups (1/10, 1/5, and 1/3 MAD of lacunary WD POM) were in a state of acidosis. Although there was no significant difference in the pH values between the groups, the difference was found in the values of $pCO_2$, $HCO_3^-$, base excess of blood, and anion gap (Table 1). The $pCO_2$ values were above the normal range and higher in all treated groups relative to the control group (within the normal range). This suggests that the respiratory system was trying to compensate for the acidosis in the control group, whereas in the treated groups, it failed to do so [24]. The $pCO_2$ values in the control group were in the normal range, which implies that low pH in this group was a consequence of metabolic acidosis [24,25]. Moreover, the metabolic acidosis in the control group was confirmed with the values of base excess of blood that were below the lower limit of the normal range [24,25]. The statistically higher values of anion gap in the control group than in the 1/3 and 1/5 MAD groups additionally confirmed the development of metabolic acidosis in the control group [24,25]. The results of the blood gas analysis indicate that animals in the control group developed metabolic acidosis. Since anesthetics are known to be associated with metabolic acidosis due to an increase in lactate level [26], the most probable cause of metabolic acidosis in the control group was anesthesia. On the contrary, in the treated groups, the low values of pH together with high values of $pCO_2$, as well as the values of base excess of blood and anion gap (Table 1), indicate the presence of mixed metabolic and respiratory acidosis. The significantly higher $HCO_3^-$ values in the treated groups point out that respiratory acidosis persists for a longer period [24,25]. All of this suggests that the metabolic acidosis observed both in the treated groups and in the control group is a consequence of anesthesia, whereas respiratory acidosis is a consequence of mono-WD POM-induced toxic effects on the lung. The concentration-dependent effect in blood gas analysis was noted in the treated groups, with the most pronounced effect in animals that received the highest concentration (1/3 MAD) of mono-WD POM.

CO-oximetry is used for the diagnosis of intoxication that is manifested with pulmonary symptoms. Oxyhemoglobin and methemoglobin levels were significantly lower, whereas deoxyhemoglobin levels were higher in all treated groups (Table 2). The oxyhemoglobin level was below the lower limit of the normal range, methemoglobin was in the normal range, and deoxyhemoglobin was significantly above the normal range (about six times) in groups receiving mono-WD POM. The fact that oxyhemoglobin is a form of hemoglobin with oxygen and deoxyhemoglobin is a form of hemoglobin without oxygen indicates that the animals in the treated groups were in hypoxia and that the oxygenation of blood in the lungs was poor [27]. An additional confirmation of poor blood oxygenation is the low values of oxygen saturation and oxygen content of hemoglobin, which were below the lower limit of the normal range for all three treated groups of experimental animals. Accordingly, the results of CO-oxymetry analysis indicate that the intravenous application of mono-WD POM in the applied doses could result in lung damage and respiratory acidosis. This assumption is in accordance with our previous study [9], as well as with literature reports on tungsten-associated lung pathologies [28,29].

The glucose and lactate levels (Figures 2 and 3) were above the upper limit of the normal range in both control and all three treated groups. Significantly higher values of glucose, relative to the control group, were detected only in the 1/3 MAD group, whereas the lactate levels were not significantly different between the experimental groups. The elevation of glucose levels is most likely a consequence of surgical intervention and anesthesia. The surgical intervention, performed to expose organs and collect blood samples, triggered a stress induced-hyperglycemia, which is a common response to this type of intervention [30–32]. Similar results were obtained in our previous study that followed the same protocol [9]. In this study, the rise of glucose levels, as a consequence of anesthesia and surgical intervention, was also recorded in both control and WD-POM treated groups [9]. The significantly higher glucose values in the 1/3 MAD group point out that this POM-based compound interferes in some way with the organism's response to stress. The higher lactate levels were also a result of stress, which initially increased glucose levels and accelerated metabolism, causing a rise in lactate levels. Furthermore, the rise in lactate level is common during anesthesia and surgical interventions [33–35] and explains isolated metabolic acidosis in the control group. It is also an explanation for a metabolic part of mixed respiratory and metabolic acidosis in all three treated groups, meaning that mono-WD POM induced only respiratory acidosis. Respiratory acidosis, developed as a consequence of the intravenous administration of POM-based compound, was also found in our previous study [9], additionally confirming the toxic effect of WD polyoxotungstates on the lungs.

The concentration of electrolytes measured in this study (Table 3) indicates possible kidney problems in mono-WD POM-treated animals, mainly due to overall high (above the upper limit of the normal range) $Mg^{2+}$ levels that were significantly different only when the control group was compared to 1/3 MAD group. $Ca^{2+}$ level was also significantly higher in the 1/3 MAD group, but in the other groups, $Ca^{2+}$ values were within the normal range. Hypermagnesemia is commonly seen in kidney disease and is a consequence of decreased glomerular filtration rates [36]. Elevated calcium levels are common in acute kidney disease (AKD). They are usually the reason for AKD development, not a consequence of AKD [37]. Hypercalcemia occurs when there is a higher rate of absorption from the intestine or when the bones release more calcium into the blood [38]. Most probably in our case, mono-WD POM affected the bones and increased calcium release.

Considering the role of the kidney in the excretion of different compounds from the organism, as well as the fact that the use of various drugs and xenobiotics can damage the kidney and affect its function, functional and histological analysis was performed to evaluate the nephrotoxic potential of mono-WD POM. Creatinine values (Figure 4), measured from the blood two weeks after the intravenous administration of saline and mono-WD POM, were in the physiological range for all experimental groups except for 1/3 MAD, with no statistical difference between the groups. On the other hand, BUN levels (Figure 5) were in the physiological range only in the control group. A dose-dependent increase in BUN concentrations, as well as a significant difference in BUN values, was observed when comparing the control group with the 1/3 and 1/5 MAD groups.

Since a rise in creatinine and BUN levels is commonly seen in acute kidney injury (AKI) [39], based on the obtained results, it can be assumed that mono-WD POM caused AKI. Nevertheless, the increase in creatinine can be postponed, especially in the early stages. This is because serum creatinine level is also dependent on non-renal factors [39]. In our study, the most prominent effects were observed in animals treated with 1/3 MAD of mono-WD POM in which both creatinine and BUN levels were high, which indicates that the degree of AKI was dose-dependent. Since AKI develops most commonly as a consequence of acute tubular necrosis [40], the development of acute tubular necrosis in animals treated with mono-WD POM might be an explanation for the development of AKI. Additional confirmation for this hypothesis is the detected hypermagnesemia, which is a common phenomenon in tubular necrosis [41]. These findings can be ascribed to the highest deposition of tungsten in the kidneys (Figure 10) and the necrosis of tubular cells confirmed by histopathological analysis of kidney tissue (Figure 6).

The results of the biodistribution study (Figure 10) are in agreement with the highest tungsten quantity found in kidneys as the main excretory organ, as it was also reported for the parent compound (WD-POM) of mono-WD POM [9] and previously published data for polyoxotungstate, $Cs_2K_4Na[SiW_9Nb_3O_{40}]$ [23]. Approximately three times less levels of tungsten detected in the rat liver samples (0.2242 *vs.* 0.6744 ppm for 1/3 MAD) (Figure 10) are in line with the reported tissue distribution of the parent molecule (WD-POM) and consequently induced morphological irregularities in liver sections (Figure 7). Overall, a dose-response tendency of tungsten deposition was observed in all investigated tissues and corresponded to morphological changes revealed by light microscopy and TEM. For example, no histological changes, fibrosis, and prominent fibrosis of the lung tissue were observed for 1/10, 1/5, and 1/3 MAD of mono-WD POM, respectively (Figure 8).

Gadolinium-based contrast agents clinically approved in the EU and USA were reported to exhibit three types of biodistribution: extracellular with renal or mixed renal/hepatobiliary elimination and intravascular or blood pool agents with renal elimination after intravenous administration. These hydrophilic, water-soluble metal complexes can quickly be distributed to the extracellular space and be eliminated mainly via the kidneys [42–44]. In accordance, mono-WD POM, as a metal-containing cluster soluble in water with the highest accumulation in kidney tissue (Figure 10), is expected to belong to extracellular fluid agents that are cleared via the kidneys by glomerular filtration and were proven as very safe after an intravenous administration at high doses [44]. Furthermore, literature data on gadolinium deposition in rats and mice after the intravenous injection of the approved gadolinium-based contrasts and "free gadolinium" (gadolinium acetate/chloride) revealed remarkable gadolinium levels in the liver and bone as the main deposition organs [45–47]. The results of this tissue distribution study (Figure 10) also detected statistically significant tungsten concentrations in the liver and femur of *Wistar* rats two weeks post-intravenously applied 1/10, 1/5, and 1/3 MAD of mono-WD POM. Similarly, the biodistribution profile of iohexol, the standard iodine-based contrast agent for CT, revealed its existence in mice's hearts, livers, and kidneys for seven days after intravenous application [48].

In summary, mono-WD POM demonstrated superior contrast performance compared to its parent WD-POM, Keggine structure WSiA, and the conventionally used CT contrast agent, iohexol. All administrated doses of mono-WD POM did not induce mortality of the treated animals, indicating a better safety profile in comparison with the parent WD-POM. The dose-dependent side effects were confirmed by biochemical and histological analysis, whereas the most prominent changes were found in kidney tissue in which the highest tungsten amount was accumulated. The observed biochemical changes, particularly the rise in $Mg^{2+}$ and BUN levels in the 1/3 MAD group, confirmed metabolic acidosis and kidney function alterations. Accordingly, mono-WD POM merits further in vivo research on imaging efficacy and possible toxicity, which may facilitate the development of inexpensive, safe, and innovative CT contrast agents for clinical application. Finally, future research directions could include the latest potential strategies like deep learning to improve novel contrast agents' effectiveness and potential innovations in imaging and biomaterial technologies [49–51].

## 4. Materials and Methods

### 4.1. Chemicals

Monolacunary Wells-Dawson POM, $\alpha_2$-$K_{10}P_2W_{17}O_{61}\cdot 20H_2O$ (mono-WD POM) was synthesized according to a previously described procedure [52] and dissolved in a saline solution for further experiments. The investigated $\alpha_2$-$K_{10}P_2W_{17}O_{61}\cdot 20H_2O$ is directly obtained by the removal of one WO octahedron from the cap position of the $\alpha$–$[P_2W_{18}O_{62}]^{6-}$ structure. The structure of the synthesized mono-WD POM was validated by $^{31}P$ NMR spectroscopy, showing characteristic peaks at $\delta = -7.2$ and $-14.4$ ppm, belonging to two non-equivalent phosphorus atoms in the lacunary structure [53]. 12-tungstosilicic acid, $H_4SiW_{12}O_{40}$ (WSiA) was purchased from Sigma–Aldrich, Munich, Germany.

A stock solution of mono-WD POM (64 mg/mL) was prepared by heating at 50 °C and intensive stirring. The stock solution was diluted with saline up to particular concentrations in order to achieve the desired application doses: 1/10 MAD (96 mg/kg), 1/5 MAD (192 mg/kg), and 1/3 MAD (320 mg/kg). The animals were treated immediately after the solutions were prepared.

### 4.2. In Vitro CT Imaging

The in vitro measurement of X-ray attenuation was performed in plastic test tubes using a clinical CT GE Medical Systems Revolution Evo 128-Detector CT scanner (the X-ray source current—400 mA, the X-ray source voltage—80 kV). The phantom images were acquired for two polyoxotungstates differing in shape and structure: Keggin structure 12-tungstosilicic acid, $H_4SiW_{12}O_{40}$ (WSiA) and monolacunary Wells–Dawson POM, $\alpha_2$-$K_{10}P_2W_{17}O_{61}\cdot 20H_2O$ (mono-WD POM). The selected concentrations of the tested solutions were (in mM): 3.125, 6.25, 12.5, 25, 50, and 100. Water was used as a reference for Hounsfield units (HU) of 0.0.

### 4.3. Ethical Approval

The ethical approval for methodological procedures performed in this study was given by the Faculty of Medicine of the University of Belgrade Ethical Commission for Experimental Animal Welfare Protection, (N°6447/1-2020). All procedures performed in this study followed Guidelines from the European Convention for the Protection of Vertebrate Animals Used for Experimental and Other Scientific Purposes. Good Laboratory practice was fully applied as well. The ARRIVE (Animal Research: Reporting of In Vivo Experiments) guidelines 2.0 were used for reporting in this study [54].

### 4.4. Experimental Animals

Healthy, male, 5-week-old *Wistar albino* rats weighing 180–220 g were selected for these experiments. Rats are the preferred species for conducting toxicological studies in accordance with the guidelines outlined by the Organization for Economic Cooperation and Development (OECD) [55,56] Moreover, previous research has demonstrated the efficacy and reliability of this model [9,57]. The selected animals were housed in polycarbonate cages (3–4 animals per one cage). The husbandry conditions were standard for this type of experiment. More precisely, the animals had *ad libitum* access to standard rodent food and tap water. A temperature of 23 ± 2 °C, humidity of 55 ± 10%, and a 12/12 dark/light cycle were maintained during the whole study period.

*4.5. Experimental Design*

Since the in vivo toxicity of mono-WD POM was tested for the first time in this study, the recommended protocol [55] was applied for the selection of doses and the follow-up period. First, experimental rats received single doses of 5, 50, 300, and 960 mg/kg *per os*, and their general habits were monitored. The animals were observed individually during the period of 14 days. A dose of 2000 mg/kg, which is recommended by the protocol, was not applied because the maximum solubility of the studied mono-WD POM was 64 mg/mL, which allowed a maximum dose of 960 mg/kg to be administered. Considering the limited solubility of the studied mono-WD POM (maximum solubility—64 mg/mL), a maximum-administered dose (MAD) of 960 mg/kg was used for an acute oral toxicity evaluation [58] instead of the maximum-tolerated dose (MTD) [9,55,56]. Four experimental groups were created for this study: the control group (received saline intravenously) and three treated groups created based on the proportion of the intravenously administered MAD dose of mono-WD POM [9,57,59]. The first treated group received 1/3 of MAD (320 mg/kg), the second group received 1/5 of MAD (192 mg/kg), and the third group received 1/10 of MAD (96 mg/kg).

At the beginning of the experiment, the animals were brought from a central housing facility and were given five days to acclimate to laboratory housing conditions. Twenty-four hours before saline or mono-WD POM application, food was withdrawn while leaving free access to tap water. Both saline and mono-WD POM were administered into the lateral rat tail vein in a single dose that did not exceed 1 mL. An intravenous approach was used because mono-WD POM is expected to be used as an intravenous contrast agent. Instantly, after the administration of saline/mono-WD POM, the animals were given access to food. For the next 14 days, after saline/mono-WD POM administration, animals were twice daily observed for any changes in action that deviated from their usual behavior, physical changes, or for the presence of visible toxicity symptoms. The detailed study design is depicted in Scheme 2.

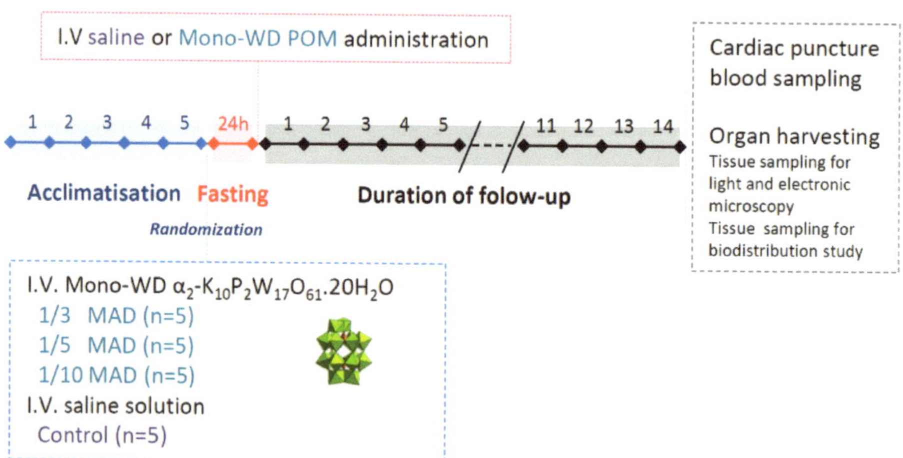


**Scheme 2.** Experimental design for in vivo acute toxicity evaluation of monolacunary Wells–Dawson polyoxotungstate (mono-WD POM).

Before the study started, the exact number of animals needed was calculated to follow the 3R rule (replacement, reduction, and refinement of experimental animals). For the calculation of the sample size, power analysis was performed, and the obtained results were checked with the resource equation method [60]. In the power analysis, the effect size and standard deviation were taken from an initial study [16], the type I error was set at 5%, power was 80%, a two-tailed test was used, and an expected attrition of sample size was set at 20%. The calculated number of animals was twenty (five in each group).

To ensure that each animal has an equal probability of receiving a particular treatment (saline or the selected proportions of MAD) and to achieve an equal number of animals in each group, a randomization procedure was performed during the allocation of animals to the control and treated groups. For randomization, the random numbers were generated using the standard = *RAND ()* function in Microsoft Excel(v2401) ®.

### 4.6. Blood Sampling and Organ Harvesting

After a 14-day follow-up period, the animals were weighed, and the weight was used to determine the dose of anesthetic urethane. In the morning of the last day of the experimental protocol, urethane was prepared and administrated as a single dose. The anesthetic was given intraperitoneally at a dose of 125 mg per 100 g of body weight. After the induction of anesthesia and a stabilization process, which took 90 min, the depth of anesthesia was determined with a needle poking the tail and tweezers pinching the space between the toes on the back paw. The lack of response to tail poking with a needle was considered superficial anesthesia, while the cessation of pedal withdrawal reflex after pinching was considered a state of deep surgical anesthesia.

After reaching a state of surgical anesthesia, the rat's chest and abdomen were opened to expose the animal's organs. For arterial blood gas analysis, a heparinized syringe was used. Arterial blood was sampled by entering the needle into the exposed left heart. After blood collection, to expose the blood to heparin and prevent unvented clotting, the syringe was thoroughly mixed. Once the blood was collected, the animals were decapitated, and the following organs were harvested from each *Wistar albino* rat: kidney, liver, lung, heart, and femur. The organs harvested for light microscopy were fixed in 4% buffered formaldehyde. For electronic microscopy, the organs were fixed in 3% glutaraldehyde. The tissue samples (kidney, liver, lung, heart, and femur) were taken for the biodistribution study and kept at −80 °C. All blood samples for gas analysis and tissue samples for histological and biodistribution studies were coded. The codes were revealed after completing all analyses to enable statistical calculations.

### 4.7. Laboratory Blood Analysis

The Profile Prime Plus® Critical Care Analyzer (Nova Biomedical Co., Waltham, MA, USA) was used for blood gas analysis (pH, $pCO_2$, $HCO_3^-$, base excess of blood, and anion gap), CO–oximetry (total hemoglobin, oxyhemoglobin, carboxyhemoglobin, methemoglobin, deoxyhemoglobin, oxygen saturation, oxygen content of hemoglobin, and oxygen capacity of hemoglobin) analysis, and for measuring electrolytes ($Na^+$, $K^+$, $Cl^-$, $Ca^{2+}$, $Mg^{2+}$, $Ca^{2+}$ normalized to pH 7.4, and $Mg^{2+}$ normalized to pH 7.4), namely, glucose, lactate, creatinine, and blood urea nitrogen (BUN) levels. These analyses were performed immediately after the blood was collected.

### 4.8. Histological Analysis

For histological analysis by light microscopy, tissue samples (kidney, liver, lung, and heart) were first fixed and then embedded in paraffin. Then, the tissues were cut into 5 µm-thick sections and stained with Hematoxylin and Eosin (HE). HE-stained micrographs were taken using a light microscope (Leica DMLS) with a digital camera (Leica DFC295).

TEM analysis was carried out with the tissue samples that were first fixed with 3% glutaraldehyde in cacodylate buffer. After fixation in 1% $OsO_4$ and dehydration in graded alcohols, the samples were fixed in Epoxy medium (Sigma-Aldrich, 45345). Then, thin tissue sections were prepared and mounted on copper grids (Sigma–Aldrich, G4901). After staining with uranyl acetate and lead citrate, the samples were analyzed under an electron microscope (Morgagni 268D, FEI, Hillsboro, OR, USA).

### 4.9. Biodistribution Study In Vivo

0.1 g of tissue sample (kidney, liver, lung, heart, or femur), 4 mL $HNO_3$, 5 mL HCl, and 2 mL $HBF_4$ were added to a Teflon vessel. The pre-set 'Easy Prep *i*wave Zeolite' protocol was applied for digesting the samples using a MARS6 microwave (15 min ramp time to 210 °C, 15 min hold at 210 °C, 30 min cool down before opening the vessel). Then, the digested sample solutions were diluted (Milli-Q water) to 2% $HNO_3$ (final volume of 10 mL).

The digested tissue samples were analyzed by inductively coupled plasma–optical emission spectroscopy (ICP-OES), using a PerkinElmer optical emission spectrometerAvio 500 instrument. A calibration curve was prepared with a series of standard tungsten solutions in the concentration range of 0.0001–50 ppm.

### 4.10. Statistical Analysis

For statistical analysis, the Graph Pad Prism (Graph Pad Software, San Diego, CA, USA) software was used. The difference between the four groups was calculated with one-way analysis of variance (ANOVA) followed by a Tukey post hoc test and was considered significant if $p < 0.05$. The obtained results are expressed as mean ± standard deviation (SD). The presented *p*-values in both the text and tables of this paper indicate significance determined through one-way analysis of variance. The distinctions between the control and experimental groups (1/10 MAD, 1/5 MAD, and 1/3 MAD) reflect the outcomes of post hoc Tukey analysis and are denoted in tables and graphs by the symbols: *, †, and ‡.

**Author Contributions:** Conceptualization, M.B.Č., T.N.P.-V. and D.K.; data curation, J.L., K.D. and B.R.; formal analysis, M.S., M.B.Č., M.Ć. and T.K.-S.; investigation, M.S., J.L., A.M. and N.D.S.; methodology, A.M., N.D.S., K.D, B.R., M.Ć. and T.K.-S.; project administration, D.K.; writing—original draft, M.S., M.B.Č. and D.K.; writing—review and editing, T.N.P.-V. All authors have read and agreed to the published version of the manuscript.

**Funding:** This study was funded by the Ministry of Science, Technological Development, and Innovation of the Republic of Serbia (contracts no. 451-03-9/2023-14/200110 and 451-03-66/2024-03/200017), the Science Fund of the Republic of Serbia (POMCACT project no. 6526393), and the Serbian Academy of Sciences and Arts (SASA, grant No. F-35). N.D.S. (1267623N) thanks the FWO for a senior post-doctoral fellowship.

**Institutional Review Board Statement:** The animal study protocol was approved by the Faculty of Medicine of the University of Belgrade Ethical Commission for Experimental Animal Welfare Protection (N°6447/1-2020).

**Informed Consent Statement:** Not applicable.

**Data Availability Statement:** The data presented in this study are available upon request from the corresponding author.

**Conflicts of Interest:** The authors declare no conflicts of interest.

## References

1. Hisieh, J. *Computed Tomography: Principles, Design, Artifacts, and Recent Advances*, 3rd ed.; SPIE Press: Bellingham, WA, USA, 2015.
2. Solomon, J.; Marin, D.; Roy Choudhury, K.; Patel, B.; Samei, E. Effect of Radiation Dose Reduction and Reconstruction Algorithm on Image Noise, Contrast, Resolution, and Detectability of Subtle Hypoattenuating Liver Lesions at Multidetector CT: Filtered Back Projection versus a Commercial Model-based Iterative Reconstruction Algorithm. *Radiology* **2017**, *284*, 777–787. [PubMed]
3. De Bournonville, S.; Vangrunderbeeck, S.; Kerckhofs, G. Contrast-Enhanced MicroCT for Virtual 3D Anatomical Pathology of Biological Tissues: A Literature Review. *Contrast. Media. Mol. Imaging* **2019**, *2019*, 8617406. [CrossRef]
4. Cha, M.J.; Kang, D.Y.; Lee, W.; Yoon, S.H.; Choi, Y.H.; Byun, J.S.; Lee, J.; Kim, Y.H.; Choo, K.S.; Cho, B.S.; et al. Hypersensitivity Reactions to Iodinated Contrast Media: A Multicenter Study of 196 081 Patients. *Radiology* **2019**, *293*, 117–124. [CrossRef]
5. Morcos, R.; Kucharik, M.; Bansal, P.; Al Taii, H.; Manam, R.; Casale, J.; Khalili, H.; Maini, B. Contrast-Induced Acute Kidney Injury: Review and Practical Update. *Clin. Med. Insights. Cardiol.* **2019**, *13*, 1179546819878680. [CrossRef] [PubMed]
6. Huang, L.; Luo, Y.; Chen, Z.L.; Yang, Z.Y.; Wu, Y. Thyroid dysfunction associated with iodine-contrast media: A real-world pharmacovigilance study based on the FDA adverse event reporting system. *Heliyon* **2023**, *9*, e21694. [CrossRef]
7. Shin, H.; Taghavifar, S.; Salehi, S.; Joyce, P.; Gholamrezanezhad, A. Current comments on contrast media administration in patients with renal insufficiency. *Clin. Imaging* **2021**, *69*, 37–44. [CrossRef]

8. Kumar, P.P.P.; Mahajan, R. Gold Polymer Nanomaterials: A Promising Approach for Enhanced Biomolecular Imaging. *Nanotheranostics* **2024**, *8*, 64–89. [CrossRef]
9. Stojanović, M.; Lalatović, J.; Milosavljević, A.; Savić, N.; Simms, C.; Radosavljević, B.; Ćetković, M.; Kravić Stevović, T.; Mrda, D.; Čolović, M.B.; et al. In vivo toxicity evaluation of a polyoxotungstate nanocluster as a promising contrast agent for computed tomography. *Sci. Rep.* **2023**, *13*, 9140. [CrossRef]
10. De Clercq, K.; Persoons, E.; Napso, T.; Luyten, C.; Parac-Vogt, T.N.; Sferruzzi-Perri, A.N.; Kerckhofs, G.; Vriens, J. High-resolution contrast-enhanced microCT reveals the true three-dimensional morphology of the murine placenta. *Proc. Natl. Acad. Sci. USA* **2019**, *116*, 13927–13936. [CrossRef] [PubMed]
11. Kerckhofs, G.; Stegen, S.; van Gastel, N.; Sap, A.; Falgayrac, G.; Penel, G.; Durand, M.; Luyten, F.P.; Geris, L.; Vandamme, K.; et al. Simultaneous three-dimensional visualization of mineralized and soft skeletal tissues by a novel microCT contrast agent with polyoxometalate structure. *Biomaterials* **2018**, *159*, 1–12. [CrossRef]
12. De Bournonville, S.; Vangrunderbeeck, S.; Ly, H.G.T.; Geeroms, C.; De Borggraeve, W.M.; Parac-Vogt, T.N.; Kerckhofs, G. Exploring polyoxometalates as non-destructive staining agents for contrast-enhanced microfocus computed tomography of biological tissues. *Acta. Biomater.* **2020**, *105*, 253–262. [CrossRef]
13. Zhang, S.; Li, M.; Zhang, Y.; Wang, R.; Song, Y.; Zhao, W.; Lin, S. A supramolecular complex based on a Gd-containing polyoxometalate and food-borne peptide for MRI/CT imaging and NIR-triggered photothermal therapy. *Dalton. Trans.* **2021**, *50*, 8076–8083. [CrossRef]
14. Dong, Y.C.; Hajfathalian, M.; Maidment, P.S.N.; Hsu, J.C.; Naha, P.C.; Si-Mohamed, S.; Breuilly, M.; Kim, J.; Chhour, P.; Douek, P. Effect of Gold Nanoparticle Size on Their Properties as Contrast Agents for Computed Tomography. *Sci. Rep.* **2019**, *9*, 14912. [CrossRef] [PubMed]
15. Carvalho, F.; Aureliano, M. Polyoxometalates impact as anticancer agents. *Int. J. Mol. Sci.* **2023**, *24*, 5043. [CrossRef]
16. Dinčić, M.; Čolović, M.B.; Sarić Matutinović, M.; Ćetković, M.; Kravić Stevović, T.; Mougharbel, A.S.; Todorović, J.; Ignjatović, S.; Radosavljević, B.; Milisavljević, M.; et al. In vivo toxicity evaluation of two polyoxotungstates with potential antidiabetic activity using Wistar rats as a model system. *RSC Adv.* **2020**, *10*, 2846–2855. [CrossRef] [PubMed]
17. Čolović, M.B.; Lacković, M.; Lalatović, J.; Mougharbel, A.S.; Kortz, U.; Krstić, D.Z. Polyoxometalates in biomedicine: Update and overview. *Curr. Med. Chem.* **2020**, *27*, 362–379. [CrossRef] [PubMed]
18. Krinke, G.J. The laboratory rat. In *Handbook of Experimental Animals*, 1st ed.; Elsevier: Amsterdam, The Netherlands, 2000.
19. Čolović, M.B.; Medić, B.; Ćetković, M.; Kravić Stevović, T.; Stojanović, M.; Ayass, W.W.; Mougharbel, A.S.; Radenković, M.; Prostran, M.; Kortz, U.; et al. Toxicity evaluation of two polyoxotungstates with anti-acetylcholinesterase activity. *Toxicol. Appl. Pharmacol.* **2017**, *333*, 68–75. [CrossRef]
20. Kim, S.J.; Xu, W.; Ahmad, M.W.; Baeck, J.S.; Chang, Y.; Bae, J.E.; Chae, K.S.; Kim, T.J.; Park, J.A.; Lee, G.H. Synthesis of nanoparticle CT contrast agents: In vitro and in vivo studies. *Sci. Technol. Adv. Mat.* **2015**, *16*, 055003. [CrossRef]
21. Yang, Z.; Wang, J.; Liu, S.; Sun, F.; Miao, J.; Xu, E.; Tao, L.; Wang, Y.; Ai, S.; Guan, W. Tumor-targeting $W_{18}O_{49}$ nanoparticles for dual-modality imaging and guided heat-shock-response-inhibited photothermal therapy in gastric cancer. *Part. Part. Syst. Char.* **2019**, *36*, 1900124. [CrossRef]
22. Herman, T.F.; Santos, C. First Pass Effect. In *StatPearls*; StatPearls Publishing: Treasure Island, FL, USA, 2023.
23. Wang, J.; Qu, X.; Qi, Y.; Li, J.; Song, X.; Li, L.; Yin, D.; Xu, K.; Li, J. Pharmacokinetics of anti-HBV polyoxometalate in rats. *PLoS ONE* **2014**, *9*, e98292. [CrossRef]
24. Burns, G.P. Arterial blood gases made easy. *Clin. Med.* **2014**, *14*, 66–68. [CrossRef] [PubMed]
25. Verma, A.K.; Roach, P. The interpretation of arterial blood gases. *Aust. Prescr.* **2010**, *33*, 124–129. [CrossRef]
26. Choi, Y.J.; Kim, M.C.; Lim, Y.J.; Yoon, S.Z.; Yoon, S.M.; Yoon, H.R. Propofol infusion associated metabolic acidosis in patients undergoing neurosurgical anesthesia: A retrospective study. *J. Korean Neurosurg. Soc.* **2014**, *56*, 135–140. [CrossRef] [PubMed]
27. McGee, S. *Cyanosis in Evidence-Based Physical Diagnosis*, 2nd ed.; McGee, S., Ed.; Saunders Elsevier: St Louis, MO, USA, 2007; pp. 85–89.
28. Tanaka, J.; Moriyama, H.; Terada, M.; Takada, T.; Suzuki, E.; Narita, I.; Kawabata, Y.; Yamaguchi, T.; Hebisawa, A.; Sakai, F.; et al. An observational study of giant cell interstitial pneumonia and lung fibrosis in hard metal lung disease. *BMJ Open* **2014**, *4*, e004407. [CrossRef]
29. Miller, K.; McVeigh, C.M.; Barr, E.B.; Herbert, G.W.; Jacquez, Q.; Hunter, R.; Medina, S.; Lucas, S.N.; Ali, A.S.; Campen, M.J.; et al. Inhalation of tungsten metal particulates alters the lung and bone microenvironments following acute exposure. *Toxicol. Sci.* **2021**, *184*, 286–299. [CrossRef]
30. Maitra, S.; Kirtania, J.; Pal, S.; Bhattacharjee, S.; Layek, A.; Ray, S. Intraoperative blood glucose levels in nondiabetic patients undergoing elective major surgery under general anaesthesia receiving different crystalloid solutions for maintenance fluid. *Anesth. Essays Res.* **2013**, *7*, 183–188. [CrossRef]
31. Nair, B.G.; Horibe, M.; Neradilek, M.B.; Newman, S.F.; Peterson, G.N. The Effect of intraoperative blood glucose management on postoperative blood glucose levels in noncardiac surgery patients. *Anesth. Analg.* **2016**, *122*, 893–902. [CrossRef]
32. El-Radaideh, K.; Alhowary, A.A.; Alsawalmeh, M.; Abokmael, A.; Odat, H.; Sindiani, A. Effect of spinal anesthesia versus general anesthesia on blood glucose concentration in patients undergoing elective cesarean section surgery: A prospective comparative study. *Anesthesiol. Res. Pract.* **2019**, *2019*, 7585043. [CrossRef] [PubMed]

33. Govender, P.; Tosh, W.; Burt, C.; Falter, F. Evaluation of increase in intraoperative lactate level as a predictor of outcome in adults after cardiac surgery. *J. Cardiothorac. Vasc. Anesth.* **2020**, *34*, 877–884. [CrossRef]
34. Cotter, E.K.; Kidd, B.; Flynn, B.C. Elevation of intraoperative lactate levels during cardiac surgery: Is there power in this prognostication? *J. Cardiothorac. Vasc. Anesth.* **2020**, *34*, 885–887. [CrossRef]
35. Klee, P.; Rimensberger, P.C.; Karam, O. Association between lactates, blood glucose, and systemic oxygen delivery in children after cardiopulmonary bypass. *Front. Pediatr.* **2020**, *8*, 332. [CrossRef]
36. Azem, R.; Daou, R.; Bassil, E.; Anvari, E.M.; Taliercio, J.J.; Arrigain, S.; Schold, J.D.; Vachharajani, T.; Nally, J.; Na Khoul, G.N. Serum magnesium, mortality and disease progression in chronic kidney disease. *BMC Nephrol.* **2020**, *21*, 49.
37. Moysés-Neto, M.; Guimarães, F.M.; Ayoub, F.H.; Vieira-Neto, O.M.; Costa, J.A.; Dantas, M. Acute renal failure and hypercalcemia. *Ren Fail.* **2006**, *28*, 153–159. [CrossRef]
38. Sadiq, N.M.; Naganathan, S.; Badireddy, M. Hypercalcemia. In *StatPearls*; StatPearls Publishing: Treasure Island, FL, USA, 2023. Available online: https://www.ncbi.nlm.nih.gov/books/NBK430714 (accessed on 17 December 2023).
39. Edelstein, C.L. Biomarkers of acute kidney injury. *Adv. Chronic Kidney Dis.* **2008**, *15*, 222–234. [CrossRef] [PubMed]
40. Hanif, M.O.; Bali, A.; Ramphul, K. Acute renal tubular necrosis. In *StatPearls*; StatPearls Publishing: Treasure Island, FL, USA, 2023. Available online: https://www.ncbi.nlm.nih.gov/books/NBK507815/ (accessed on 20 December 2023).
41. Root, A.W.; Diamond, F.B. Disorders of mineral homeostasis in children and adolescents. In *Pediatric Endocrinology*; Saunders: Philadelphia, PA, USA, 2014; pp. 734–845.
42. Tombach, B.; Bremer, C.; Reimer, P.; Schaefer, R.M.; Ebert, W.; Geens, V.; Heindel, W. Pharmacokinetics of 1M gadobutrol in patients with chronic renal failure. *Invest. Radiol.* **2000**, *35*, 35–40. [CrossRef] [PubMed]
43. Baker, J.F.; Kratz, L.C.; Stevens, G.R.; Wible, J.H., Jr. Pharmacokinetics and safety of the MRI contrast agent gadoversetamide injection (OptiMARK) in healthy pediatric subjects. *Invest. Radiol.* **2004**, *39*, 334–339. [CrossRef]
44. Aime, S.; Caravan, P. Biodistribution of gadolinium-based contrast agents, including gadolinium deposition. *J. Magn. Reson. Imaging* **2009**, *30*, 1259–1267. [CrossRef]
45. Wedeking, P.; Kumar, K.; Tweedle, M.F. Dissociation of gadolinium chelates in mice: Relationship to chemical characteristics. *Magn. Reson. Imaging.* **1992**, *10*, 641–648. [CrossRef] [PubMed]
46. Wedeking, P.; Kumar, K.; Tweedle, M.F. Dose-dependent biodistribution of [153Gd]Gd(acetate)n in mice. *Nucl. Med. Biol.* **1993**, *20*, 679–691. [CrossRef]
47. Tweedle, M.F.; Wedeking, P.; Kumar, K. Biodistribution of radiolabeled, formulated gadopentetate, gadoteridol, gadoterate, and gadodiamide in mice and rats. *Invest. Radiol.* **1995**, *30*, 372–380. [CrossRef]
48. Zheng, J.; Liu, J.; Dunne, M.; Jaffray, D.A.; Allen, C. In vivo performance of a liposomal vascular contrast agent for CT and MR-based image guidance applications. *Pharm. Res.* **2007**, *24*, 1193–1201. [CrossRef]
49. Joshi, A.A.; Aziz, R.M. Deep learning approach for brain tumor classification using metaheuristic optimization with gene expression data. *Int. J. Imaging Syst. Tech.* **2023**, *34*, e23007. [CrossRef]
50. Pan, Y.; Abazari, R.; Tahir, B.; Sanati, S.; Zheng, Y.; Tahir, M.; Gao, J. Iron-based metal–organic frameworks and their derived materials for photocatalytic and photoelectrocatalytic reactions. *Coordin. Chem. Rev.* **2024**, *499*, 215538. [CrossRef]
51. Wang, L.; Dai, P.; Ma, H.; Sun, T.; Peng, J. Advancing biomedical applications of polyoxometalate-based metal–organic frameworks: From design to therapeutic potential. *Inorg. Chem. Front.* **2024**. [CrossRef]
52. Ginsberg, A.P. *Inorganic Syntheses*; John Wiley and Sons: New York, NY, USA, 1990; Volume 27.
53. Brianda, L.E.; Thomas, H.J.; Baronetti, G.T. Thermal stability and catalytic activity of Wells-Dawson tungsten heteropoly salts. *Appl. Catal. A-Gen.* **2000**, *201*, 191–202. [CrossRef]
54. Percie du Sert, N.; Hurst, V.; Ahluwalia, A.; Alam, S.; Avey, M.T.; Baker, M.; Browne, W.J.; Clark, A.; Cuthill, I.C.; Dirnagl, U.; et al. The ARRIVE guidelines 2.0: Updated guidelines for reporting animal research. *PLoS Biol.* **2020**, *18*, e3000410.
55. OECD. *Guidance for the Testing of Chemicals, Acute Oral Toxicity—Fixed Dose Procedure*; No 420; OECD: Paris, France, 2001.
56. OECD. *Guidance Document on the Recognition, Assessment and Use of Clinical Signs as Humane Endpoints for Experimental Animals Used in Safety Evaluation. Environmental Health and Safety Monograph Series on Testing and Assessment*; No 19; OECD: Paris, France, 2000.
57. Abdalla, Y.O.A.; Nyamathulla, S.; Shamsuddin, N.; Arshad, N.M.; Mun, K.S.; Awang, K.; Nagoor, N.H. Acute and 28-day sub-acute intravenous toxicity studies of 1′-S-1′-acetoxychavicol acetate in rats. *Toxicol. Appl. Pharmacol.* **2018**, *356*, 204–213. [CrossRef] [PubMed]
58. Le Tourneau, C.; Stathis, A.; Vidal, L.; Moore, M.J.; Siu, L.L. Choice of starting dose for molecularly targeted agents evaluated in first-in-human phase I cancer clinical trials. *J. Clin. Oncol.* **2010**, *28*, 1401–1407. [CrossRef] [PubMed]
59. Mahdian-Shakib, A.; Hashemzadeh, M.S.; Anissian, A.; Oraei, M.; Mirshafiey, A. Evaluation of the acute and 28-day sub-acute intravenous toxicity of α-l-guluronic acid (ALG.; G2013) in mice. *Drug Chem. Toxicol.* **2022**, *45*, 151–160. [CrossRef] [PubMed]
60. Charan, J.; Kantharia, N.D. How to calculate sample size in animal studies? *J. Pharmacol. Pharmacother.* **2013**, *4*, 303–306. [CrossRef] [PubMed]

International Journal of *Molecular Sciences*

MDPI

*Article*

# Zinc(II) Iminopyridine Complexes as Antibacterial Agents: A Structure-to-Activity Study

**Silvia de la Mata Moratilla [1], Sandra Casado Angulo [1], Natalia Gómez-Casanova [2], José Luis Copa-Patiño [2], Irene Heredero-Bermejo [2,*], Francisco Javier de la Mata [1,3,4] and Sandra García-Gallego [1,3,4,*]**

1 University of Alcalá, Faculty of Sciences, Department of Organic and Inorganic Chemistry and Research Institute in Chemistry "Andrés M. del Río" (IQAR), 28805 Alcalá de Henares, Spain; silvia.matam@edu.uah.es (S.d.l.M.M.); sandra.casado@edu.uah.es (S.C.A.); javier.delamata@uah.es (F.J.d.l.M.)

2 University of Alcalá, Faculty of Pharmacy, Department of Biomedicine and Biotechnology, 28805 Alcalá de Henares, Spain; natalia.gomezc@uah.es (N.G.-C.); josel.copa@uah.es (J.L.C.-P.)

3 Networking Research Center on Bioengineering, Biomaterials and Nanomedicine (CIBER-BBN), 28029 Madrid, Spain

4 Institute Ramón y Cajal for Health Research (IRYCIS), 28034 Madrid, Spain

* Correspondence: irene.heredero@uah.es (I.H.-B.); sandra.garciagallego@uah.es (S.G.-G.)

**Abstract:** Antibiotic resistance is currently a global health emergency. Metallodrugs, especially metal coordination complexes, comprise a broad variety of candidates to combat antibacterial infections. In this work, we designed a new family of Schiff base zinc(II) complexes with iminopyridine as an organic ligand and different inorganic ligands: chloride, nitrate, and acetate. The antibacterial effect of the Zn(II) complexes was studied against planktonic bacterial cells of *Staphylococcus aureus* (Gram-positive) and *Escherichia coli* (Gram-negative) strains. The results showed a moderate biocide activity in both types of planktonic bacteria, which arises from the metal complexation to the Schiff base ligand. Importantly, we confirmed the crucial effect of the metal, with Zn(II) improving the activity of Cu(II) counterparts previously reported. On the other hand, the impact of the inorganic ligands was not significant for the antibacterial effect but was relevant for the complex solubility. Finally, as proof of concept of topical antibacterial formulation, we formulated an emulsion containing the most lipophilic Zn(II) complex and confirmed a sustained release for 24 h in a vertical cell diffusion assay. The promising activity of iminopyridine Zn(II) complexes is potentially worth exploring in more detailed studies.

**Keywords:** Schiff base; iminopyridine; zinc; metal complex; bacteria; antibiotic; *Staphylococcus aureus*; *Escherichia coli*

## 1. Introduction

Antibiotic resistance is a worldwide health emergency. Resistance to all antibiotics in clinical use has been detected, so next-generation antimicrobial therapies must be developed [1]. An encouraging approach is the use of metallodrugs [2]. These compounds have been employed for antibacterial treatment since ancient times. It is well known that first-row transition metals (manganese, iron, cobalt, nickel, copper, and zinc) are the key metals required by most organisms. Among them, zinc is unique since it possesses a filled d-orbital and does not undergo redox cycling. Furthermore, zinc is crucial in normal host immune function, as macrophages adopt different strategies of zinc starvation and zinc toxicity to kill the bacteria they phagocytize [3]. Inspired by this mechanism, different Zn metallodrugs have been designed, such as antibacterial ZnO nanoparticles [4].

Unlike solvated metal ions or metal nanoparticles, organometallic and metal coordination complexes present a well-defined arrangement of ligands around the metal center, thus expanding their versatility and applications [5,6]. The outstanding tunable properties of metal complexes arise from the vast variety of coordination numbers, geometries, structural diversity, and kinetics of ligand exchange. This structural versatility translates into a range of different mechanisms of action, such as redox reactions, ligand exchange processes, the generation of Reactive Oxygen Species (ROS), and the competitive inhibition of enzymes.

Schiff base ligands are a promising family of organic ligands in such metallodrugs [7]. The metal chelation to the Schiff base increases the lipophilicity of the ligand due to the delocalization of $\pi$-electrons in the entire chelate system, and thus increases the penetration through cell membranes. A broad list of Schiff base metal complexes mainly based on Cu(II), Ru(II), Ni(II), or Zn(II) show antibacterial activity [8,9]. It is important to mention that Zn(II) complexes frequently exhibit superior antimicrobial activity than other metals, like their Cu(II) counterparts. For example, a benzimidazole Schiff base Zn(II) complex exhibited an inhibition zone of 18.9 mm in *E. coli*, while for the Cu(II) analog, the zone was only 11.8 mm [10]. Another organometallic Schiff base Zn(II) complex produced inhibition zones of 26 mm and 17 mm against *E. coli* and *S. aureus*, respectively, while no activity was reported for the Cu(II) counterpart [11]. This promising activity, together with the biocompatibility and lower cost of zinc salts, enhances the potential of these metallodrugs.

Silane-containing Schiff base ligands further tune the lipophilic character and thus the biological activity of metal complexes. İspir et al. synthesized Zn(II) and Cd(II) complexes comprising 3-iminopropyltrimethoxysilane and 3-iminopropyltriethoxysilane chains, which exhibited certain antibacterial and antifungal activities [12]. In our previous works, we demonstrated the antibacterial activity of carbosilane iminopyridine complexes of Cu(II) and Ru(II) against *S. aureus* and *E. coli* [13]. All complexes exhibited better activity than the corresponding metal salts precursors, confirming the impact of the metal coordination to the iminopyridine ligand. Furthermore, relevant differences were found in the Minimum Inhibitory Concentration (MIC) and the Minimum Bactericide Concentration (MBC) depending on the type of metal or the counterions in the complex. Monometallic complexes were employed to predict the behavior of first- and second-generation carbosilane metallodendrimers, where the multivalent properties significantly improved the antibacterial effect [14]. This multivalency, characteristic of dendritic materials, is responsible for their outstanding properties for the prevention, treatment, and diagnosis of infectious diseases [15]. However, the combination of dendritic scaffolds with metal ions further expands the versatility and efficacy of the candidates by merging the activity of both components.

Considering the well-documented antibacterial activity of Schiff base Zn(II) complexes, in this study, we focused on the design of Zn(II) carbosilane iminopyridine complexes with three different inorganic ligands: chloride, nitrate, and acetate. These complexes exhibited moderate antibacterial activity against Gram-positive and Gram-negative bacteria and could be formulated as an emulsion for antibacterial topical uses. We confirmed the impact of the type of metal center and the ligands on the structure, as well as the antibacterial activity of these metallodrugs, finding that Zn(II) iminopyridine complexes appear as promising candidates for antibacterial therapy.

## 2. Results

### 2.1. Synthesis of Zn(II) Iminopyridine Complexes

The synthesis of the iminopyridine ligand (**I**) was carried out as described in [16]. This ligand was employed to synthesize three different Zn(II) complexes with several inorganic ligands: chloride, nitrate, and acetate (Figure 1). As Zn(II) precursors, $ZnCl_2$, $Zn(NO_3)_2{\cdot}6H_2O$ and $Zn(CH_3CO_2)_2{\cdot}2H_2O$ were used. In all the reactions, 1:1 stoichiometry (ligand:Zn) was employed. The ligand and the metal salt were dissolved in MeOH, mixed, and reacted for 12 h at room temperature. The complexes **Ia**, **Ib**, and **Ic** were isolated as brown oils with high yield after the evaporation of the solvent.

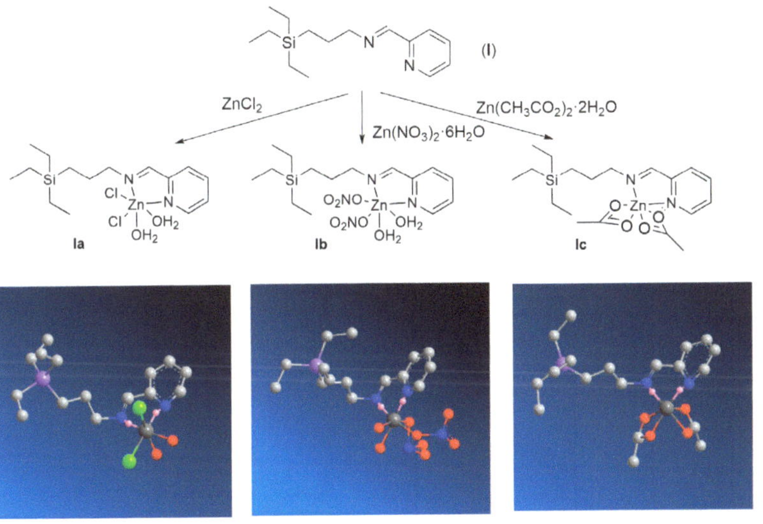


**Figure 1.** Synthetic protocol to prepare the Zn(II) iminopyridine complexes **Ia–c** and proposed structures. The reactions proceeded in MeOH for 12 h at room temperature, using 1:1 (M:L) stoichiometry. Insert: Snapshots from 3D spatial arrangement of complexes **Ia–c** in solid state after energy minimization. Hydrogen atoms have been hidden for simplicity. [Chem3D 22.0.0 software: job 1 (minimize energy to minimum RMS gradient of 0.010)].

### *2.2. Structural Characterization of Zn(II) Complexes: Impact of the Metal Coordination and the Nature of the Inorganic Ligands*

To evaluate the impact of structural parameters, Zn(II) complexes **Ia–c** were comparatively studied through $^{1}$H-NMR and $^{13}$C-NMR (Figures 2 and S1). Unlike the Cu(II) counterparts, the diamagnetic properties of Zn(II) complexes facilitated this study, which offered useful insight into the impact of the different structural parameters. The signals were compared to the precursor ligand **I**. In $^{1}$H-NMR spectra, it can be observed that all the signals close to the metal center appeared at lower field, compared to the precursor **I**. The signal corresponding to the methylene group connected to the N atom appeared at 3.66 ppm in the ligand, and it was shifted to 3.95 (**Ia**), 3.88 (**Ib**), and 3.76 (**Ic**). For the imine group, the signal was shifted from 8.37 (**I**) to 8.51 (**Ia**), 8.48 (**Ib**), and 8.34 (**Ic**). Additionally, relevant differences were observed in the shifting for the aromatic protons (Figure 2), probably as a result of the distortion of the aromatic ring through the chelation of the metal ion. The overall shifting can be explained by the removal of electronic density from the ligand, caused by the metal atom, as well as the chelating effect. A similar effect was observed in $^{13}$C-NMR (Figure S1). The imine carbon appeared in the range of 159.6–161.5 ppm for the new metal complexes.

Among the three Zn(II) complexes, relevant differences were also observed in NMR spectra, which can be ascribed to several reasons: The first reason is the different nature of the inorganic ligands. All of them are σ,π-donating ligands whose base strength decreases in the following order: $CH_3CO_2^- > NO_3^- > Cl^-$. As Zn(II) is a borderline Lewis acid, it will find higher affinity towards chloride ligands. The second reason is the different coordination modes of the ligands. All the complexes probably present the same coordination index (6) in solid state, which may vary in solution. However, the acetate ligands appear as bidentate ligands, as later confirmed through FTIR. This result is in line with the zeta potential values (Table 1). The Zn(II) complexes exhibited slightly negative values, and the negative charge decreases when increasing the metal–ligand affinity, reaching values close to zero for the chloride **Ia**. Again, Zn(II) complexes behave differently to the Cu(II) and Ru(II) counterparts previously reported, with clearly positive values. The fully filled $d^{10}$

electronic configuration of Zn(II) liable with its redox inertness may be behind this behavior. Other Schiff base complexes of $d^{10}$ metals (Zn(II), Cd(II) and Hg(II)) have been reported as antibacterial agents [17]. To predict the composition of our complexes, we compared to similar Zn(II) complexes bearing bidentate ligands with the iminopyridine moiety, such as 2-pyrilidineaniline [18] or 4-(4-aminophenoxy)-N-(1-(pyridin-2-yl)ethylidene)aniline [19]. In these examples, the authors confirmed a $ZnLCl_2(H_2O)_2$ composition. However, in our complexes, the analysis of the $^1$H-NMR spectra in $CDCl_3$ showed broad signals which, overall, integrated for around one water molecule in each complex. The exchange with the deuterated solvent might be responsible for this underestimation. Other analytical techniques were used to gain further insight into the composition of the complexes.

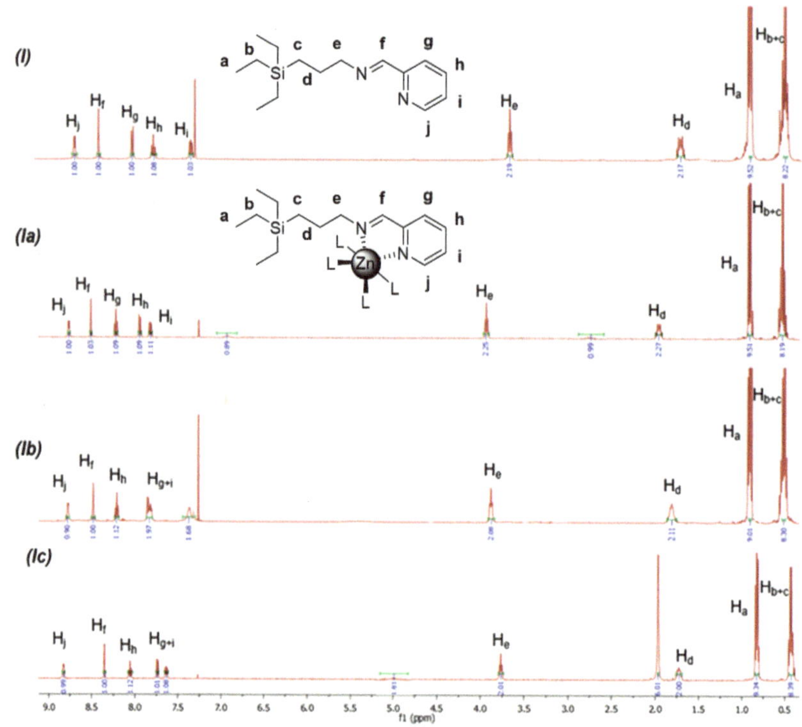


**Figure 2.** Comparative $^1$H-NMR spectra of compounds **I**, **Ia–c** in $CDCl_3$.

**Table 1.** Minimum Inhibitory Concentration (MIC) and Minimum Bactericidal Concentration (MBC) effect of Schiff base complexes in planktonic cells and comparative values of Z-potential.

| Compound | Zeta Potential, [mV] | *S. aureus* MIC [mg/L] | MBC [mg/L] | *E. coli* MIC [mg/L] | MBC [mg/L] |
|---|---|---|---|---|---|
| G0[NCPh(*o*-N)ZnCl$_2$·2H$_2$O] (**Ia**) | −0.33 ± 0.06 | 64 | 64 | 128 | 128 |
| G0[NCPh(*o*-N)Zn(NO$_3$)$_2$·2H$_2$O] (**Ib**) | −6.28 ± 0.33 | 64 | 64 | 128 | 128 |
| G0[NCPh(*o*-N)Zn(O$_2$CCH$_3$)$_2$] (**Ic**) | −8.43 ± 1.53 | 64 | 64 | 64–128 | 64–128 |
| G0[NCPh(*o*-N)CuCl$_2$·H$_2$O] [a] | 10.45 ± 1.25 | 64 | 64 | 256 | 256 |
| G0[NCPh(*o*-N)Cu(NO$_3$)$_2$·H$_2$O] [a] | 14.79 ± 1.92 | 128 | 128 | 128 | 256 |
| G0[NCPh(*o*-N)Ru(Cp)(PTA)]Cl [a] | 18.70 ± 4.21 | 16 | 16 | 64 | 64 |

[a] previously published results [13].

These complexes, in solid-state form, were explored through FT-IR (Figures S2–S4). The samples were dried under vacuum overnight to eliminate non-coordinated water and facilitate the interpretation of the spectra. The precursor ligand presented a characteristic band around 1650 $cm^{-1}$, assigned to the imine C=N stretching. For the chloride complex **Ia**, the spectrum showed a shifting of this band to ~1600 $cm^{-1}$. Additionally, a band in the 3400–3100 $cm^{-1}$ range confirms the presence of water molecules. This corresponds to coordinated water, as two bands around 880–980 $cm^{-1}$ can be observed due to $\nu(H_2O)$ rocking and the wagging modes of vibrations [20]. For the nitrate complex **Ib**, the C=N band was also observed around 1650 $cm^{-1}$. The characteristic stretching of monodentate nitrate ligands were observed at 1480, 1300, and 1010 $cm^{-1}$, [21], and the presence of coordinated water molecules was observed in the ranges of 3500–3000 and 820–920 $cm^{-1}$. Finally, for the acetate complex **Ic**, two broad peaks at 1400 and 1600 $cm^{-1}$, assigned to the symmetric and asymmetric stretching of COO, were observed, which overlap with the peak from C=N stretching. In this case, no water molecules were observed. To explore the coordination mode of acetate, we evaluated the divergence of $\nu_{asym}(COO)$ and $\nu_{sym}(COO)$. The precursor $Zn(acet)_2{\cdot}2H_2O$ is described as octahedral, with two chelating acetate groups and two water molecules, and $\Delta\nu$ = 145 $cm^{-1}$ [22]. For **Ic**, we found a similar $\Delta\nu$, so we proposed a similar bidentate coordination of the ligands in the complex as a consequence of the elimination of the labile water molecules in the octahedral environment of the precursor that were replaced by the iminopyridine ligand. It has been described that $\Delta\nu$ follows the following trend: chelating < bridging < ionic < monodentate [23]. In our case, sodium acetate exhibits $\Delta\nu$ = 164 $cm^{-1}$, so chelating coordination mode occurs when $\Delta\nu \ll 164$ $cm^{-1}$, and bidentate bridging coordination occurs when $\Delta\nu \leq 164$ $cm^{-1}$. In complex **Ic**, the divergence $\Delta\nu$ = 138 $cm^{-1}$, so we can propose a chelating mode of the acetates. The Zn-O stretching appeared in the 500–570 $cm^{-1}$ range, and the Zn-N stretching appeared in the 400–470 $cm^{-1}$ range [24]. To further confirm the composition of our compounds, we performed an elemental analysis and ICP-OES. The hygroscopicity of our compounds was responsible for the slight deviations from the values calculated in elemental analysis. For example, for complex **Ic**, the experimental analysis matched the calculated values, considering the presence of one non-coordinated water molecule. This molecule was not observed in the FTIR analysis, when the sample was tested just directly after drying. ICP-OES revealed a higher content of Zn compared to the calculated values. Such deviations, with no clear explanation, highlighted the need for the optimization of the analysis parameters and sample preparation to be able to accurately quantify the amount of zinc in our complexes.

Additionally, we studied the stability of our complexes. We dissolved our complexes in water and analyzed the potential changes in the $^1$H-NMR spectra at t = 0 and t = 7 days. Samples were evaporated and redissolved in MeOD for analysis. After one week in water solution, we observed no changes in the $^1$H-NMR spectra (Figures S8–S10). These results are in agreement with former studies on the stability of Ru(II) complexes of the chelating ligand **I** [16]. The authors did not observe any appreciable changes in NMR spectra in $D_2O$, PBS-$D_2O$, or neat DMSO-$d^6$, even after 72 h. However, similar monodentate complexes showed the displacement of the ligand from the metal coordination sphere at high DMSO ratios.

Furthermore, the thermal stability was explored through TGA (Figures S5–S7). The thermogravimetric study indicated that chloride complex **Ia** presented a small humidity loss around 85 °C and required temperatures around 250 °C to lose the two water molecules (8.3% weight loss), confirming their coordination to the central metal atom. For the nitrate complex **Ib**, the 7.4% loss, corresponding to two water molecules, was observed at above 125 °C, and the subsequent loss of the nitrate ligands was observed at above 190 °C. Again, this confirms the presence of coordinated water molecules in the coordination sphere of the central metal ion. These water molecules are more strongly bonded to the metal ions and, therefore, eliminated at higher temperatures. For the acetate complex **Ic**, the loss of non-coordinated water occurred at around 85 °C, and that for the acetate ligands occurred at above 190 °C. Overall, we observed a higher thermal stability for the chloride derivative compared to the nitrate and acetate counterparts. However, the three complexes were stable under the conditions required for the biological evaluation.

Other similar ligands, but with a tridentate NNO nature, like 2-[[(2-pyridinylmethyl)im ino]methyl]phenol (HSALIMP), generated two types of complexes: the dimeric [Zn(SALIM P)$NO_3$]$_2$ and the monomeric [Zn(HSALIMP)$Cl_2$]·$H_2O$ [25]. In these examples, the ligand acted as a tridentate in the presence of the moderately coordinating nitrate or as a bidentate in the presence of the anionic chloride. In order to evaluate the nuclearity of our complexes, we performed High-Resolution Mass Spectrometry (Figure S11). Unfortunately, we could not identify the molecular peak of any of our complexes. However, the results pointed to mononuclear species, and only in the case of complex **Ic** may the presence of small peaks above m/z 700 indicate the presence of minor binuclear species. These binuclear species could be explained by bridging acetate ligands, but only as minor species, according to the FTIR spectra.

### 2.3. Antibacterial Activity

#### 2.3.1. Biocidal Activity

The tested compounds showed antibacterial activity against *S. aureus* and *E. coli*. The MIC values of the compounds were 64 mg/L for *S. aureus* and 128 mg/L for *E. coli*, except for **Ic**, with a range of 64–128 mg/L (Table 1). This confirmed that they are bactericidal compounds whose MBC values correspond with the MIC values. The biocidal activity was compared with the precursor ligand **I** as well as with the precursor salts ($ZnCl_2$, $Zn(NO_3)_2{\cdot}6H_2O$, $Zn(CH_3CO_2)_2{\cdot}2H_2O$), with all of them exhibiting MIC and MBC values higher than 512 mg/L.

#### 2.3.2. Kinetic Studies

Kinetic studies showed that bacteria treated with compounds **Ib** and **Ic** required 2 h more than the control to grow for the *S. aureus* strain and 1 h for the *E. coli* strain at one concentration below their MIC values. However, the chloride complex **Ia** inhibited the growth of both bacteria up to 3 h later than the control at one concentration below the MIC value. This confirmed the impact of the type of inorganic ligand. In the studies against *E. coli*, the bacterial population treated with this compound was unable to reach a viability of 50% at a concentration below its MIC (64 mg/L) at 20 h, while in *S. aureus*, a viability of 90.9% was recorded at a concentration below its MIC (32 mg/L) at the same time.

### 2.4. Evaluation of Topical Antibacterial Emulsion

According to the zeta potential values, the chloride complex **Ia** exhibited more lipophilic properties. This could be interesting for topical antibacterial action, where the complex must cross the skin barriers. Furthermore, kinetic studies revealed a prolonged inhibitory effect of the chloride derivative compared to the nitrate and acetate complexes. As proof of concept, compound **Ia** was formulated in a W/O emulsion (at 4%) comprising glycerin, milliQ water, PEG-400, and Emolivan™ base (see the Section 4 Materials and Methods). A thorough quality check confirmed the correct formation of the emulsion.

Subsequently, the release of **Ia** from the formulation was studied through a vertical diffusion cell HDT-1000 using Stat-M membranes at 32 °C to simulate skin conditions (Figure 3). The emulsion (total 55 mg, containing 2.2 mg of **Ia**) was deposited on the membrane in the donor chamber, and PBS was used in the collecting chamber. Aliquots of 0.2 mL were collected at different times, filtered through a 0.2 µm filter, and evaluated through HPLC. As represented in Figure 3B, the drug release reached a plateau after 24 h, which matches the loss of the brownish color in the emulsion. This assay confirmed the potential of Zn(II) iminopyridine complexes as antibacterial candidates in topical formulations, enabling a sustained release.

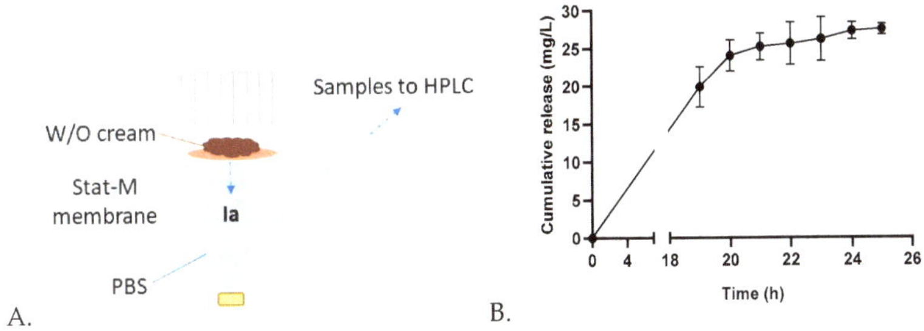


**Figure 3.** (**A**) Scheme depicting the vertical cell for the diffusion assay. (**B**) Cumulative release of compound **Ia** from a 4% W/O cream formulation at 32 °C. In the donor chamber, 55 mg of the cream was applied on the membrane. The collector chamber was filled with PBS. Average results from two independent experiments are represented in the graph.

## 3. Discussion

Schiff base metal complexes are promising therapeutic metallodrugs with applications in different fields, such as cancer therapy, or they could be used as antimicrobial agents [7–9]. In our previous studies, we explored the activity of Schiff base complexes comprising a carbosilane iminopyridine ligand, as well as different metal ions such as Cu(II) and Ru(II) [13]. These complexes exhibited promising antitumor and antimicrobial activity and, more importantly, could emulate the behavior of their multivalent counterparts: metallodendrimers. The monometallic complexes enabled a simplified evaluation of the impact of different structural parameters, like the type of metal ion and the nature of the inorganic ligands. We concluded that these parameters are crucial for the biological activity of these metallodrugs.

In line with our former studies, in this work we designed a family of Zn(II) iminopyridine-containing metallodrugs and evaluated the influence of these structural parameters on their antimicrobial activity (Figure 1). These complexes were prepared through a straightforward reaction and isolated with high yields. Their structural characterization through NMR, FTIR, elemental analysis, ICP, TGA, and HRMS offered relevant information about the purity, structure, and stability of the complexes. The new Zn(II) compounds probably exhibit an octahedral coordination in solid state and are stable in water solution for long periods of time and at temperatures below 120 °C. As expected, the metal ion coordination to the iminopyridine ligand induced a shift in the $^{1}H$ and $^{13}C$ NMR signals, potentially explained by the "heavy atom-light atom" (HALA) effect [26]. For example, Barone et al. explored the effect of different metal ions (Ni(II), Pd(II), Pt(II), and Zn(II)) on the $^{1}H$-NMR shift of Schiff base complexes [27]. They observed that the nearest protons to the metal center are the most deshielded, and the covalent nature of the M-O and M-N bonds significantly affected such deshielding. In our case, we observed a deshielding of $H_e$ and $H_f$ in the trend **Ia** > **Ib** > **Ic**, even reaching a small shielding of $H_f$(**Ic**). We might interpretate this result as indicative of the more covalent nature of the Zn-N bond for the chloride complex, followed by the nitrate and the acetate systems.

Unexpectedly, the Zn(II) complexes exhibited slightly negative Z-potential values, and their charge increased in the following order: **Ia** < **Ib** < **Ic** (Table 1). This indicated that the most colloidally unstable and lipophilic compound was the chloride derivative, also confirmed by the strong interaction with the chloride and water ligands. Compared to the Cu(II) and Ru(II) counterparts, which showed strong cationic properties that could disrupt bacterial membrane through electrostatic interactions, the new Zn(II) derivatives seem to behave differently. A redox mechanism is also ruled out due to the $d^{10}$ electronic configuration of zinc.

Similar Zn(II) Schiff base complexes have been described in the literature, with these ones mainly being used as catalysts in esterification reactions. Di Serio et al. modulated the Lewis acidity of zinc by altering the inorganic ligands in the iminopyridine complexes with different coordination abilities: $Cl^- > AcO^- > CF_3CO_2^- > CF_3SO_3^-$ [28]. Through NMR and conductivity measurements, they confirmed that chloride and acetate were non-conductive, while the rest were readily displaced by solvent molecules affording cationic species. Our complexes exhibited a similar behavior, as shown through DLS. The same authors described a complex analogous to **Ic**, bearing a triethoxosilyl ligand instead of a triethylsilyl [29]. In these two cases, the authors depicted a tetrahedral environment around the Zn atom, but few characterization details were provided. Sharma et al. bound zinc acetate to silica-supported magnetic nanoparticles decorated with a similar ligand [30]. The C=N peak shifted from 1647 to 1636 $cm^{-1}$ after metalation, but no studies were carried out on the coordination sphere of the metal atom. A more detailed characterization was found for a Zn(II) bis(imino)pyridine complex, where single-crystal X-ray diffraction confirmed the monomeric nature and trigonal bipyramidal geometry around the zinc center, with two monodentate acetate ligands [31]. The monodentate disposition of the acetate ligands is favored due to the presence of the tridentate Schiff base ligand, unlike the complex herein described.

The antibacterial effect of the Zn(II) complexes was evaluated by measuring the inhibitory (MIC) and bactericide (MBC) properties against Gram-positive and Gram-negative bacteria. From the results summarized in Table 1, we can draw the following conclusions:

(1) The metal complexation to the ligand, which was clearly demonstrated through the shifting of the signals in NMR and FTIR, is crucial for the antibacterial effect. The precursor ligand and the metal salts used as controls are inactive, but the chelating effect of the iminopyridine ligand stabilizes the metallodrug and increases its lipophilicity. Furthermore, it rules out the possibility that the antibacterial activity is related to metal release, as we previously demonstrated with similar complexes [13].

(2) Carbosilane metallodrugs are promising broad-spectrum antibiotics exhibiting activity against both *S. aureus* (Gram-positive) and *E. coli* (Gram-negative) bacteria. As expected, all compounds were more active against *S. aureus* than against *E. coli*. This effect is frequent because Gram-negative microorganisms are slightly more resistant to treatments than Gram-positive ones due to the outer membrane on the cell wall of Gram-negative bacteria [32].

(3) The nature of the inorganic ligand (chloride, nitrate, or acetate) produces a different shifting in the NMR signals, arising from the electronic withdrawal and the chelating effect. This is relevant for the solubility of the complexes (with the chloride system being the most lipophilic) but not significant for the antibacterial effect in these Zn(II) complexes. Similar MIC and MBC values were obtained for **Ia–c**, and just a slightly better activity was observed for the chloride complex **Ia** in the kinetic studies.

(4) The water solubility of complexes **Ia–c** is a relevant advantage for antibacterial applications. A great number of Schiff base metal complexes described in the literature exhibit very poor or no solubility in water; thus, the testing of their antibacterial activity through the inhibition zone assay is required. For example, Naureen et al. studied two different Zn(II) Schiff base complexes bearing monodentate acetate ligands, which exhibited 11 and 19 mm inhibition zones for *E. coli* and 11 and 15 mm inhibition zones for *S. aureus* [33]. Other examples are the Zn(II) and Cu(II) complexes from the ligand (E)-2-(((7-chloro-2-hydroxyquinolin-3-yl) methylene) amino)-3-phenylpropanoic acid [23]. At 100 μg/mL, they showed diameters of 21.08 and 6.08 (in *E. coli*) and 25.05 and 6.58 mm (in *S. aureus*), respectively. The different techniques used for the evaluation of the antibacterial activity hinders the accurate comparison of complexes **Ia–c** with others described in the literature. However, they can be compared to the Cu(II) and Ru(II) complexes bearing the same ligand [13].

(5) The type of metal ion is crucial for the antimicrobial activity. The chloride Zn(II) complex **Ia** and its Cu(II) counterpart exhibited similar activity in *S. aureus*, but the Zn(II) complex was slightly more active against *E. coli*. The nitrate Zn(II) complex **Ib**, though, improves the activity against both bacteria compared to its Cu(II) counterpart. This behavior is in line with former studies that confirm the higher antibacterial activity of Zn(II) Schiff

base complexes compared to Cu(II) compounds [13]. Overall, we could establish the following trend regarding the antibacterial activity of carbosilane iminopyridine systems: Ru(II) complexes > Zn(II) complexes > Cu(II) complexes.

(6) The Z-potential values observed with these complexes rule out a main antibacterial mechanism due to the cationic charge. As we previously reported, the antibacterial properties must rely on other mechanisms related to the presence of metal ions. Zinc is a redox-inactive metal required for catalytic activity and/or structural stability for thousands of proteins. Bacterial pathogens have developed mechanisms to keep zinc homeostasis, relying on a complex network of metal transporters and zinc buffering systems [3]. For example, to overcome Zn intoxication, *E. coli* possesses a metalloregulator named ZntR, which causes DNA conformational changes, leading to the expression of the P-type ATPase ZntA. This mechanism, together with the expression of cation diffusion family transporters and the use of glutathione as a buffering agent, alleviates Zn toxicity. For *S. aureus*, bacteria produce the metallophore staphylopine with broad metal-chelating abilities, which regulates Zn homeostasis. The use of the metallodrugs designed in this study can affect Zn homeostasis, potentially explaining their antibacterial activity.

As proof of concept, we formulated the most lipophilic complex **Ia** as an emulsion and evaluated the topical diffusion behavior using membranes that simulated skin. The results indicated a complete release of the compound in the first 24 h, confirming the potential use of these metallodrugs in topical antimicrobial formulations.

Overall, we concluded that the promising results obtained with the Zn(II) complexes encourage us to design multivalent platforms such as carbosilane metallodendrimers to significantly improve the antimicrobial activity of the monometallic complexes and to understand the impact of the presence of several Zn(II) complexes in a single scaffold.

## 4. Materials and Methods

### 4.1. *Synthesis and Characterization of Zn(II) Iminopyridine Complexes*

Solvents were purified from appropriate drying agents. Chemicals were purchased from commercial sources and used as received. Elemental analyses were performed on a LECO CHNS-932. NMR spectra were recorded on a Varian Unity VXR-300 Hz and 500 Hz instruments (Varian, Inc., Palo Alto, CA, USA) in $CDCl_3$. Chemical shifts are given in ppm. $^1H$ and $^{13}C$ resonances were measured relative to internal deuterated solvents peaks. $\{^1H\text{-}^{13}C\}$-HSQC-2D-NMR experiments were carried out to support the assignment of the signals. FT-IR spectra were recorded using a Cary 630 FTIR Agilent instrument (Agilent, Santa Clara, CA, USA) equipped with an attenuated total reflectance (ATR) accessory. Agilent Microlab PC software v5.1 was used for spectra acquisition, which was carried out from 4000 to 400 $cm^{-1}$. ICP studies were performed in an ICP-OES 720 Agilent (Agilent, Santa Clara, CA, USA) automatic injector SPS3 at $\lambda$ 202.548 and 213.857 nm. The samples were digested in $HNO_3$ (6 mL, ultrapure ppb trace quality, Scharlab, Barcelona, Spain) in a microwave digestor Milestone Ethos Easy (Sorisole, BG, Italy) at 230 °C for 20 min and then diluted to 25 mL with milliQ water. The calibration curve contained 0, 400, 1000, and 2000 ppb of Zn. TGA evaluation was performed in a TGA55 (TA Instruments, New Castle, DE, USA) in the range of 25–270 °C, with an increase of 10 °C/min. Spectra were analyzed with software Trios V5.6.0.87 (TA Instruments, New Castle, DE, USA). High-Resolution Mass Spectrometry studies were performed in a high-resolution Triple TOF 5600 plus (Sciex, Framingham, MA, USA).

General synthetic protocol: A MeOH solution of the metal salt (1 eq., in 4 mL) was added over a MeOH of the ligand **I** (1 eq., in 4 mL). After 24 h stirring at r.t., the solvent was evaporated, and the product was isolated as dark brown oil with high yield.

#### 4.1.1. G0[NCPh(o-N)$ZnCl_2 \cdot 2H_2O$] (**Ia**)

**Ia** was prepared according to the general protocol using the following reagents: ligand **I** (1.27 mmol) and $ZnCl_2$ (1.27 mmol). Yield: 90%.

$^1$H-NMR: δ 8.42 (N=CH); 8.81, 8.20, 7.84 (4H, $CH^{py}$); 3.95 (t, 2H, -$CH_2$N); 1.98 (m, 2H, $SiCH_2CH_2CH_2N$); 0.92 (m, 9H, $SiCH_2CH_3$); 0.54 (m, 8H, $SiCH_2CH_3$ & $SiCH_2CH_2CH_2N$). $^{13}$C-NMR {$^1$H}: δ 161.5 (N=CH); 146.5 ($C^{py}$); 149.4, 142.1, 129.4, 128.2 ($CH^{py}$); 61.8 ($SiCH_2CH_2CH_2N$); 24.7 ($SiCH_2CH_2CH_2N$); 8.9 ($SiCH_2CH_2CH_2N$); 7.3 ($SiCH_2CH_3$), 2.7 ($SiCH_2CH_3$). Elemental analysis (%): $C_{15}H_{30}Cl_2N_2O_2SiZn$ (434.8 g/mol) Calc.: C, 41.44%; H, 6.96%; N, 6.44%. Exp.: C, 41.77%; H, 6.40%; N, 6.55%. FT-IR: ν(C=N): 1643.8 cm$^{-1}$. ICP (Zn%): Calc. 15.0%, Exp. 19.1%.

4.1.2. G0[NCPh(o-N)Zn$(NO_3)_2 \cdot 2H_2O$] (**Ib**)

**Ib** was prepared according to the general protocol using the following reagents: ligand **I** (0.487 mmol) and $Zn(NO_3)_2 \cdot 6H_2O$ (0.487 mmol). Yield: 93%.
$^1$H-NMR: δ 8.35 (N=CH); 8.79, 8.19, 7.80 (4H, $CH^{py}$); 3.88 (t, 2H, -$CH_2$N); 1.81 (m, 2H, $SiCH_2CH_2CH_2N$); 0.92 (m, 9H, $SiCH_2CH_3$); 0.54 (m, 8H, $SiCH_2CH_3$ & $SiCH_2CH_2CH_2N$). $^{13}$C-NMR {$^1$H}: δ 161.3 (N=CH); 146.1 ($C^{py}$); 149.1, 141.1, 129.0, 127.4 ($CH^{py}$); 61.3 ($SiCH_2CH_2CH_2N$); 24.0 ($SiCH_2CH_2CH_2N$); 8.5 ($SiCH_2CH_2CH_2N$); 7.1 ($SiCH_2CH_3$), 2.8 ($SiCH_2CH_3$). Elemental analysis (%): $C_{15}H_{30}N_4O_8SiZn$ (487.9 g/mol). Calc.: C, 36.93%; H, 6.20%; N, 11.48%. Exp.: C, 36.51%; H, 6.11%; N, 10.96%. FT-IR: ν(C=N): 1645.6 cm$^{-1}$. ICP (Zn%): Calc. 13.4%, Exp. 14.1%.

4.1.3. G0[NCPh(o-N)Zn$(OOCCH_3)_2$] (**Ic**)

**Ic** was prepared according to the general protocol using the following reagents: ligand **I** (0.497 mmol) and $Zn(O_2CCH_3)_2 \cdot 2H_2O$ (0.497 mmol). Yield: 98%.
$^1$H-NMR: δ 8.34 (N=CH); 8.82, 8.05, 7.73, 7.63 (4H, $CH^{py}$); 3.76 (t, 2H, -$CH_2$N); 1.91 (6H, $CH_3CO_2$); 1.73 (m, 2H, $SiCH_2CH_2CH_2N$); 0.88 (m, 9H, $SiCH_2CH_3$); 0.40 (m, 8H, $SiCH_2CH_3$ & $SiCH_2CH_2CH_2N$). $^{13}$C-NMR {$^1$H}: δ 179.8 ($CH_3CO_2$); 159.6 (N=CH); 146.6 ($C^{py}$); 149.9, 140.8, 128.4, 126.6 ($CH^{py}$); 61.8 ($SiCH_2CH_2CH_2N$); 24.2 ($SiCH_2CH_2CH_2N$); 22.5 ($CH_3CO_2$); 8.9 ($SiCH_2CH_2CH_2N$); 7.2 ($SiCH_2CH_3$); 3.0 ($SiCH_2CH_3$). Elemental analysis (%): $C_{19}H_{32}N_2O_4SiZn$ (445.9 g/mol). Calc.: C, 51.17%; H, 7.23%; N, 6.28%. Calc.(+$H_2O$): C, 49.19%; H, 7.39%; N, 6.04%. Exp.: C, 49.17%; H, 7.27%; N, 5.92%. Calc.: FT-IR: ν(C=N) & ν(C=O): 1593.4 cm$^{-1}$. ICP (Zn%): Calc. 14.1%, Exp. 20.0%.

### 4.2. Formulation in W/O Emulssion

The cream was formulated using the following components: active principle (**Ia**) (40 mg), glycerin (50 mg), milliQ water (50 mg), polyethylene(glycol)-400 (60 mg), and Emolivan base (800 mg). In a flask, **Ia**, glycerin, PEG-400, and water were mixed. In another flask, Emolivan base was added and heated to 85 °C until fusion. The first mixture is heated to 85 °C and then added drop-wise to the Emolivan base while stirring with a glass pipette. The mixture was stirred until it reached room temperature. Subsequently, quality control was performed. The organoleptic control indicated that a viscous brown emulsion had been obtained. From the external/internal phase control, the test with Sudan III and methylene blue confirmed the formation of a W/O emulsion.

### 4.3. Vertical Cell Diffusion Assay

The assays were conducted in a HDT 1000 vertical diffusion cell tester from Copley Scientific using a vertical Franz skin cell with a nominal volume of the acceptor compartment of 12 mL and a diameter of 15 mm. For the donor compartment, the **Ia**-loaded emulsion was initially set. The experiments were conducted in triplicate and carried out at 37 °C and 480 rpm for 24 h. Samples were taken at different time points and quantified through HPLC. The extracted volume was continuously replaced to keep the volume constant. The assay was performed in aqueous media as the acceptor phase, mimicking physiological conditions corresponding to buffer solution (PBS) at pH 7.4. A transdermal diffusion test model Strat-M® membrane was used. These membranes have two different layers, simulating human epidermis and dermis; thus, the permeation rates are slower than with other types of membranes.

Drug (compound **Ia**) release was quantified using an HPLC Agilent 1200 Series (Agilent, Santa Clara, CA, USA) with a DAD detector at 258 nm using water/acetonitrile (9:1) as the mobile phase. The calibration curve was performed with increasing amounts of the active principle **Ia**. Additionally, a blank emulsion was prepared without the active principle and evaluated through HPLC after dissolution in PBS.

### 4.4. Zeta Potential Evaluation

Zeta potential was measured using a Photon Correlation spectrometer Zetasizer Nano ZS (Malvern Instruments, Malvern, UK). The Helmholtz–Smoluchowski equation was used to calculate the final value. Five measurements in seven cycles of each sample were made. Compounds were measured in distilled water at a concentration of 30 μM. The data were analyzed using Zetasizer Software (version 7.11, Malvern Instruments Ltd., Malvern, UK).

### 4.5. Bacterial Strains

The bacterial strains used in this study were *Escherichia coli* (CECT 515, Gram-negative) and *Staphylococcus aureus* (CECT 240, Gram-positive), provided by the Spanish Type Culture Collection (CECT) in lyophilized form. The strains were kept at −20 °C and grown on Plate Count Agar (PCA).

### 4.6. In Vitro Antibacterial Activity Tests

The assay was based on the ISO 20776-1:2006 protocol [34]. First, colonies were transferred and grown in Mueller Hinton Broth medium (Scharlab S.L., Spain) and subjected to agitation (150 rpm) at 37 °C for 24 h. Then, the bacteria were incubated with the three compounds. The range of concentrations tested was from 0.5 to 512 mg/L. The three compounds were dissolved in water. Assays were carried out in sterile 96-well plates. Different controls were added, such as inoculum without compound, compound without inoculum, and culture medium without neither inoculum nor compound. Assays were run on technical triplicates. The treated plates were incubated at 37 °C for 24 h. Absorbance was measured at 630 nm in an Ultra Microplate reader (BIO-TEK Instruments, model ELx808, Winooski, VT, USA) after the incubation period. The results were collected to obtain the Minimum Inhibitory Concentration (MIC) values. Finally, well content was resuspended, and 5 μL of each well was deposited onto a PCA Petri dish and incubated at 37 °C for 24 h to obtain the Minimum Bactericidal Concentration (MBC) values.

### 4.7. Kinetic Studies

*S. aureus* and *E. coli* were grown as described in Section 4.6. Kinetic studies were performed by measuring absorbance values at 630 nm every hour up to 20 h. Microplates were incubated at 37 °C. All assays were performed in triplicate. The data at each time-point were analyzed by comparing the values of the different concentration gradients with the value of the control at the corresponding time. The cell viability percentage was calculated for each compound concentration and incubation time.

**Supplementary Materials:** The following supporting information can be downloaded at: https://www.mdpi.com/article/10.3390/ijms25074011/s1.

**Author Contributions:** Conceptualization, S.G.-G.; formal analysis, S.d.l.M.M., S.G.-G., I.H.-B. and N.G.-C.; investigation, S.d.l.M.M., S.C.A. and N.G.-C.; resources, S.G.-G., F.J.d.l.M., J.L.C.-P. and I.H.-B.; data curation, S.d.l.M.M., S.C.A., N.G.-C. and I.H.-B.; writing—original draft preparation, S.G.-G.; writing—review and editing, S.d.l.M.M., S.G.-G., I.H.-B., N.G.-C. and F.J.d.l.M.; supervision, S.G.-G., F.J.d.l.M., J.L.C.-P. and I.H.-B.; funding acquisition, S.G.-G., F.J.d.l.M. and I.H.-B. All authors have read and agreed to the published version of the manuscript.

**Funding:** The authors acknowledge the funding received from Ministerio de Ciencia e Innovación (PID2020112924RB-100; PID2020-113274RA-I00/AEI/10.13039/501100011033), Comunidad de Madrid and University of Alcalá (projects CM/BG/2021-01 and CM/JIN/2021-003). S.G.-G. thanks the Ministry of Universities for a Beatriz Galindo research grant (BG20/00231). CIBER-BBN is an initiative funded by the VI National R&D&i Plan 2008–2011, Iniciativa Ingenio 2010, Consolider Program, CIBER Actions, and financed by the Instituto de Salud Carlos III, with assistance from the European Regional Development Fund. S.d.l.M. thanks the University of Alcalá for the research grants ref. EPU-INV-UAH/2021/002 and UAH Introduction to Research 2023.

**Institutional Review Board Statement:** Not applicable.

**Informed Consent Statement:** Not applicable.

**Data Availability Statement:** Data are contained within the article or Supplementary Materials.

**Conflicts of Interest:** The authors declare no conflicts of interest.

## References


1. Darby, E.M.; Trampari, E.; Siasat, P.; Gaya, M.S.; Alav, I.; Webber, M.A.; Blair, J.M.A. Molecular mechanisms of antibiotic resistance revisited. *Nat. Rev. Microbiol.* **2023**, *21*, 280–295. [CrossRef] [PubMed]
2. Frei, A.; Verderosa, A.D.; Elliott, A.G.; Zuegg, J.; Blaskovich, M.A.T. Metals to combat antimicrobial resistance. *Nat. Rev. Chem.* **2023**, *7*, 202–224. [CrossRef]
3. Lonergan, Z.R.; Skaar, E.P. Nutrient Zinc at the host-Pathogen interface. *Trends Biochem. Sci.* **2019**, *44*, 1041–1056. [CrossRef] [PubMed]
4. Li, Y.; Liao, C.; Tjong, S.C. Recent advances in Zinc Oxide nanostructures with antimicrobial activities. *Int. J. Mol. Sci.* **2020**, *21*, 8836. [CrossRef] [PubMed]
5. Simpson, D.H.; Scott, P. Chapter Seven-Antimicrobial Metallodrugs. In *Inorganic and Organometallic Transition Metal Complexes with Biological Molecules and Living Cells*; Lo, K.K.-W., Ed.; Academic Press: Cambridge, MA, USA, 2017; pp. 205–243.
6. Waters, J.E.; Stevens-Cullinane, L.; Siebenmann, L.; Hess, J. Recent advances in the development of metal complexes as antibacterial agents with metal-specific modes of action. *Curr. Opin. Microbiol.* **2023**, *75*, 102347. [CrossRef]
7. Shekhar, S.; Khan, A.M.; Sharma, S.; Sharma, B.; Sarkar, A. Schiff base metallodrugs in antimicrobial and anticancer chemotherapy applications: A comprehensive review. *Emergent Mater.* **2022**, *5*, 279–293. [CrossRef]
8. Malik, M.A.; Dar, O.A.; Gull, P.; Wani, M.Y.; Hashmi, A.A. Heterocyclic Schiff base transition metal complexes in antimicrobial and anticancer chemotherapy. *Med. Chem. Commun.* **2018**, *9*, 409–436. [CrossRef] [PubMed]
9. Sinicropi, M.S.; Ceramella, J.; Iacopetta, D.; Catalano, A.; Mariconda, A.; Rosano, C.; Saturnino, C.; El-Kashef, H.; Longo, P. Metal Complexes with Schiff Bases: Data Collection and Recent Studies on Biological Activities. *Int. J. Mol. Sci.* **2022**, *23*, 14840. [CrossRef]
10. Mahmood, K.; Hashmi, W.; Ismail, H.; Mirza, B.; Twamley, B.; Akhter, Z.; Rozas, I.; Baker, R.J. Synthesis, DNA binding and antibacterial activity of metal(II) complexes of a benzimidazole Schiff base. *Polyhedron* **2019**, *157*, 326–334. [CrossRef]
11. Mahmoud, W.H.; Deghadi, R.G.; Mohamed, G.G. Metal complexes of novel Schiff base derived from iron sandwiched organometallic and 4-nitro-1,2-phenylenediamine: Synthesis, characterization, DFT studies, antimicrobial activities and molecular docking. *Appl. Organomet. Chem.* **2018**, *32*, e4289. [CrossRef]
12. İspir, E.; Kurtoğlu, M.; Toroğlu, S. The $d^{10}$ metal chelates derived from Schiff base ligands having silane: Synthesis, characterization, and antimicrobial studies of Cadmium(II) and Zinc(II) complexes. *Synth. React. Inorg. Met.-Org. Nano-Met. Chem.* **2006**, *36*, 627–631. [CrossRef]
13. Llamazares, C.; Sanz Del Olmo, N.; Ortega, P.; Gómez, R.; Soliveri, J.; de la Mata, F.J.; García-Gallego, S.; Copa-Patiño, J.L. Antibacterial effect of carbosilane metallodendrimers in Planktonic cells of Gram-positive and Gram-negative bacteria and *Staphylococcus aureus* Biofilm. *Biomolecules* **2019**, *9*, 405. [CrossRef] [PubMed]
14. Sanz del Olmo, N.; Carloni, R.; Ortega, P.; García-Gallego, S.; de la Mata, F.J. Chapter One-Metallodendrimers as a promising tool in the biomedical field: An overview. In *Advances in Organometallic Chemistry*; Pérez, P.J., Ed.; Academic Press: Cambridge, MA, USA, 2020; Volume 74, pp. 1–52.
15. Ortega, M.Á.; Guzmán Merino, A.; Fraile-Martínez, O.; Recio-Ruiz, J.; Pekarek, L.; GGuijarro, L.; García-Honduvilla, N.; Álvarez-Mon, M.; Buján, J.; García-Gallego, S. Dendrimers and dendritic materials: From laboratory to medical practice in infectious diseases. *Pharmaceutics* **2020**, *12*, 874. [CrossRef] [PubMed]
16. Maroto-Díaz, M.; Elie, B.T.; Gómez-Sal, P.; Pérez-Serrano, J.; Gómez, R.; Contel, M.; De La Mata, F.J. Synthesis and anticancer activity of carbosilane metallodendrimers based on arene ruthenium(II) complexes. *Dalton Trans.* **2016**, *45*, 7049–7066. [CrossRef] [PubMed]
17. Tyula, Y.A.; Goudarziafshar, H.; Yousefi, S.; Dušek, M.; Eigner, V. Template synthesis, characterization and antibacterial activity of d10 ($Zn^{2+}$, $Cd^{2+}$, $Hg^{2+}$) Schiff base complexes: A novel supramolecular $Cd^{2+}$ complex with two 1D helical chains, and its Hirshfeld surface analysis. *J. Mol. Struct.* **2023**, *1272*, 134051. [CrossRef]

18. Capitan, F.; Salinas, F.; Capitan-Vallvey, L.F. Study of the complexes formed with 2-pyrilidineaniline and copper(II), silver(I), zinc(II), cadmium(II) and mercury(II). *An. Quim.)* **1978**, *74*, 265–271.
19. Deghadi, R.G.; Elsharkawy, A.E.; Ashmawy, A.M.; Mohamed, G.G. Antibacterial and anticorrosion behavior of bioactive complexes of selected transition metal ions with new 2-acetylpyridine Schiff base. *Appl. Organomet. Chem.* **2022**, *36*, e6579. [CrossRef]
20. Mahmoud, W.H.; Mohamed, G.G.; El-Sayed, O.Y. Coordination compounds of some transition metal ions with new Schiff base ligand derived from dibenzoyl methane. Structural characterization, thermal behavior, molecular structure, antimicrobial, anticancer activity and molecular docking studies. *Appl. Organomet. Chem.* **2017**, *32*, e4051. [CrossRef]
21. Hadjiivanov, K.I. Infrared spectra of surface nitrates: Revision of the current opinions based on the case study of ceria. *J. Catalysis.* **2021**, *394*, 245–258.
22. Edwards, D.A.; Hayward, R.N. Transition metal acetates. *Can. J. Chem.* **1968**, *46*, 3443–3446. [CrossRef]
23. Kumar, U.; Thomas, J.; Thirupathi, N. Factors Dictating the Nuclearity/Aggregation and Acetate Coordination Modes of Lutidine-Coordinated Zinc(II)Acetate Complexes. *Inorg. Chem.* **2010**, *49*, 62–72. [CrossRef] [PubMed]
24. Dinku, D.; Demissie, T.B.; Beas, I.N.; Eswaramoorthy, R.; Abdi, B.; Desalegn, T. Antimicrobial activities and docking studies of new Schiff base ligand and its Cu (II), Zn (II) and Ni (II) Complexes: Synthesis and Characterization. *Inorg. Chem. Commun.* **2024**, *160*, 111903. [CrossRef]
25. Tandon, S.S.; Chander, S.; Thompson, L.K. Ligating properties of tridentate Schiff base ligands, 2-[[(2-pyridinylmethyl)imino]methyl] phenol (HSALIMP) and 2-[[[2-(2-pyridinyl)ethyl]imino]methyl]phenol (HSALIEP) with zinc(II), cadmium(II), nickel(II) and manganese(III) ions. X-ray crystal structures of the $[Zn(SALIEP)(NO_3)]_2$ dimer, $[Mn(SALIEP)_2](ClO_4)$, and $[Zn(AMP)_2(NO_3)_2]$. *Inorg. Chim. Acta* **2000**, *300–302*, 683–692.
26. Vicha, J.; Novotny, J.; Komorovsky, S.; Straka, M.; Kaupp, M.; Marek, R. Relativistic Heavy-Neighbor-Atom Effects on NMR Shifts: Concepts and Trends Across the Periodic Table. *Chem. Rev.* **2020**, *120*, 7065–7103. [CrossRef] [PubMed]
27. Butera, V.; D'Anna, L.; Rubino, S.; Bonsignore, R.; Spinello, A.; Terenzi, A.; Barone, G. How the Metal Ion Affects the 1H NMR Chemical Shift Values of Schiff Base Metal Complexes: Rationalization by DFT Calculations. *J. Phys. Chem. A.* **2023**, *127*, 9283–9290. [CrossRef] [PubMed]
28. Di Serio, M.; Carotenuto, G.; Cucciolito, M.E.; Lega, M.; Ruffo, F.; Tesser, R.; Trifuoggi, M. Shiff base complexes of zinc(II) as catalysts for biodiesel production. *J. Mol. Cat. A* **2012**, *353–354*, 106–110. [CrossRef]
29. Benessere, V.; Cucciolito, M.E.; Esposito, R.; Lega, M.; Turco, R.; Ruffo, F.; Di Serio, M. A novel and robust homogeneous supported catalyst for biodiesel production. *Fuel* **2016**, *171*, 1–4. [CrossRef]
30. Sharma, R.K.; Monga, Y. Silica encapsulated magnetic nanoparticles-supported Zn(II) nanocatalyst: A versatile integration of excellent reactivity and selectivity for the synthesis of azoxyarenes, combined with facile catalyst recovery and recyclability. *Appl. Catal. A.* **2013**, *454*, 1–10. [CrossRef]
31. Lakshkar, R.R.; Yadav, S.; Thirumoorthi, R.; Ray, S.; Dash, C. Zinc-bis(imino)pyridine Complexes as Catalysts for Intermolecular Amidation of Benzylic C(sp3)−HBond. *Chem. Select.* **2024**, *9*, e202302804. [CrossRef]
32. Breijyeh, Z.; Jubeh, B.; Karaman, R. Resistance of Gram-negative bacteria to current antibacterial agents and approaches to resolve it. *Molecules* **2020**, *25*, 1340. [CrossRef]
33. Naureen, B.; Miana, G.A.; Shahid, K.; Asghar, M.; Tanveer, S.; Sarwar, A. Iron (III) and zinc (II) monodentate Schiff base metal complexes: Synthesis, characterisation and biological activities. *J. Mol. Structure* **2021**, *1231*, 129946. [CrossRef]
34. Gómez-Casanova, N.; Torres-Cano, A.; Elias-Rodriguez, A.X.; Lozano, T.; Ortega, P.; Gómez, R.; Pérez-Serrano, J.; Copa-Patiño, J.L.; Heredero-Bermejo, I. Inhibition of Candida glabrata biofilm by combined effect of dendritic compounds and amphotericin. *Pharmaceutics* **2022**, *14*, 1604. [CrossRef] [PubMed]

International Journal of
*Molecular Sciences*

MDPI

*Article*

# UV-C Seed Surface Sterilization and Fe, Zn, Mg, Cr Biofortification of Wheat Sprouts as an Effective Strategy of Bioelement Supplementation

**Katarzyna Czarnek [1,*], Małgorzata Tatarczak-Michalewska [2], Piotr Dreher [3], Vishnu D. Rajput [4], Grzegorz Wójcik [5], Anna Gierut-Kot [6], Agnieszka Szopa [7] and Eliza Blicharska [2,*]**

1 Institute of Medical Science, Faculty of Medical, The John Paul II Catholic University of Lublin, Konstantynów 1 H Str., 20-708 Lublin, Poland

2 Department of Pathobiochemistry and Interdisciplinary Applications of Ion Chromatography, Biomedical Sciences, Medical University of Lublin, 1 Chodźki Str., 20-093 Lublin, Poland; malgorzatatatarczakmichalewska@umlub.pl

3 Chair and Department of Public Health, Medical University of Lublin, 1 Chodźki Str., 20-093 Lublin, Poland; piotr.dreher@umlub.pl

4 Academy of Biology and Biotechnology, Southern Federal University, 344090 Rostov-on-Don, Russia; rajput.vishnu@gmail.com

5 Department of Inorganic Chemistry, Institute of Chemical Sciences, Faculty of Chemistry, Maria Curie-Skłodowska University, 20-031 Lublin, Poland; g.wojcik@poczta.umcs.lublin.pl

6 Intermag sp. z o.o. R+D Department, Al. 1000-Lecia 15G, 32-300 Olkusz, Poland; el2000gato@gmail.com

7 Chair and Department of Pharmaceutical Botany, Jagiellonian University Medical College, Medyczna 9 Str., 30-688 Kraków, Poland; a.szopa@uj.edu.pl

* Correspondence: katarzyna.czarnek@kul.pl (K.C.); eliza.blicharska@umlub.pl (E.B.)

**Abstract:** Metalloenzymes play an important role in the regulation of many biological functions. An effective way to prevent deficiencies of essential minerals in human diets is the biofortification of plant materials. The process of enriching crop sprouts under hydroponic conditions is the easiest and cheapest to conduct and control. In this study, the sprouts of the wheat (*Triticum aestivum* L.) varieties Arkadia and Tonacja underwent biofortification with Fe, Zn, Mg, and Cr solutions in hydroponic media at four concentrations (0, 50, 100, and 200 $\mu g\ g^{-1}$) over four and seven days. Moreover, this study is the first to combine sprout biofortification with UV-C ($\lambda$ = 254 nm) radiation treatment for seed surface sterilization. The results showed that UV-C radiation was effective in suppressing seed germination contamination by microorganisms. The seed germination energy was slightly affected by UV-C radiation but remained at a high level (79–95%). The influence of this non-chemical sterilization process on seeds was tested in an innovative manner using a scanning electron microscope (SEM) and EXAKT thin-section cutting. The applied sterilization process reduced neither the growth and development of sprouts nor nutrient bioassimilation. In general, wheat sprouts easily accumulate Fe, Zn, Mg, and Cr during the applied growth period. A very strong correlation between the ion concentration in the media and microelement assimilation in the plant tissues ($R^2 > 0.9$) was detected. The results of the quantitative ion assays performed with atomic absorption spectrometry (AAS) using the flame atomization method were correlated with the morphological evaluation of sprouts in order to determine the optimum concentration of individual elements in the hydroponic solution. The best conditions were indicated for 7-day cultivation in 100 $\mu g\ g^{-1}$ of solutions with Fe (218% and 322% better nutrient accumulation in comparison to the control condition) and Zn (19 and 29 times richer in zinc concentration compared to the sprouts without supplementation). The maximum plant product biofortification with magnesium did not exceed 40% in intensity compared to the control sample. The best-developed sprouts were grown in the solution with 50 $\mu g\ g^{-1}$ of Cr. In contrast, the concentration of 200 $\mu g\ g^{-1}$ was clearly toxic to the wheat sprouts.

**Keywords:** iron; zinc; magnesium; chromium; trace elements; SEM; biofortification; atomic absorption spectrometry; wheat sprouts; UV seed sterilization; nutrient deficiency

## 1. Introduction

Trace elements function in the human body mainly as catalysts for cellular enzymes to perform various functions, the most important of which is their structural role in specific metalloenzymes. These metals remain a constant component of such enzyme complexes, and their amount depends on the structure of the enzyme involved. The specificity of enzymatic reactions often requires the use of specific ions by the complex, which cannot be exchanged for other ions or will have an inhibitory effect on a given enzymatic reaction [1,2].

Iron is the most abundant metal in the human body. Body Fe content is approximately 3–4 g, which corresponds to a concentration of 40–50 mg of Fe per kilogram of body weight [3]. In the human body, this element occurs in the form of complexes with proteins (hemoproteins), among others, as a component of heme, heme enzymes, and non-heme proteins such as transferrins, ferritins, and flavin enzymes. It is necessary for the synthesis of hemoglobin, myoglobin, heme enzymes, and enzymes involved in oxidation and reduction processes [4,5]. Zinc is another metal present in the human body and is the second most abundant (about 2.5 g) trace element after Fe. Zn is an essential trace element that functions as a cofactor for certain enzymes involved in metabolism and cell growth; it is found in nearly 300 specific enzymes [6–8]. Chromium is found primarily in two forms: trivalent, which is biologically active and found in food, and hexavalent, a toxic form that results from industrial pollution. Cr (III) has been postulated to be involved in regulating carbohydrate, lipid, and protein metabolism by enhancing insulin efficacy [9,10]. $Mg^{2+}$ is the second richest intracellular cation after $K^+$ and is a cofactor in more than 325 enzyme systems in cells. Mg is used in many biological functions, where it functions as a cofactor in enzyme systems that regulate diverse biochemical reactions in the body, including protein synthesis, muscle and nerve functions, blood glucose control, and blood pressure regulation [1,11].

Micronutrients play a central role in metabolism and the maintenance of tissue functions [12]. In contrast to macronutrients (fat, carbohydrate, and protein), they cannot substitute for one another and cannot be synthesized within the body. Consequently, we depend on the delivery of all essential micronutrients via our diet [13].

Nowadays, plant products tend to be less nutrient-dense in comparison to plant products from previous decades. This is partly due to the progressive process of soil depletion as well as the introduction of highly productive plant varieties into agriculture [14]. These plants guarantee a high yield but are also characterized by low nutrient content. Therefore, the diet of an average human lacks vitamins and minerals such as iron, zinc, selenium, iodine, magnesium, calcium, and other nutrients. Macro- and micronutrient deficiencies affect a significant part of the world population. Growing concerns about the widespread impacts of micronutrient deficiencies have been magnified by the poor nutritional quality of crops that are less resistant to drought and other stresses, which are likely to be further exacerbated by continuing climate change [15,16]. It is estimated that 60–80%, 30%, 30%, and 15% of people suffer from deficiencies of Fe, I, Zn, and Se, respectively [17,18]. According to the WHO and FAO [19], we have to deal with so-called "hidden hunger" when food quality does not meet our nutrient requirements. The problem is so common that about two billion people are estimated to suffer from vitamin and mineral deficiencies.

An average human diet lacking in vitamins, minerals, or other nutrients poses a real challenge. The resolution is supposed to be found in plant product biofortification. Biofortification is understood as a process that increases the uptake and accumulation of mineral nutrients in agricultural products through the selection of agricultural practices, plant breeding, or/and product modification via genetic engineering [14,20–22]. The results should provide health benefits as well as pose a minimal risk for consumers [23]. Through biofortification, the nutrient content increases during crop growth rather than during crop processing. Commonly used practices are based on selecting fertilization and allowing the uptake of nutrients from the soil, or foliar fertilization [14].

Recently, a new approach for increasing micronutrient content in plants has been genetic engineering, which has been shown to be a feasible and cost-effective alternative to traditional fortification programs. In transgenic methods of fortification, new cultivars with desired traits can be developed by transferring new genes, overexpressing the genes already present, or blocking genes that provide inhibitor synthesis. These methods aim to transport and distribute micronutrients between tissues, thereby increasing their concentration in the edible parts of crops, and facilitate the efficiency and productivity of biochemical pathways involved in their synthesis. Transgenic biofortification requires time, effort, and investment to optimize the process. For that reason, this process requires making calculations as to whether it will pay off in long-term applications. There are also time-consuming and costly legal regulations regarding the commercial distribution of biofortified food crops, including difficulties in obtaining government approval, concerns and demands of anti-GMO activists for more tests before distribution, and low social acceptance. These main disadvantages of plant genetic engineering must be taken into account [24–28].

Plant products that are already rich in essential nutrients include crop sprouts. They are abundant in valuable components such as vitamins, amino acids, enzymes, dietary fibers, flavonoids, or antioxidants. Their high nutritional value classifies them as being highly desired in the human diet [29]. The biggest advantage of introducing sprouts to the human diet is the relatively short process of preparation as well as the possibility of biofortification.

The effectiveness of biofortification of edible seeds has been studied by several research teams. In the research by Park et al. [30] the effect of iron supplementation on alfalfa, broccoli and radish sprouts was analyzed. Iron concentration was supplemented by soaking seeds in Fe(III)-EDTA or Fe(III)-citrate solution at a concentration of 2.5, 5.0, or 10 mM. The soaking treatment significantly increased the iron concentration in 5-day-old alfalfa sprouts by up to 1.8 times the concentration observed in the controls. In broccoli and radish sprouts, the increase in Fe concentration was insignificant [30]. In other studies, the iron content increased by 1.1–15.6 times via soaking brown rice grains in an $FeSO_4$ solution before germination compared to germinating the grains [31]. Se biofortification during sprouting could represent a valid strategy to improve Se concentration in tartary buckwheat sprouts [32] and brown rice [33]. In wheat, Se-enriched kernels can be obtained with 35 mg $L^{-1}$ $Na_2SeO_3$ in a germination medium for 24 h at 25 °C [34].

The enrichment of plant sprouts with nutrients in hydroponic media does not require any special conditions [23,35]. However, there are still possibilities for the introduction of new materials for vessels for hydroponic cultivation [36] or new solutions for maintaining seed microbiological quality. Recently, researchers have conducted hydroponic cultivation of soybean, lettuce, garden cress, mung bean, broccoli, and onion sprouts or seedlings via biofortification [23,35,37–39]. These researchers noticed that enrichment with one element may affect the physiological and biochemical parameters of sprouts, including the assimilation of other micro- and macro-nutrients. Depending on the content of iodine in hydroponic media, the biofortified lettuce seedlings contained more potassium and less sodium and manganese than the control seedlings [35]. In the case of Zn, this element significantly influenced the Fe and Ca contents in soybean sprouts but did not affect the contents of Mn, Cu, and Mg [39].

Since common wheat (*Triticum aestivum* L.) is the most popular crop in the world and wheat sprouts show relatively high carbohydrate and amino acid values [29], we attempted to enhance the macro- and micro-nutrient densities of this food product. The aim of the present work was an evaluation of the ability to accumulate Fe, Zn, Mg, and Cr in germinated wheat seeds of two varieties, Tonacja and Arkadia, in hydroponic conditions. This work also aimed to assess the impact of seed surface sterilization via UV-C light on biofortification effectiveness and seed properties using a scanning electron microscope (SEM). The study performed using SEM provides information about the elemental composition of the samples in the microregion. Thus, it is possible to determine in which part of the plant the most biofortified element is deposited. The analysis of elemental composition in the microregion included a qualitative analysis of the test samples. The quantitative analysis of selected micronutrients in the plant tissues was performed using atomic absorption spectrometry (AAS) with flame atomization.

## 2. Results and Discussion

### 2.1. Sample Preparation and Macroscopic Analysis of Wheat Seeds

Before sprouting, seeds are soaked and then maintained in a humid environment that is favorable for sprouting. Bacteria and associated biofilms grow well under conditions with enough moisture and nutrients. Thus, the sprouting stage has been categorized as a major source of bacterial contamination in sprouts because bacteria present in seeds can become internalized during the process of sprouting if they are not inactivated [40]. Since sprouts are consumed with minimal processing, it is extremely important to use effective decontamination methods that will ensure the elimination of pathogenic microbes and the safety of sprouts.

In our research, wheat seeds of the Tonacja and Arkadia varieties were sterilized with UV-C radiation in order to maintain microbiological purity. UV lamps (254 nm) were used in the tests, which are standard equipment in most laboratories. The minimum radiation dose (1.5 kJ $m^{-2}$) was used in the experiments to inhibit the spread of yeasts/molds. This dose was experimentally optimized. Lower doses of UV-C did not inhibit infections caused by undesirable microorganisms during the experiment. The proposed UV treatment of seeds improves on current technology used for the production of hydroponic crops by taking into account modern requirements for energy savings and obtaining environmentally friendly products.

The influence of this non-chemical sterilization process on seeds was tested in an innovated manner via a scanning electron microscope and EXAKT thin-section cutting. The EXAKT system was developed in collaboration with Prof. Donath at the Institute for Pathology at the University of Hamburg in 1987, and since then, it has been widely used. The precise cutting leaves a perfect surface between different tissue structures. Burr-free, plane-parallel surfaces and thicknesses of down to 100 μm allow many slices to be cut in order to obtain as much structural information as possible from the original sample. In addition to organic specimen testing, the EXAKT system also finds application in the development of novel combinations of natural and synthetic materials used in industry or in recycling.

In this study, this method was adjusted for macroscopic analysis of plant seeds. It successfully allowed the observation of wheat grains of the Tonacja variety. The outer and internal structures were well captured. It allowed a visual separation of the outer layer, including the pericarp, brush, crease, starchy endosperm with color gradient, pigment stand, and embryo with scutellum, radicle, and coleoptile (Figure 1A,B).

Between the two tested specimens, there were no significant changes in morphology. This indicated that wheat seed sterilization for 30 min using UV-C radiation was not a too harsh treatment (Figure 1A).

The presented approach appeared to be suitable for following the morphological changes caused by the seed germination process and sprout formation as well. Figure 2 presents the hypocotyl development.

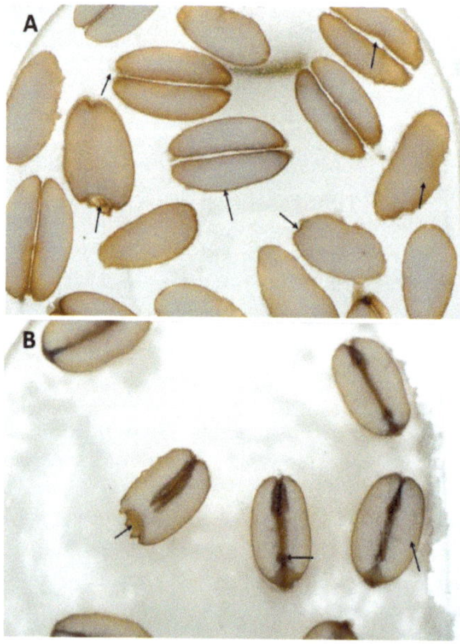


**Figure 1.** Macroscopic analysis of wheat seeds of the Tonacja variety treated (**A**) and not treated (**B**) with UV-C radiation as a factor for improving microbiological quality.

**Figure 2.** Morphological changes in wheat seeds caused by germination mechanisms.

*2.2. Seed Germination Energy*

It has been proven that wheat germination efficiency is affected by the presence of microbiological and mineral components in hydroponic media [41]. In this study, the type of ion enrichment did not affect the seed germination energy (Table 1). An exception is Cr biofortification, which leads to a reduction in germination energy (79–88%) in comparison to the control samples (94–100%). Cr is responsible for reactive oxygen species production. It results in lipid, protein, and nucleic acid degradation in plants. That may lead to seed germination, photosynthesis, or water balance dysfunction [42–44].

The more prominent differences in seed germination energy (Table 1) are mainly associated with the plant material treatment used. Previous experiments have shown that UV-C radiation stimulates the germination of edible seeds, including wheat seeds [45]. In this experiment, UV-C radiation reduced the seed germination energy and noticeably suppressed yeast/mold spread (data not shown). This result leads to the conclusion that microflora may promote germination of wheat seeds in the early stages of infection (within four days). The results reported by other authors reveal diverse relations between wheat seed germination energy and the type of microflora in the environment, including a nonsignificant difference in germination between healthy and infected seeds due to infection by *Alternaria alternate* [46], negative effects induced by fungal seed-borne pathogens (*Alternaria tenuis, Aspergillus niger, Fusarium moniliforme, Curvuluria lunata,* and *Stemphylium herhurum*) [47], and positive effect caused by bacteria [41].

**Table 1.** Wheat seed germination energy in the hydroponic media with the presence of Fe, Zn, Cr, and Mg at a concentration of 100 μg $g^{-1}$: (A) Tonacja and (B) Arcadia. Data are presented as mean ± SD.

| Ion Concentration in Media (100 μg $g^{-1}$) | Seed Germination Energy (%) | | |
|---|---|---|---|
| | Variety A (Non-UV) | Variety A (UV) | Variety B (UV) |
| 0 (control) | 100.0 ± 0.0 | 94.3 ± 0.6 | 95.3 ± 1.5 |
| $Fe^{3+}$ | 98.3 ± 0.6 | 91.0 ± 0.0 | 91.0 ± 0.0 |
| $Zn^{2+}$ | 97.0 ± 0.0 | 89.7 ± 1.2 | 92.0 ± 0.0 |
| $Mg^{2+}$ | 97.7 ± 1.5 | 89.3 ± 0.6 | 90.3 ± 0.6 |
| $Cr^{3+}$ | 88.0 ± 1.7 | 80.7 ± 1.5 | 79.3 ± 1.5 |

It is difficult to clearly conclude whether some changes may be induced primarily by the lack/presence of microflora or by UV-C radiation itself. It should be noted that long-term exposure to UV light shortens the germination process and subsequently deforms the seedlings [48]. Permanent exposure may interrupt germination altogether. After 30 min of seed treatment with UV-C (λ = 254), we noticed any physiologically incorrect plant growth (Tables 2 and 3).

In this experiment, no significant differences in germination energy between the UV-exposed seeds of the wheat varieties Tonacja and Arcadia were detected. Both show exceptionally efficient seed germination energy.

**Table 2.** Morphological assessment of wheat sprouts in the hydroponic media with the presence of Fe, Zn, Cr, and Mg at four concentrations (0, 50, 100, and 200 μg $g^{-1}$): (A) Tonacja and (B) Arcadia, with −/+ indicating physiologically correct growth, - indicating weaker growth or slightly deformation, -- indicating substantial deformation or inhibition of seed germination, + indicating faster growth, and ++ indicating substantial longer sprouts.

| Ion Concentration in Media (μg $g^{-1}$) | | Morphological Features of Sprout | | |
|---|---|---|---|---|
| | | Variety A (Non-UV) | Variety A (UV) | Variety B (UV) |
| Fe | 0 (control) | −/+ | −/+ | −/+ |
| | 50 | −/+ | −/+ | −/+ |
| | 100 | + | + | + |
| | 200 | - | - | - |
| Zn | 0 (control) | −/+ | −/+ | −/+ |
| | 50 | −/+ | + | −/+ |
| | 100 | + | ++ | + |
| | 200 | −/+ | −/+ | - |
| Mg | 0 (control) | −/+ | −/+ | −/+ |
| | 50 | −/+ | −/+ | −/+ |
| | 100 | + | + | + |
| | 200 | + | + | + |
| Cr | 0 (control) | −/+ | −/+ | −/+ |
| | 50 | −/+ | −/+ | + |
| | 100 | −/+ | −/+ | + |
| | 200 | -- | -- | -- |

**Table 3.** Morphological appearance of the *Triticum aestivum* variety Tonacja treated with UV-C and grown in hydroponics supplemented with different concentrations of bioelements.

| Elements | Concentrations of Elements (μg $g^{-1}$) | | |
|---|---|---|---|
| | 50 | 100 | 200 |
| Iron (Fe) |  |  |  |
| Zinc (Zn) |  |  |  |
| Magnesium (Mg) |  |  |  |
| Chromium (Cr) |  |  |  |
| Control [1] | |  | |

[1] Control—medium without bioelement supplementation.

### 2.3. The Influence of Seed UV-C Sterilization on Biofortification Efficiency

The microbiological quality of seeds during the germination process must be maintained. This is especially important when the plant material is susceptible to fungal diseases, as in the case of common wheat [49]. We used a non-chemical treatment for seed surface sterilization by UV-C radiation and evaluated its impact on biofortification efficiency.

To confirm the effectiveness of biofortification and to demonstrate that the quantified content of trace elements does not come from grain surface contamination but from an effective enrichment process, a microscopic analysis was performed, and selected samples with the best morphological characteristics were tested. The samples for the SEM examination were biofortified wheat varieties (100 μg $g^{-1}$) and a control sample. The analysis with a scanning electron microscope allowed us to obtain high-resolution images and maps of the distribution of elements. Examples of the SEM analysis results obtained for the control sample and the Mg-biofortified sample are presented in Table 4.

**Table 4.** SEM analysis results obtained for the control sample and the sample of the *Triticum aestivum* variety Tonacja biofortified with Mg.

| Mg Concentration in Media | BSE Micrograph with Graphic Arrangement of Measurement Points Made by SEM | Element Distribution Maps |
|---|---|---|
| 0 μg $g^{-1}$ (Control) | 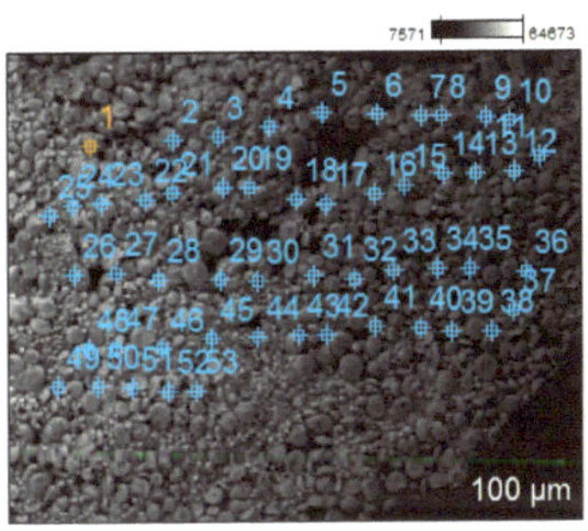  | 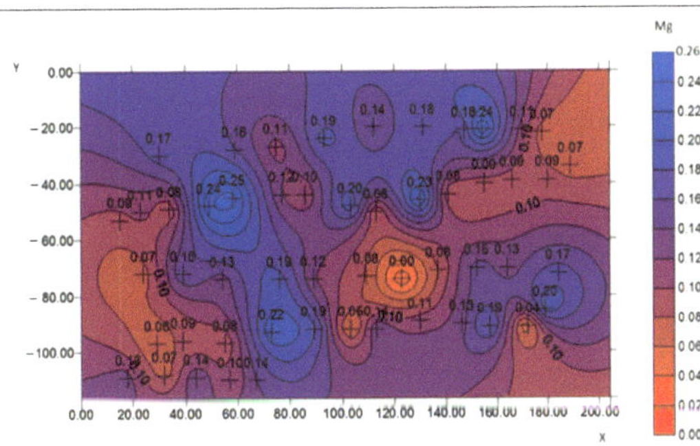  |
| 100 μg $g^{-1}$ | 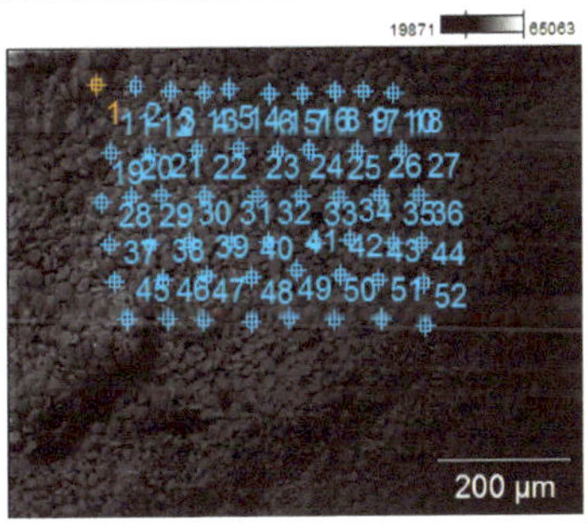  | 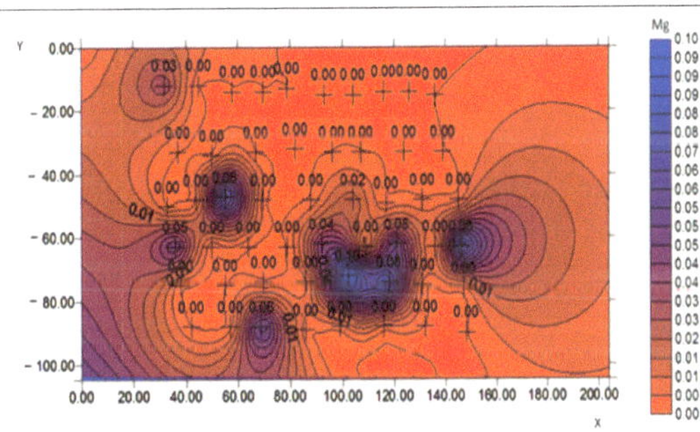  |

The quantitative analysis of selected micronutrients in plant tissues was performed using atomic absorption spectrometry with flame atomization. In this experiment, we observed slightly better assimilation of Cr and Zn from the hydroponic media into the sprout tissues of the sample treated with UV-C radiation (Figure 3). Only medium supplementation with chromium (100 μg $g^{-1}$) significantly improved the element concentration in sprouts. Without medium enrichment, Cr is detectable only in wheat seeds infected by microorganisms [50]. Based on this study, a lack of seed surface sterilization (the presence of microflora) results in insignificantly lower Cr assimilation in sprout tissues. UV-C sterilization had no influence on Fe and Mg concentrations in the tested sprouts. The study by Blicharska et al. [23] reported that there is a slight correlation between the occurrence of microbiological contamination of sprouts and the presence of a particular type of ion in the hydroponic media or tested plant species.

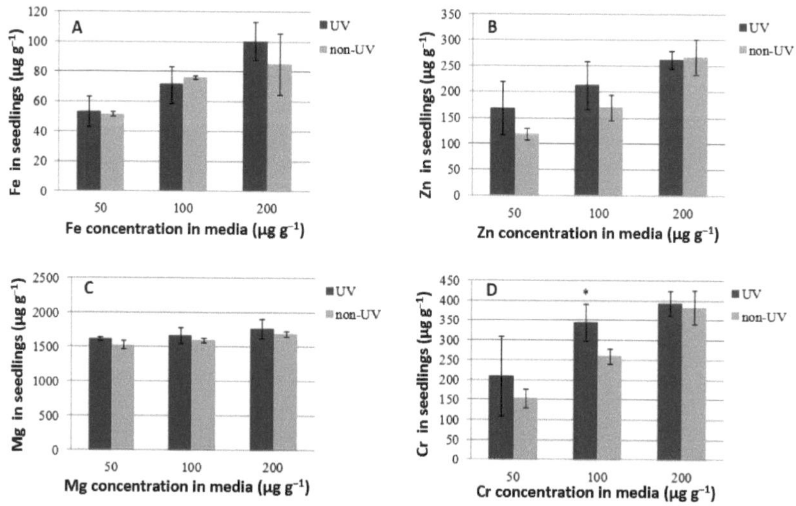


**Figure 3.** The influence of seed UV-C sterilization on micronutrient concentration in the wheat sprouts of the Tonacja variety. * refers to a result that is significantly different at the level of $p < 0.05$ according to Tukey's test. Data are presented as mean ± SD.

This study is the first to feature UV-C irradiation as a factor for seed surface sterilization and to correlate it with the level of nutrient uptake by sprouts or seedlings. Thus far, seed treatment with low doses of UV has proven to induce various morphological and physiological changes in the later stages of plant growth. This method improves plant productivity and yield quality [51–53]. It activates mechanisms of biotic and abiotic stress adaptation in plants by, e.g., stimulating the synthesis of antioxidants in leaves [52,54,55]. It may affect photosynthesis and the content of chlorophyll in leaves [51,56]. Pre-sowing UV-C irradiation of wheat seeds may result in seed coat thicknesses that relate to better resistance against pathogens [51]. Most of the studies have focused on post-sowing UV-A/B radiation [57]. It is considered a stress factor connected with stratospheric ozone layer depletion. Under natural conditions, ozone loss may affect plant growth, including the uptake of nutrients by wheat seedlings [58].

### 2.4. Wheat Biofortification with Nutrients

As shown in Table 5, the ion type and concentration in the media affect the final concentrations of Fe, Zn, Mg, and Cr in the wheat sprouts. We correlated this result with a visual assessment of the plant materials (Table 2) in order to select the optimum conditions for wheat biofortification with macro- and micro-nutrients. Even if vegetables or cereals are able to assimilate a great amount of particular nutrients, the final, maximum concentration may be toxic to seedlings/sprouts.

**Iron.** In comparison to other nutrients, the correlation between the ion concentration in the media and the accumulation of this element in the plant material is the weakest. The linear relationship becomes even weaker over time (Figure 4). For the Tonacja variety, the coefficient of determination decreases from $R^2 = 0.8933$ to $R^2 = 0.6479$, and for the Arkadia variety, it decreases from $R^2 = 0.9576$ to $R^2 = 0.8531$. However, this result is still promising. The morphological assessment revealed that the most healthy-looking sprouts were grown in the solution with a concentration of 100 µg $g^{-1}$.

**Table 5.** Nutrient concentrations in the dry weight of the wheat sprouts of Tonacja (A) and Arkadia (B) varieties (μg g$^{-1}$) with the presence of Fe, Zn, Cr, and Mg in the hydroponic media at four concentrations (0, 50, 100, and 200 μg g$^{-1}$). Different letters within each column indicate significant differences at $p < 0.05$ based on Tukey's test. Data are presented as mean ± SD.

| Ion Concentration in Media (μg g$^{-1}$) | | Ion Concentration in Sprouts (μg g$^{-1}$) | | | |
|---|---|---|---|---|---|
| | | 4 Days | | 7 Days | |
| | | Variety A | Variety B | Variety A | Variety B |
| Fe | 0 (control) | 34.96 ± 3.96 c | 35.65 ± 0.85 d | 44.77 ± 2.30 c | 37.55 ± 2.43 c |
| | 50 | 51.61 ± 1.36 bc | 53.03 ±2.83 c | 58.36 ± 3.02 c | 133.40 ± 8.37 b |
| | 100 | 75.89 ± 1.15 ab | 89.13 ± 9.79 b | 142.41 ± 3.91 b | 158.40 ± 9.26 a |
| | 200 | 85.20 ± 20.5 a | 114.96 ± 3.44 a | 163.27 ± 9.55 a | 163.47 ± 7.25 a |
| Zn | 0 (control) | 22.30 ± 2.09 c | 16.11 ± 0.92 c | 19.54 ± 0.64 d | 16.72 ± 0.17 d |
| | 50 | 118.79 ±10.48 b | 114.29 ± 1.96 b | 208.64 ± 45.71 c | 141.14 ±13.10 c |
| | 100 | 169.99 ±24.30 b | 117.44 ± 12.76 b | 301.32 ± 11.64 b | 208.92 ± 24.52 b |
| | 200 | 267.31 ±34.61 a | 305.12 ± 64.95 a | 560.53 ± 16.75 a | 320.06 ± 34.15 a |
| Mg | 0 (control) | 1486.32 ± 7.92 c | 1464.48 ± 70.78 c | 1527.41 ± 60.30 d | 1473.89 ± 31.30 c |
| | 50 | 1527.18 ±55.21 bc | 1657.53 ± 72.16 b | 1715.02 ± 13.35 c | 1625.96 ±57.53 c |
| | 100 | 1599.22 ±33.41 ab | 1717.31 ± 42.40 ab | 1928.44 ± 34.53 b | 1851.03 ± 92.55 b |
| | 200 | 1693.68 ± 39.14 a | 1813.90 ± 31.11 a | 2117.65 ± 111.59 a | 2015.14 ± 32.52 a |
| Cr | 0 (control) | 0.00 ± 0.00 d | 0.00 ± 0.00 c | 0.00 ± 0.00 c | 0.00 ± 0.00 d |
| | 50 | 179.79 ± 22.79 c | 130.42 ± 13.88 b | 203.60 ± 7.26 b | 221.64 ± 33.74 c |
| | 100 | 260.58 ± 19.13 b | 331.35 ± 10.82 a | 483.30 ± 28.46 a | 346.85 ± 63.34 b |
| | 200 | 383.80 ± 43.85 a | 399.12 ± 79.01 a | 463.60 ± 31.99 a | 538.19 ± 10.71 a |

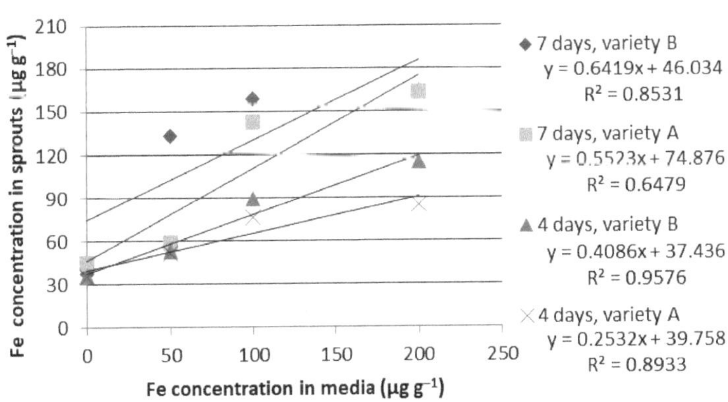


**Figure 4.** Relationship between iron (Fe) concentrations in the growth media and in the dry weight of the wheat sprouts of the Tonacja (A) and Arkadia (B) varieties (μg g$^{-1}$).

The concentration of 200 μg g$^{-1}$ significantly reduced wheat development, which expressed shorter hypocotyl and physical sprout deformation (Table 2). After four days of hydroponic cultivation in the 100 μg g$^{-1}$ solution, there was an increase in 117% and 150% in iron accumulation for the Tonacja and Arkadia varieties, respectively. The 7-day cultivation resulted in 218% and 322% better nutrient accumulation in comparison to the control condition, respectively. Further supplementation (200 μg g$^{-1}$) significantly increased biofortification effectiveness only for the Arkadia variety on the fourth (114.96 μg g$^{-1}$) and seventh days (163.47 μg $g^{-1}$) of cultivation in comparison to the lower dose (100 μg $g^{-1}$). The primary Fe concentration in these wheat grains was at a similar level (35–44 μg $g^{-1}$) to that in most wheat crops [47].

Several methods to correct Fe deficiency, such as broadcast application, foliar spray, and seed priming (soaking seeds in a nutrient-rich solution before sowing), have been used [59].

A previous study showed that foliar application of Fe at each of the different growth stages of wheat significantly increased grain yield and the concentration of Fe in grains. Foliar feeding involves applying a fertilizer spray in a liquid state over leaves. A single micronutrient solution or a mixture of such solutions in combination with the target concentration of salt is applied as a spray on leaves, where it is absorbed via stomata and epidermis. In particular cases, foliar feeding is shown to be more efficient when compared to soil applications for the effective uptake of nutrients. In one study, foliar sprays of Fe ($FeSO_4 \cdot 2H_2O$) were applied at different growth stages of wheat, starting from the maximum tillering, flower initiation, milk, and dough stages [60]. A similar strategy has been used with other edible grains [61].

However, in the case of biofortification of sprouts (taking into account the specificity of the products and the method of their cultivation), the strategy of Fe supplementation by soaking seeds in appropriate solutions seems to be more appropriate. Seed priming involves the controlled hydration of seeds that permits them to perform their pre-germination metabolic events without radical emergence. This treatment activates enzymatic and metabolic processes, which may enhance the plant's capacity for nutrient uptake and their subsequent translocation [59].

**Zinc.** The Zn concentration in wheat sprouts was very strongly correlated with the ion concentration in the media. Regardless of the cultivation time or plant variety, the determination coefficient was higher than $R^2$ = 0.940 (Figure 5). This nutrient assimilation was also the most effective. A similar effect of biofortification of various vegetable sprouts was previously noticed by Blicharska et al. [23].

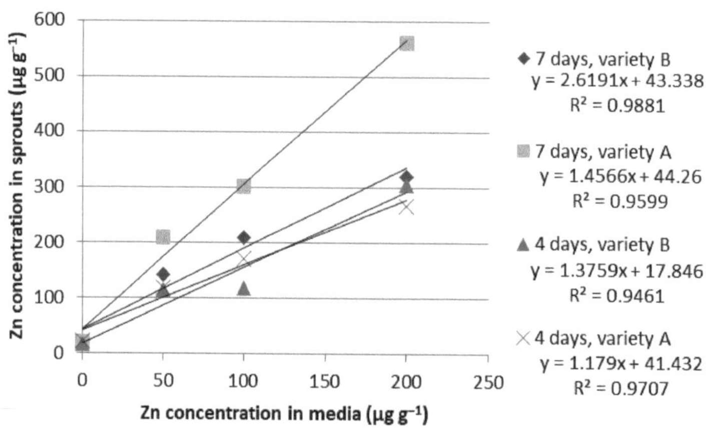


**Figure 5.** Relationship between zinc (Zn) concentrations in the growth media and in the dry weight of the wheat sprouts of the Tonacja (A) and Arkadia (B) varieties (μg $g^{-1}$).

Even the smallest dose of 50 μg $g^{-1}$ caused an approximately 5–7 fold increase in the Zn sprout content during the first step of the experiment. The final Zn concentration after seven days was approximately 8–11 times higher in comparison to the control (0 μg $g^{-1}$). The enrichment of the hydroponic media with a solution of 100 μg $g^{-1}$ initially induced an 8–11 fold increase in Zn concentration in the plant tissues; afterward, it reached a 12–15 fold higher level in comparison to the control condition. The maximum wheat biofortification was observed with a solution of 200 μg $g^{-1}$. At the end of hydroponic cultivation (seven days), the sprouts of the Arkadia (320.06 μg $g^{-1}$) and Tonacja (560.53 μg $g^{-1}$) varieties were 19 and 29 times richer in zinc compared to the sprouts without supplementation (17–20 μg $g^{-1}$) (Table 5). The morphological response of the sprouts was strongly correlated with the type of plant. The sprouts of the Tonacja variety were characterized by very stable growth compared to the Arkadia variety. These sprouts were longer and better developed. We noticed that the sprouts, especially of the Tonacja variety, were in visibly better condition when grown in the hydroponic media with a 100 μg $g^{-1}$ concentration. The solution of 200 μg $g^{-1}$ was associated with uneven growth of sprouts and inhibition of root elongation (Table 2). These results indicate that the Tonacja variety is more resistant to the toxic effects of this element at high concentrations.

There is evidence in the literature demonstrating that foliar-applied Zn can be absorbed by the leaf epidermis and then remobilized and transferred into rice grains through the phloem [62]. In recent years, several studies have been conducted to adjust the time of foliar Zn application in cereal crops [63–65]. It is now well established that foliar Zn application after the flowering stage (e.g., at the early milk and dough stages) more distinctly increases the Zn concentration in grains [64]. However, the process of enriching of crop sprouts under hydroponic conditions is the easiest and cheapest to conduct and control.

**Magnesium.** There was a strong correlation between the Mg concentrations in the media and in the wheat sprouts (Figure 6). It was most pronounced for the seeds of the Tonacja variety cultivated by the fourth day ($R^2$ = 0.9923) and weakest for the seeds of the Arkadia variety tested at the same time ($R^2$ = 0.8708). The enrichment of the hydroponic media with the ion solution at a concentration of 50 μg $g^{-1}$ did not significantly affect the biofortification level, regardless of the sampling time. Using solutions of 100 and 200 μg $g^{-1}$ in the first stage of the experiment (four days) resulted in a similar level of Mg concentration in the sprouts of the Tonacja (1599–1693 μg $g^{-1}$) and Arkadia varieties (1717–1814 μg $g^{-1}$). After seven days, we noticed significant improvement in Mg assimilation with increasing ion concentration in the media from 1928 μg $g^{-1}$ to 2118 μg $g^{-1}$ for the Tonacja variety and from 1851 μg $g^{-1}$ to 2015 μg $g^{-1}$ for the Arkadia variety. Overall, the maximum plant product biofortification with magnesium did not exceed 40% in intensity when compared to the control sample (Table 5). The relatively low percentage of magnesium absorption is reflected in the morphological condition of the sprouts (Table 2). Both the Tonacja and Arkadia varieties showed significantly better sprout growth with an increased concentration of Mg in the hydroponic solution. At concentrations of 100 and 200 μg $g^{-1}$, we observed long sprouts with a well-developed root system. However, it could be observed that the roots formed in the medium supplemented with 200 μg $g^{-1}$ of magnesium are slightly shorter in comparison to the roots grown in a solution containing 100 μg $g^{-1}$ of magnesium.

**Chromium.** According to Singh et al. [50] and our results, Cr does not occur naturally in wheat grains. This element is highly toxic to wheat plants. It interferes with various metabolic processes and leads to the inhibition of plant growth and development [43]. Our results show that wheat assimilates chromium very effectively. Chromium concentrations after four days of cultivation reached 179.79, 260.58, and 383.80 μg $g^{-1}$ for the Tonacja variety and 130.42, 331.35, and 399.12 for the Arkadia variety when grown in the ion solutions of 50, 100, and 200 μg $g^{-1}$, respectively. The maximum wheat biofortification with Cr was achieved after seven days of cultivation. The sprouts of the Tonacja variety reached the highest micronutrient concentration (483.30–463.60 μg $g^{-1}$) in the presence of 100 and 200 μg $g^{-1}$ of ion solution. In contrast, the Arkadia variety assimilated 538.19 μg $g^{-1}$ of Cr when cultivated in the hydroponic medium supplemented with 200 μg $g^{-1}$ of ion solution (Table 5). In this case, we also noted a strong linear correlation between the ion concentration in the media and the accumulation of Cr in the plant material ($R^2$ = 0.9569) (Figure 7). The morphological assessment revealed that the Arkadia variety of wheat has a greater tolerance for the toxic properties of this element. In the presence of Cr at concentrations of 50 and 100 μg $g^{-1}$, the sprouts of the Arkadia variety were more healthy-looking. They were characterized by better developed hypocotyls and root systems.

Regardless of the plant type, the best-developed sprouts were grown in the solution with 50 μg $g^{-1}$ of chromium. In contrast, the concentration of 200 μg $g^{-1}$ was clearly toxic to the wheat sprouts. The sprouts were significantly shorter. Root growth was also inhibited. For many seeds, a complete inhibition of germination was noticed (Table 2).

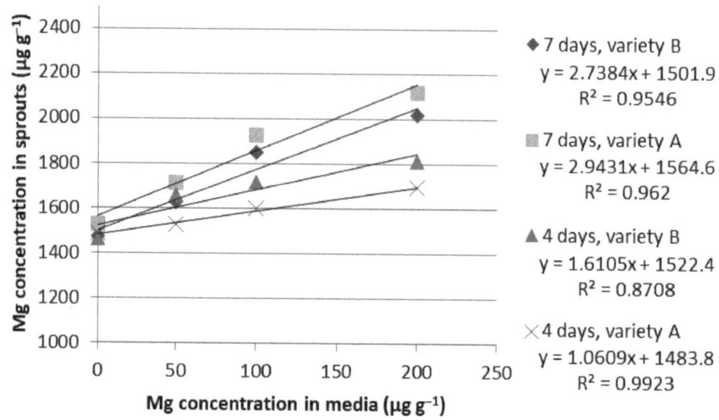


**Figure 6.** Relationship between magnesium (Mg) concentrations in the growth media and in the dry weight of the wheat sprouts of the Tonacja (A) and Arkadia (B) varieties (μg $g^{-1}$).

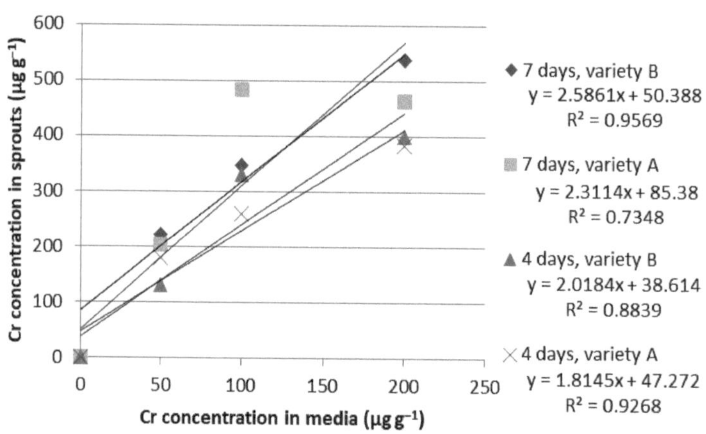


**Figure 7.** Relationship between chromium (Cr) concentrations in the growth media and in the dry weight of the wheat sprouts of the Tonacja (A) and Arkadia (B) varieties (μg $g^{-1}$).

Since wheat is the most popular crop in the world, there is a high interest in increasing its nutritional value [66]. The grains of cereal crops like wheat contain various nutrients, such as phytic acid and tannins. Phytic acid forms a complex with mineral elements and reduces Fe and Zn bioavailability. This leads to a deficiency of these nutrients in the human body via their uptake through diet [67,68]. Overall, the percent bioavailability of Fe and Zn in seeds and grains of full-grown cereal crops is as low as 5% and 25%, respectively [69]. In this experiment, we obtained significantly better results when testing their bioavailability in wheat sprouts. The maximum assimilation of Fe and Zn reached 322% and 1442%, respectively, when the optimal ion concentration was applied (100 μg $g^{-1}$). About 90% of Mg from hydroponic media is stored in leaves, not grains [70]. This confirms that sprouts are a fast, easily available, and relatively safe source of nutrient supplementation [38].

## 3. Materials and Methods

### *3.1. Ion Solutions and AAS Standards*

In order to prepare the solutions of Zn, Cr, Fe, and Mg, the following chemicals were used: zinc nitrate hexahydrate ($Zn(NO_3)_2 \cdot 6H_2O$), chromium (III) sulfate hydrate ($Cr_2(SO_4)_3 \cdot 18H_2O$), iron (III) chloride hexahydrate ($FeCl_3 \cdot 6H_2O$), and magnesium nitrate hexahydrate ($Mg(NO_3)_2 \cdot 6H_2O$). For every individual metal, concentrations of 50, 100, and 200 $\mu g\ g^{-1}$ were prepared in solutions with a volume of 200 mL. The prepared solutions were sterilized under a UV-C lamp for 30 min before their use.

The AAS ion standards for zinc, chromium, iron, and magnesium at a concentration of 1000 $\mu g\ L^{-1}$ (Merck, Darmstadt, Germany) were acquired. For the quantitative analysis, solutions at varying concentrations, depending on the primary metal concentration in the plant material, were prepared via the dilution of appropriate standards. The following dilutions ($\mu g\ g^{-1}$) of $Fe^{3+}$ (0.3, 2.5, 5.0, 7.5, and 10.0), $Zn^{2+}$ (0.1, 0.3, 0.5, 0.8, and 1.0), $Mg^{2+}$ (0.5, 1.2, 1.8, 2.4, and 3.0), and $Cr^{3+}$ (0.1, 0.2, 0.3, 0.4, and 0.5) were prepared.

### *3.2. Plant Material*

The plant material selection was made carefully. For the purpose of this experiment, two varieties of common wheat, *Triticum aestivum* var. Tonacja (variety A) and *Triticum aestivum* var. Arkadia (variety B), with short germination times and high nutrient values, were chosen. The research material was obtained from the research farm in Czesławice (Poland). Both varieties were reported to have great popularity and wide application. The seeds were pure and free of preservatives.

### *3.3. UV-C Exposure*

Sprouts are desired to be incorporated into human diet routines, and as with other edible materials, they need to meet some quality standards. A portion of the wheat seeds underwent a non-chemical sterilization process with a UV-C lamp ($\lambda$ = 254 nm) for 30 min in order to maintain microbiological quality and safety. The UV lamp was TUV 15W/G15 T8 type (Philips, Eindhoven, The Netherlands), and 15 W power was used as the irradiation source. The distance between the lamp and the samples was 25 cm. The intensity of energy irradiation reached 0.85 W $m^{-2}$. The measurements of UV-C radiation dose were conducted using a UVC-2GRM radiometer (Sonopan, Białystok, Poland).

### *3.4. Sample Preparation and Analysis of Wheat Seeds via Scanning Electron Microscopy (SEM)*

In order to trace the features of wheat seeds affected by UV-C light exposure, a new approach was introduced. In this study, the well-known method of scanning electron microscopy (SEM), together with sample preparation, was altered for this specific plant material. The wheat seeds were submerged in plastic resin and subsequently cut into thin layers. In this process, the optimization of light polymerization was challenging because many factors play a role in the preservation of organic material structures when they are embedded in resin. To avoid blistering and cracking, the curing time, degree of curing, and temperature development must be controlled and adapted to the type and origin of the material.

In this study, the polymerization of the wheat seeds of the Tonacja variety (treated and not treated with UV-C radiation) in the Technovit® 2000 LC resin provided by Heraeus Kulze (Wehrheim, Germany) was carried out using Light Polymerizer Exakt 520 (Exakt, Hamburg-Norderstedt, Germany). The following conditions were incorporated into the polymerization process: an exposure time of 45 min and the use of blue light (the bulb type was 9W/71).

Furthermore, the high-precision EXAKT thin-section cutting and grinding system was introduced. It was used for the preparation of the sections, with a thickness down to 10 $\mu$m to retain the native morphology of the samples. The method of thin-section cutting is especially important in the study of the surface of biomaterials since the tested area can be prepared while preserving the overall structure, and light microscopic techniques can be used directly, e.g., SEM. The use of the saw-like EXAKT 312 allows the quick and precise preparation of the native condition of a sample and creates a precondition for further diagnostics.

In the present experiment, the resin blocks were cut using a Diamond Bend Saw EXAKT 312 (Exakt, Hamburg-Norderstedt, Germany) with a diamond band. The cutting time was 60 s, and the depth of intersection was 2 mm. The type of diamond band used to cut the samples was Exakt 0.3/D151.

The samples were prepared with the courtesy of the company AMP Medical, which is the exclusive distributor of Exakt machines in Poland.

The microarea studies were carried out using a Hitachi SU6600 (Hitachi, Tokyo, Japan) scanning electron microscope. The samples were tested without sputtering under low vacuum conditions (5–10 Pa). As part of the analysis of the samples, microphotographic documentation was made using the backscattered electron (BSE) technique; then, using an EDS adapter with a silicon detector provided by Thermo (Waltham, MA, USA), the chemical composition of the tested samples was analyzed. The analysis was carried out at a voltage of 25 kV, and the time of data collection was 90 s. The working distance was about 10 mm, and the magnification was about 100–500 times.

### 3.5. Hydroponic Cultivation

The experiment was performed under hydroponic conditions in a thermostatic chamber. Plates that were 80 mm in diameter were coated with wet filter paper. The plates with seeds were incubated at room temperature (21–23 °C), with a relative humidity of 60–70% and natural light exposure (day/night). The appropriate humidity level and medium concentration were maintained by systematic refilling. An equal number of seeds (100) per plate was placed directly on the surface of filter papers soaked with ion solutions at four different concentrations: 0 (control), 50, 100, and 200 $\mu g\ g^{-1}$ for each nutrient. The incubation was performed between 4 and 7 days. Additional protocols, including the cultivation of unsterilized wheat seeds of variety A for 4 days in the four hydroponic media with the four elements at a concentration of 100 $\mu g\ g^{-1}$, were performed. The experiments were carried out in three independent replications for each treatment. During the experiment, sprout development as well as microbiological infection spread were precisely monitored.

### 3.6. Seed Germination Energy

The seed germination energy was determined on the basis of the International Seed Testing Association protocol (ISTA, Essen, Germany) [71]. For this purpose, we used the seeds of varieties A and B that were treated with UV-C radiation (UV-treated) and the seeds of variety A that were free from any surface sterilization (non-UV-treated). The germinated seeds were counted at 24, 48, and 96 h after sowing in the hydroponic media with 100 $\mu g\ g^{-1}$ of ion solutions (Fe, Zn, Mg, and Cr) and control (0 $\mu g\ g^{-1}$). The fraction of germinated seeds (number of sprouts) after 4 days of germination was defined as the germination energy $G$. The threshold was at least 2 mm for the long radicle [71]. The germination energy was expressed as a fraction of the germinated seeds $G$ after a defined time $t$ and calculated using the following equation:

$$G = \frac{n}{n_T} \times 100\% \tag{1}$$

where $n$ is the number of seeds germinated at time $t$, and $n_T$ is the total number of sown seeds.

### 3.7. Collection and Mineralization of Sprout Samples

After 4 and 7 days of incubation, the sprouts were washed several times with deionized water. This allowed the removal of metal ions absorbed on their surface. The samples were dried at 60–65 °C to an equal weight. The next step was sample mineralization with the use of a Mars 5 microwave-accelerated reaction system with XP1500 vessels (vol. 50 milliliters, max. 55 bar) obtained from CEM (Kamp-Lintfort, Germany). A total of 0.5 g of dried material was placed into an extraction vessel. The sample digestion was performed in the presence of 3 mL of 69% nitric acid and 7 mL of deionized water. The mineralization procedure included a temperature–time ramp for 10 min with a final temperature of 180 °C that was held for 20 min and a cooling step lasting 15 min. The plant material was then suspended in deionized water to a volume of 25 mL.

### 3.8. The AAS Conditions

The accumulation of micronutrients in the plant material was tested using high-resolution continuum source flame atomic absorption spectrometry (Analytik Jena AG, Germany) in the presence of acetylene as the gas fuel and air as an oxidant. The optimal wavelength (nm) for each element was as follows: Fe at 248.3, Zn at 231.9, Mg at 285.2, and Cr at 357.9. The reference solution was prepared with 0.5% nitric acid and 0.5 mL of 19% KCl in a volume of 100 mL. The signals reported as the peak areas were analyzed using the Aspect CS 2.0.0 software (Analytik Jena AG, Jena, Germany) and calculated as the ion content (μg) per sprout of dry matter (g). The calculation included the material volume after mineralization and the dilution used for a particular sample.

### 3.9. Statistical Analysis

One-way analysis of variance (ANOVA) was performed using the Statistica 10.0 software, with a level of significance of $p < 0.05$. Homogeneous groups were determined using Tukey's test.

## 4. Conclusions

The obtained results confirmed that both tested common wheat (*Triticum aestivum* L.) varieties (Arkadia and Tonacja) have a high ability to accumulate Fe, Zn, Mg, and Cr during the germination process. The highest feasibility was indicated for Zn biofortification ($R^2 > 0.940$). For other elements, a very strong correlation between the ion concentration in the media and their bioassimilation in the plant tissues was also noticed ($R^2 > 0.90$). The ion concentration of 100 μg $g^{-1}$ was assumed to be the most appropriate for effective bioaccumulation of ions in the wheat sprouts. After seven days of hydroponic cultivation in the 100 μg $g^{-1}$ solution, there was an increase in iron accumulation of 218% and 322% in the plant materials of the Tonacja and Arkadia varieties, respectively. The final Zn concentration after seven days was approximately 8–11 times higher in comparison to the control. The ion dose of 200 μg $g^{-1}$ showed high potential in producing nutrient-rich sprouts; however, it reduced sprout biomass growth.

The applied UV-C seed sterilization was proven to be an effective factor in preventing infections during seed germination. UV-C irradiation reduced the seed germination energy by 5–9%, depending on the type of ion present in the hydroponic media. However, it did not have an adverse effect on the morphological features of wheat grains or future sprouts, as well as on nutrient assimilation, and it could even significantly improve the biofortification of Cr.

In summary, this study showed that the conditions applied could be an effective tool for obtaining fortified food in a cheap and quick way.

**Author Contributions:** Conceptualization, E.B., K.C., M.T.-M. and A.S.; methodology, K.C., E.B., M.T.-M. and A.S.; validation, A.G.-K. and P.D.; formal analysis, E.B., K.C., G.W. and M.T.-M.; investigation, K.C., E.B., A.S. and V.D.R.; resources, K.C. and E.B.; writing—original draft preparation, E.B., K.C., A.G.-K. and M.T.-M.; writing—review and editing, E.B., A.S., K.C. and M.T.-M.; visualization, V.D.R., G.W. and P.D.; supervision, E.B. and K.C.; project administration, E.B. and K.C.; funding acquisition, K.C. and E.B. All authors have read and agreed to the published version of the manuscript.

**Funding:** This research received no external funding.

**Institutional Review Board Statement:** Not applicable.

**Informed Consent Statement:** Not applicable.

**Data Availability Statement:** Not applicable.

**Acknowledgments:** The authors wish to thank the company AMP Medical—the exclusive distributor of EXAKT machines on the territory of Poland.

**Conflicts of Interest:** The authors declare no conflict of interest.

## References

1. Sigel, A.; Sigel, H.; Sigel, R.K.O. *Neurodegenerative Diseases and Metal Ions*; John Wiley & Sons, Ltd.: Chichester, UK, 2006. [CrossRef]
2. Lee, T.G.; Park, J.-W.; Shon, H.K.; Moon, D.W.; Choi, W.W.; Li, K.; Chung, J.H. Biochemical imaging of tissues by SIMS for biomedical applications. *Appl. Surf. Sci.* **2008**, *255*, 1241–1248. [CrossRef]
3. Al-Fartusie, F.; Mohssan, S. Essential Trace Elements and Their Vital Roles in Human Body. *Indian J. Adv. Chem. Sci.* **2017**, *5*, 127–136.
4. McDowell, L.R. *Minerals in Animal and Human Nutrition*, 2nd ed.; Elsevier Science: Amsterdam, The Netherlands, 2003; p. 660.
5. Hurrell, R.F. Bioavailability of iron. *Eur. J. Clin. Nutr.* **1997**, *51*, 4–8.
6. Osredkar, J. Copper and Zinc, Biological Role and Significance of Copper/Zinc Imbalance. *J. Clin. Toxicol. S* **2011**, *3*, 2161. [CrossRef]
7. Prasad, A.S. Zinc deficiency: Has been known of for 40 years but ignored by global health organisations. *BMJ* **2003**, *326*, 409. [CrossRef] [PubMed]
8. Plum, L.M.; Rink, L.; Haase, H. The essential toxin: Impact of zinc on human health. *Int. J. Environ. Res. Public Health* **2010**, *7*, 1342–1365. [CrossRef] [PubMed]
9. Mertz, W. Chromium in human nutrition: A review. *J. Nutr.* **1993**, *123*, 626–633. [CrossRef]
10. Vincent, J.B.; Lukaski, H.C. Chromium. *Adv. Nutr.* **2018**, *9*, 505–506. [CrossRef]
11. Kostov, K.; Halacheva, L. Role of Magnesium Deficiency in Promoting Atherosclerosis, Endothelial Dysfunction, and Arterial Stiffening as Risk Factors for Hypertension. *Int. J. Mol. Sci.* **2018**, *19*, 1724. [CrossRef]
12. Shenkin, A. Micronutrients in health and disease. *Postgrad. Med. J.* **2006**, *82*, 559–567. [CrossRef]
13. Biesalski, H.K.; Tinz, J. Micronutrients in the life cycle: Requirements and sufficient supply. *NFS J.* **2018**, *11*, 1–11. [CrossRef]
14. Yin, X.; Yuan, L.; Lin, Z. Phytoremediation and Biofortification. Two Sides of One Coin. In *Phytoremediation and Biofortification*; Yin, X., Yuan, L., Eds.; Springer Briefs in Green Chemistry for Sustainability; Springer: Doetinchem, The Netherlands, 2012. [CrossRef]
15. Huey, S.L.; Krisher, J.T.; Bhargava, A.; Friesen, V.M.; Konieczynski, E.M.; Mbuya, M.N.N.; Mehta, N.H.; Monterrosa, E.; Nyangaresi, A.M.; Mehta, S. Review of the Impact Pathways of Biofortified Foods and Food Products. *Nutrients* **2022**, *14*, 1200. [CrossRef] [PubMed]
16. Beach, R.H.; Sulser, T.B.; Crimmins, A.; Cenacchi, N.; Cole, J.; Fukagawa, N.K.; Mason-D'Croz, D.; Myers, S.; Sarofim, M.C.; Smith, M.; et al. Combining the effects of increased atmospheric carbon dioxide on protein, iron, and zinc availability and projected climate change on global diets: A modelling study. *Lancet Planet Health* **2019**, *3*, e307–e317. [CrossRef]
17. Combs, G.F., Jr. Selenium in global food systems. *Brit. J. Nutr.* **2001**, *85*, 517–547. [CrossRef] [PubMed]
18. Kennedy, G.; Nantel, G.; Shetty, P. The scourge of 'hidden hunger': Global dimensions of micronutrient deficiencies. *Food Nutr. Agri.* **2003**, *32*, 8–16.
19. WHO; FAO. Better nutrition, better lives. In Proceedings of the Second International Conference on Nutrition (ICN2), Rome, Italy, 19–21 November 2014.
20. Buturi, C.V.; Mauro, R.P.; Fogliano, V.; Leonardi, C.; Giuffrida, F. Mineral Biofortification of Vegetables as a Tool to Improve Human Diet. *Foods* **2021**, *10*, 223. [CrossRef]
21. Siwela, M.; Pillay, K.; Govender, L.; Lottering, S.; Mudau, F.N.; Modi, A.T.; Mabhaudhi, T. Biofortified crops for combating hidden hunger in South Africa: Availability, acceptability, micronutrient retention and bioavailability. *Foods* **2020**, *9*, 815. [CrossRef]
22. Duborská, E.; Šebesta, M.; Matulová, M.; Zvěřina, O.; Urík, M. Current Strategies for Selenium and Iodine Biofortification in Crop Plants. *Nutrients* **2022**, *14*, 4717. [CrossRef]
23. Blicharska, E.; Flieger, J.; Oszust, K.; Frąc, M.; Świeboda, R.; Kocjan, R. High-resolution continuum source atomic absorption spectrometry with microwave-assisted extraction for the determination of metals in vegetable sprouts. *Anal. Lett.* **2015**, *48*, 2272–2287. [CrossRef]
24. Dunwell, J.M. Transgenic cereals: Current status and future prospects. *J. Cereal Sci.* **2014**, *59*, 419–434. [CrossRef]
25. Kumar, S.; Palve, A.; Joshi, C.; Srivastava, R.K. Crop biofortification for iron (Fe), zinc (Zn) and vitamin A with transgenic approaches. *Heliyon* **2019**, *5*, e01914. [CrossRef] [PubMed]
26. Van Der Straeten, D.; Bhullar, N.K.; De Steur, H.; Gruissem, W.; MacKenzie, D.; Pfeiffer, W.; Bouis, H. Multiplying the efficiency and impact of biofortification through metabolic engineering. *Nat. Commun.* **2020**, *11*, 1–10. [CrossRef] [PubMed]
27. Malik, K.A.; Maqbool, A. Transgenic crops for biofortification. *Front. Sustain. Food Syst.* **2020**, *4*, 182. [CrossRef]
28. Koç, E.; Karayiğit, B. Assessment of Biofortification Approaches Used to Improve Micronutrient-Dense Plants That Are a Sustainable Solution to Combat Hidden Hunger. *J. Soil Sci. Plant Nutr.* **2022**, *22*, 475–500. [CrossRef]

29. Lorenz, K.; D'Appolonia, B. Cereal sprouts: Composition, nutritive value, food applications. *CRC Cr. Rev. Food Sci.* **1980**, *13*, 353–385. [CrossRef]
30. Park, S.A.; Grusak, M.A.; Oh, M.M. Concentrations of minerals and phenolic compounds in three edible sprout species treated with iron-chelates during imbibition. *Hortic. Environ. Biotechnol.* **2014**, *55*, 471–478. [CrossRef]
31. Wei, Y.; Shohag, M.; Ying, F.; Yang, X.; Wu, C.; Wang, Y.; Shohag, J.I. Effect of ferrous sulfate fortification in germinated brown rice on seed iron concentration and bioavailability. *Food Chem.* **2013**, *138*, 1952–1958. [CrossRef]
32. Zhu, H. Accumulation and distribution of selenium in different parts and macromolecule of Se-enriched Tartary Buckwheat (*Fagopyrum tataricum* Gaertn.) during germination. *Int. Food Res. J.* **2014**, *21*, 991–997.
33. Liu, K.; Chen, F.; Zhao, Y.; Gu, Z.; Yang, H. Selenium accumulation in protein fractions during germination of Se-enriched brown rice and molecular weights distribution of Se-containing proteins. *Food Chem.* **2011**, *127*, 1526–1531. [CrossRef]
34. Lazo-Vélez, M.A.; Avilés-González, J.; Serna-Saldivar, S.O.; Temblador-Pérez, M.C. Optimization of wheat sprouting for production of selenium enriched kernels using response surface methodology and desirability function. *LWT-Food Sci. Technol.* **2016**, *65*, 1080–1086. [CrossRef]
35. Krzepiłko, A.; Zych-Wężyk, I.; Święciło, A.; Molas, J.; Skwaryło-Bednarz, B. Effect of iodine biofortification of lettuce seedlings on their mineral composition and biological quality. *J. Elem.* **2016**, *21*, 1071–1080. [CrossRef]
36. Klepaka, T. Construcion of axial-symmetric polymetric extrudates of complex forms. *Polimery* **2008**, *53*, 390–395. (In Polish) [CrossRef]
37. Arscott, S.; Goldman, I. Biomass effects and selenium accumulation in sprouts of three vegetable species grown in selenium-enriched conditions. *Hortscience* **2012**, *47*, 497–502. [CrossRef]
38. Zielińska-Dawidziak, M.; Siger, A. Effect of elevated accumulation of iron in ferritin on the antioxidants content in soybean sprouts. *Eur. Food Res. Technol.* **2012**, *234*, 1005–1012. [CrossRef]
39. Zou, T.; Xu, N.; Hu, G.; Pang, J.; Xu, H. Biofortification of soybean sprouts with zinc and bioaccessibility of zinc in the sprouts. *J. Sci. Food Agric.* **2014**, *94*, 3053–3060. [CrossRef]
40. Aloo, S.O.; Ofosu, F.K.; Kilonzi, S.M.; Shabbir, U.; Oh, D.H. Edible Plant Sprouts: Health Benefits, Trends, and Opportunities for Novel Exploration. *Nutrients* **2021**, *13*, 2882. [CrossRef]
41. He, W.; Yang, H.; Gu, G. Effect of Bacteria-mineral water produced from bio-reacted fowl dung on seed germination of wheat (*Triticum aestivum*) and rice (*Oryza sativa* L.). *Environ. Progr.* **2008**, *27*, 125–132. [CrossRef]
42. Tripathi, D.K.; Singh, S.; Singh, S.; Mishra, S.; Chauha, D.K.; Dubey, N.K. Micronutrients and their diverse role in agricultural crops: Advances and future prospective. *Acta Physiol. Plant.* **2015**, *37*, 139. [CrossRef]
43. Tripathi, D.K.; Singh, V.P.; Prasad, S.M.; Chauha, D.K.; Dubey, N.K.; Rai, A.K. Silicon-mediated alleviation of Cr (VI) toxicity in wheat seedlings as evidenced by chlorophyll florescence, laser induced breakdown spectroscopy and anatomical changes. *Ecotoxicol. Environ. Saf.* **2015**, *113*, 133–144. [CrossRef]
44. Sharma, P.; Jha, A.B.; Dubey, R.S.; Pessarakli, M. Reactive oxygen species, oxidative damage, and antioxidative defense mechanism in plants under stressful conditions. *J. Bot.* **2012**, 217037. [CrossRef]
45. Rupiasih, N.N.; Vidyasagar, P.B. Effect of UV-C radiation and hypergravity on germination, growth and content of chlorophyll of wheat seedlings. *AIP Conf. Proc.* **2016**, *1719*, 030035-1–030035-6. [CrossRef]
46. Partap, M.; Solanki, V.A. Impact of black point incited by Alternaria alternata on wheat trade, seed quality and seed germination. *Indian Phytopathol.* **2016**, *69*, 74–76.
47. Rajput, M.A.; Pathan, M.A.; Lodhi, A.M.; Shah, G.S.; Khanzada, K.A. Studies on seed-borne fungi of wheat in Sindh Province and their effect on seed germination. *Pak. J. Bot.* **2005**, *37*, 181–185.
48. Noble, R.E. Effects of UV-irradiation on seed germination. *Sci. Total Environ.* **2002**, *299*, 173–176. [CrossRef]
49. Kuzdraliński, A.; Kot, A.; Szczerba, H.; Nowak, M.; Muszyńska, M. A review of conventional PCR assays for the detection of selected phytopathogens of wheat. *J Mol. Microbiol. Biotechnol.* **2017**, *27*, 175–189. [CrossRef]
50. Singh, V.K.; Devi, A.; Pathania, S.; Kumar, V.; Tripathi, D.K.; Sharma, S.; Chauhan, D.K.; Singh, V.K.; Zorba, V. Spectroscopic investigation of wheat grains (*Triticum aestivum*) infected by wheat seed gall nematodes (*Anguina tritici*). *Biocat. Agri. Biotech.* **2017**, *9*, 58–66. [CrossRef]
51. Badridze, G.; Kacharava, N.; Chkhubianishvili, E.; Rapava, L.; Kikvidze, M.; Chanishvili, S.; Shakarishvili, N.; Mazanishvili, L.; Chigladze, L. Effect of UV radiation and artificial acid rain on productivity of wheat. *Russ. J. Ecol.* **2016**, *47*, 158–166. [CrossRef]
52. Thomas, D.T.T.; Puthur, J.T. UV radiation priming: A means of amplifying the inherent potential for abiotic stress tolerance in crop plants. *Environ. Exp. Bot.* **2017**, *138*, 57–66. [CrossRef]
53. Brown, J.; Lu, T.; Stevens, C.; Khan, V.; Lu, J.; Wilson, C.; Collins, D.; Wilson, M.; Igwegbe, E.; Chalutz, E.; et al. The effect of low dose ultraviolet light-C seed treatment on induced resistance in cabbage to black rot (*Xanthomonas campestris* pv. campestris). *Crop Prot.* **2001**, *20*, 873–883. [CrossRef]
54. Kacharava, N.; Chanishvili, S.; Badridze, G.; Chkhubianishvili, E.; Janukashvili, N. Effect of seed irradiationon the content of antioxidants in leaves of kidney bean, cabbage and beet cultivars. *Aust. J. Crop. Sci.* **2009**, *3*, 137–145.
55. Ouhibi, C.; Attia, H.; Rebah, F.; Msilini, N.; Chebbi, M.; Aarrouf, J.; Urban, L.; Lachaal, M. Salt stress mitigation by seed priming with UV-C in lettuce plants, growth, antioxidant activity and phenolic compounds. *Plant Phys. Biochem.* **2014**, *83*, 126–133. [CrossRef] [PubMed]

56. Shetta, N.D.; Areaf, I.M. Impact of ultraviolet-c radiation on seed germination and chlorophyll concentration of some woody trees grown in Saudi Arabia. *J. Agric. Food Env. Sci.* **2009**, *8*, 1–21.
57. Brazaitytė, A.; Viršilė, A.; Jankauskienė, J.; Sakalauskienė, S.; Samuolienė, G.; Sirtautas, R.; Novičkovas, A.; Dabašinskas, L.; Miliauskienė, J.; Vaštakaitė, V.; et al. Effect of supplemental UV-A irradiation in solid-state lighting on the growth and phytochemical content of microgreens. *Int. Agrophysics* **2015**, *29*, 13–22. [CrossRef]
58. Shukla, U.C.; Kakkar, P. Effect of dual stress of ultraviolet-B radiation and cadmium on nutrient uptake of wheat seedlings. *Commun. Soil Sci. Plant.* **2002**, *33*, 1737–1749. [CrossRef]
59. Ilyas, M.; Khan, M.J.; Murad, Z.; Satti, S.Z.; Ullah, A. Biofortification of Iron in Wheat Varieties Using Different Methods of Application. *Gesunde Pflanz.* **2022**. [CrossRef]
60. Dhaliwal, S.S.; Sadana, U.S.; Manchanda, J.S.; Dhadli, H. Biofortification of wheat grains with zinc and iron in Typic Ustochrept soils of Punjab. *Indian J. Fertil.* **2009**, *5*, 13–20.
61. Dhaliwal, S.S.; Sharma, V.; Shukla, A.K.; Verma, V.; Kaur, M.; Shivay, Y.S.; Nisar, S.; Gaber, A.; Brestic, M.; Barek, V.; et al. Biofortification—A Frontier Novel Approach to Enrich Micronutrients in Field Crops to Encounter the Nutritional Security. *Molecules* **2022**, *27*, 1340. [CrossRef]
62. Wu, C.-Y.; Lu, L.-L.; Yang, X.-E.; Feng, Y.; Wei, Y.-Y.; Hao, H.-L.; Stoffella, P.J.; He, Z.-L. Uptake, translocation, and remobilization of zinc absorbed at different growth stages by rice genotypes of different Zn densities. *J. Agric. Food Chem.* **2010**, *58*, 6767–6773. [CrossRef]
63. Cakmak, I. Enrichment of cereal grains with zinc: Agronomic or genetic biofortification? *Plant Soil.* **2008**, *302*, 1–17. [CrossRef]
64. Phattarakul, N.; Rerkasem, B.; Li, L.J.; Wu, L.H.; Zou, C.Q.; Ram, H.; Sohu, V.S.; Kang, B.S.; Surek, H.; Kalayci, M.; et al. Biofortification of rice grain with zinc through zinc fertilization in different countries. *Plant Soil.* **2012**, *361*, 131–141. [CrossRef]
65. Cakmak, I.; Kalayci, M.; Kaya, Y.; Torun, A.A.; Aydin, N.; Wang, Y.; Arisoy, Z.; Erdem, H.; Yazici, A.; Gokmen, O.; et al. Biofortification and localization of zinc in wheat grain. *J. Agric. Food Chem.* **2010**, *58*, 9092–9102. [CrossRef] [PubMed]
66. Velu, G.; Ortiz-Monasterio, I.; Cakmak, I.; Hao, Y.; Singh, R.P. Biofortification strategies to increase grain zinc and iron concentrations in wheat. *J. Cereal Sci.* **2014**, *59*, 365–372. [CrossRef]
67. Guttieri, M.J.; Peterson, K.M.; Souza, E.J. Agronomic performance of low phytic acid wheat. *Crop. Sci.* **2006**, *46*, 2623–2629. [CrossRef]
68. Liu, Z.H.; Wang, H.Y.; Zhang, G.P.; Chen, P.D.; Liu, D.J. Genotypic and spike positional difference in grain phytase activity, phytate, inorganic phosphorus, iron and zinc contents in wheat (*Triticum aestivum* L.). *J. Cereal Sci.* **2006**, *44*, 212–219. [CrossRef]
69. Bouis, H.E.; Welch, R.M. Biofortification—A sustainable agricultural strategy for reducing micronutrient malnutrition in the Global South. *Crop. Sci.* **2010**, *50*, 20–32. [CrossRef]
70. Courtois, D.; Kasternmayer, P.; Clough, J.; Vigo, M.; Sabatier, M.; Arnaud, M.J. Magnesium enrichment and distribution in plants. *Isot. Environ. Health Stud.* **2003**, *39*, 273–279. [CrossRef]
71. ISTA. International rules for seed testing. In Proceedings of the Ordinary General Meeting 2016, Tallinn, Estonia, 29 February 2016; Volume 2017.

MDPI AG
Grosspeteranlage 5
4052 Basel
Switzerland
Tel.: +41 61 683 77 34

*International Journal of Molecular Sciences* Editorial Office
E-mail: ijms@mdpi.com
www.mdpi.com/journal/ijms

www.ingramcontent.com/pod-product-compliance
Lightning Source LLC
Chambersburg PA
CBHW041353210926
51850CB00013B/3070
* 9 7 8 3 7 2 5 8 4 3 4 0 4 *